IMER OF WARRANTY

hnical descriptions, procedures, and computer programs in this book have
veloped with the greatest of care and they have been useful to the author in
range of applications; however, they are provided as is, without warranty of
d. Artech House, Inc. and the author and editors of the book titled *Design
pplications of Active Integrated Antennas* make no warranties, expressed or
d, that the equations, programs, and procedures in this book or its associated
re are free of error, or are consistent with any particular standard of mer-
bility, or will meet your requirements for any particular application. They
not be relied upon for solving a problem whose incorrect solution could result
ry to a person or loss of property. Any use of the programs or procedures in
manner is at the user's own risk. The editors, author, and publisher disclaim
bility for direct, incidental, or consequent damages resulting from use of the
ams or procedures in this book or the associated software.

For a listing of recent titles in the
Artech House *Antennas and Electromagnetics Analysis Library*,
turn to the back of this book.

Design and App
Active Integratec

DISCLA

The tec
been de
a broac
any ki
and A
implie
softwa
chanta
should
in inj
such
all lia
progr

Design and Applications of Active Integrated Antennas

Mohammad S. Sharawi

Oualid Hammi

ARTECH
HOUSE

BOSTON | LONDON
artechhouse.com

Library of Congress Cataloging-in-Publication Data
A catalog record for this book is available from the U.S. Library of Congress

British Library Cataloguing in Publication Data
A catalog record for this book is available from the British Library.

ISBN 13: 978-1-63081-358-1

Cover design by John Gomes

685 Canton Street
Norwood, MA

10 9 8 7 6 5 4 3 2 1

To my great parents, my amazing and lovely wife Rana, and my three adorable little angles, Basima, Rand, and Ibrahim

Mohammad S. Sharawi

To my loving parents, my dear and kind inspiration Ilhem, my special best buddy Zakariya, and my charismatic princess Eya

Oualid Hammi

Contents

Preface

Wireless technology has evolved dramatically in recent years. Several technologies have been invented and utilized to achieve the high data rates that we see nowadays in fourth generation (4G) systems, while others will be deployed soon to make the fifth generation (5G) a reality. Passive antennas have been used in most of the previous wireless terminal generations from first generation (1G) to 4G. Few of the recent 4G antenna systems with multiple-input-multiple-output (MIMO) capability used simple active switches to achieve a limited set of advanced features, but this has been on a restricted scale. It is expected that active integrated antennas (AIAs) will be utilized more in future wireless terminals due to their added features and enhanced performance.

The radio frequency (RF) front-end has the amplifiers as well as filters and switches to allow the wireless terminal operate in various bands. The concept of AIAs will provide the system designer more features and enhanced performance when compared to individual stand-alone connections between the front-end components (amplifier, filter, and so forth) and the antenna. Active integration between the active components (amplifier, oscillator, mixer, diodes) and the antenna can provide better power transfer, higher gains, increased efficiencies and smaller design footprints. These are all added values for using AIAs over the conventional subsystem designs of various terminal circuits.

The focus of this book is on the design and applications of AIA systems. AIAs are expected to be used in future wireless terminals to provide higher gain values, reconfigurable frequency, and pattern capabilities to select the best available spectrum for use or tilt their beams for better reception. On-chip antenna array solutions at millimeter-waves will need AIAs to compensate for the path losses at such high frequencies. All these topics, among other AIA-based designs for future technologies, will be discussed in this book. The book starts by reviewing the basic methods and techniques of passive impedance matching in Chapter 2. Lumped elements and stub matching methods are revisited in detail, as they will be used in later chapters. The chapter touches upon narrowband and wideband matching techniques highlighting their various features. A design example on impedance matching using the Microwave Office (MWO) software package is presented.

Chapter 3 covers in detail the fundamental aspects of designing RF amplifiers. Whether the amplifier is designed as a stand-alone unit in any RF-front end or as part of an AIA, the design procedures in this chapter are used in any RF amplifier design for a wireless terminal. The chapter starts by giving a detailed review on noise power analysis in two-port networks and provides the various definitions of the quantities used. The procedures for maximum gain and low noise and the trade-off between them are highlighted. Several complete examples of RF amplifier designs are provided to enhance the understanding of the procedures along with several illustrations used to show the behavior of such amplifiers under various gain and noise requirements.

Chapter 4 presents a very detailed discussion on the design of antennas for modern wireless terminals. It starts with introducing the basic definitions of antenna parameters and performance metrics. Then it discusses the most common printed antenna types used nowadays in wireless devices (i.e., dipoles, monopoles, loop, patch, and slot) along with their features and design outlines. The features and design equations of antenna arrays are provided for linear, planar, and circular configurations. The chapter then provides a detailed overview about the MIMO technology and the design and performance metrics of MIMO antenna systems. We end the chapter by several detailed design examples using computer-aided design (CAD) tools for modeling and examining the behavior of printed MIMO antenna systems using the software packages High Frequency Structure Simulator (HFSS) and Computer Simulation Technology (CST).

Chapter 5 provides a comprehensive discussion on all possible AIA types published to date. It provides detailed design procedures for each of the AIA types and gives several detailed design examples to illustrate the design process and outcome. It starts with the general performance metrics of AIAs and a detailed design example to explain them. Then it lists the various AIA types as well as their design procedures, specifically oscillator-based AIAs, amplifier-based AIAs, mixer-based AIAs, transceiver-based AIAs, frequency-reconfigurable and pattern-reconfigurable AIAs, on-chip and in-chip AIAs, and non-Foster-based AIAs. The chapter is full of design examples with details using several software packages.

The new approach of codesigning the active amplifier with the antenna without the requirements for 50Ω impedance is provided in detail in Chapter 6. The design procedure and its advantages are highlighted, and then several detailed design examples are provided with the use of design equations and software packages to illustrate the benefits and features of such a procedure. The codesign approach is applied on narrowband AIA designs, wideband AIA designs, and ultrawideband (UWB) AIA designs. A MIMO antenna is used in the UWB AIA codesign approach showing the advantages of such a method even on multi-antenna systems, which are expected to be used in upcoming 5G wireless terminals.

Four appendixes are provided with detailed step-by-step examples for the reader to experience and learn the software packages (CAD tool) features and modeling steps. In Appendix A, we provide a detailed step-by-step tutorial for using the Advanced Design System (ADS) software package through the design of an amplifier. ADS is widely used for the design of active microwave and RF circuits. In Appendix B, we provide detailed modeling and simulation steps for an RF filter using the Microwave Office (MWO) software package. This tool can also be used to design active circuits much like ADS and is widely used for system-level designs. In Appendix C, we provide a step-by-step design example using the HFSS via the design of a probe-fed printed square patch antenna. HFSS is widely used for modeling and simulating passive antenna structures and arrays. Finally, in Appendix D, we provide a detailed example of designing a microstrip-fed printed square patch antenna using CST. CST is another widely used tool for modeling and designing antenna systems but has the capability of incorporating active circuits to show their effect on the overall system performance. The appendixes will give the reader a good grasp of the tools used to design AIAs. The CD accompanying the book has the files used in these examples as well as some files for the examples appearing in the book chapters.

This book provides a complete reference on the design and applications of AIAs with numerous designs and illustrations along with design procedures for practicing and research engineers. The book provides examples based on CAD tools that will help the designers get started. It includes a comprehensive review of related and previous works on AIAs with all its types and applications as well as provides insight on future AIA systems. The book can be used in an advanced antenna design course or an advanced RF/microwave circuits design course. Its detailed examples, illustrations, and design procedures make it a must-have and valuable resource for any practicing engineer or researcher interested in this area.

Acknowledgments

We must start by thanking God (Allah) for giving us the strength to complete this timely endeavor about active integrated antenna systems. No doubt about it, our families should get the largest portion of the thanks and appreciation for their support and patience throughout the time of preparing this book. Thank you from the bottom of our hearts; we love you all.

We would like to thank the book reviewers for their valuable and constructive comments. Their detailed reviews enhanced the flow of the book contents considerably. We would also like to thank our book commissioning editor, Aileen Storry, and her assistant, Soraya Nair, for their prompt actions and valuable comments and suggestions. Thank you for forcing us to finish the book on time.

Mohammad S. Sharawi, thanks several of his past graduate students who helped in several parts of this book. Specifically, he thanks Mr. Sagar Dhar for preparing several of the examples in this book and providing some of the design procedures on which he worked hard during his thesis work extending the active integrated antenna design approach to multiple-input-multiple-output (MIMO) antenna systems. He also thanks Dr. Rifaqat Hussain, his former Ph.D. student, for providing several designs and examples specifically in MIMO reconfigurable antennas. He thanks Mr. Syed Jehangir for providing CAD tool examples and models for some of the antenna designs. Finally, he thanks King Fahd University for Petroleum and Minerals (KFUPM) and the Deanship of Scientific Research (DSR) for providing the support and facilities via research grants and funding opportunities from which several parts of this book were investigated and built.

Oualid Hammi thanks the Department of Electrical Engineering and the College of Engineering at the American University of Sharjah, for their valuable support.

CHAPTER 1

Introduction

Since the initial commercialization of the first wireless service, namely, the wireless telegraphy by Marconi in 1897, wireless systems have made their way to become essential devices in our daily life. The growth in terms of mobile devices and services provided has evolved at an unprecedented pace over the past couple of decades. We witnessed swift changes in mobile communication services as they progressed from the first generation (1G) to the fourth generation (4G), with each generation setting new goals in terms of user experience and unavoidably bringing technical challenges that engineers brightly addressed.

In this chapter, we will briefly review some aspects related to the evolution of wireless technology which will help ascertain the origins of the challenges set by the upcoming fifth generation (5G) of wireless communications. 5G is expected to mark a steep evolution and revolutionize the wireless communications landscape from an ecosystem of connected people to another of connected devices.

Since this book is devoted to active integrated antennas (AIAs), it is natural to discuss within this introductory chapter the commonly adopted architectures for wireless transmitters and receivers along with their advantages and limitations. This will allow us highlight the close interaction between amplifiers and antennas and hence the need for a codesign approach for their integration which will ultimately lead to enhanced performance metrics.

1.1 Review of Wireless Communication Technology Evolution

Wireless technology encompasses a wide range of applications in which information is transmitted and/or received through a wireless medium. Probably, the most influencing wireless technologies are those indispensable and inseparable smart mobile devices we are using on a daily basis. However, wireless technology includes other major applications in military such as radars and tactical communications; in navigation such as global positioning system (GPS) and aircraft air traffic control (ATC) radar transponders; and in broadcasting such as digital TV and satellite radio services. All these wireless technologies share the same resource, the electromagnetic (EM) spectrum. This valuable resource that is available to anyone, but regulated by government authorities, is becoming scarce especially in the sub-6-GHz range. Therefore, one of the main technical drives has always been to come up with increasingly spectrum-efficient technologies that will allow for the transmission of higher data rates within limited frequency bands. This led to the replacement of time-division multiple access (TDMA) and frequency-division multiple access

(FDMA) schemes by code-division multiple-access (CDMA)-based techniques in 2G and 3G systems and then the wide adoption of orthogonal frequency division multiple access (OFDMA) techniques for the 4G and possibly 5G systems.

Figure 1.1 illustrates the principles of TDMA, FDMA, and CDMA. This figure clearly depicts that in the TDMA technique each user has access to the entire frequency channel but uses it over a limited time interval, while in the FDMA technique the user has continuous access to a limited share of the frequency channel.

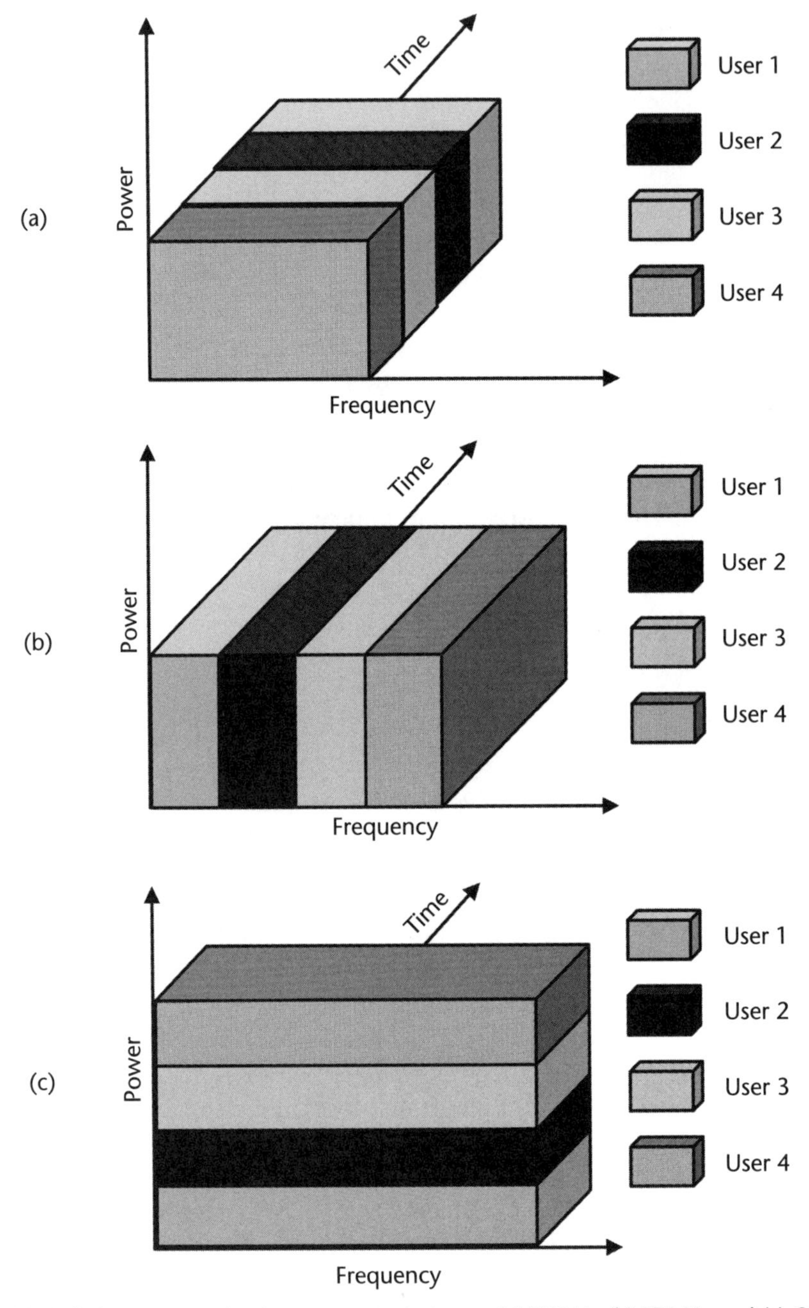

Figure 1.1 Wireless communications access techniques: (a) TDMA, (b) FDMA, and (c) CDMA.

Conversely, the CDMA technique allows all users to concurrently share the full resources (time and frequency) while separating them through the use of orthogonal codes. Similar to CDMA, OFDMA also allows for allocating all time and frequency resources to each user. However, compared to CDMA, OFDMA is more robust to severe channel conditions and narrowband interferences, has higher spectral efficiency, and is more suitable for carrier aggregation and multi-input and multi-output (MIMO) systems [1–3].

In terms of user experience and services, 1G communication systems provided basic calling features using analog technology. 2G systems started the dawn of the digital communication era in which the basic voice service was augmented with an additional service namely the short message service (SMS). The Global System for Mobile Communications (GSM) is the standard developed by the European Telecommunications Standards Institute (ETSI) for the 2G communication system. This standard later evolved into the General Packet Radio Service (GPRS) (also known as 2.5G) and Enhanced Data Rates for GSM Evolution (EDGE) (also known as 2.75G), which allowed theoretical data rates of 50 kbps and 1 Mbps, respectively. The first true mobile broadband experience was offered with 3G systems, in 2001, mainly based on CDMA technology. In fact, the EDGE data rates were bound by the channel bandwidth of GSM systems (i.e., 200 kHz) and therefore could not compete with the wideband CDMA (WCDMA) technique and its variants such as High-Speed Packet Access (HSPA), which achieved data rates of 14.4 Mbps and higher. Such data rates enabled Internet browsing, videoconferencing, and mobile TV, to cite only a few of the data-based features made possible with 3G. GSM and CDMA-based technologies later converged toward a unified OFDM-based Long Term Evolution (LTE) in 4G communication systems. 4G is mainly data oriented and aims at achieving high data rates (in the hundreds of megabits per second) with channel bandwidths varying from 1.4 MHz up to 20 MHz. Moreover, up to 1-Gbps data rates are targeted when concepts such as MIMO and carrier aggregation (CA) are applied.

The MIMO technique is based on sending parallel data streams through various transmitting antennas, and receiving them with a plurality of antennas as illustrated in Figure 1.2. MIMO techniques were proven to increase the data throughput and enhance the system robustness to multipath fading even with limited bandwidth and power levels [4–6]. However, the implementation of MIMO techniques requires some careful consideration when placing the antennas to avoid coupling and other effects [7]. CA consists of transmitting data stream through different frequency channels and allows for a significant increase in the data rate [8–11]. The number of

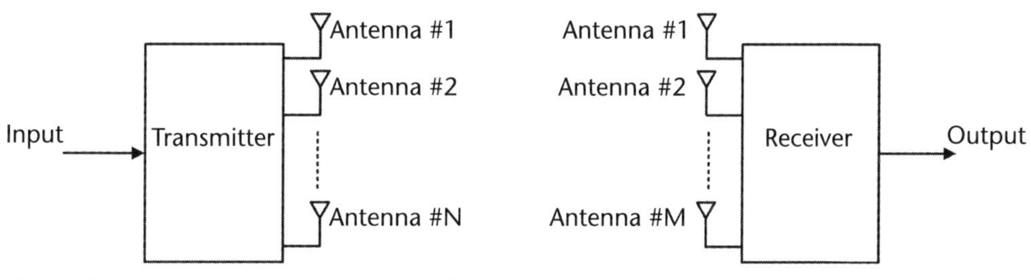

Figure 1.2 Block diagram of an $N \times M$ MIMO system.

channels can be up to 5 in the LTE-Advanced (LTE-A) standard, and each of these channels has a bandwidth of 20 MHz for a maximum total bandwidth of 100 MHz [12]. The CA can be done using contiguous carriers, intraband noncontiguous carriers or interband carriers as depicted in Figure 1.3 for the case of two carriers only. Clearly, the adoption of MIMO and carrier aggregation techniques sets stringent requirements on the radio frequency (RF) front-end where the coupling between the different antennas need to be carefully minimized, and the bandwidth of the system enhanced to support broadband and multiband signals obtained through CA.

1.2 Transmitter and Receiver Architectures

RF transmitters and receivers consist of two main subblocks: the digital signal processing (DSP) block and the RF front-end block. The arrangement of the components in the RF front-end might vary depending on the type of transmitter or receiver, and

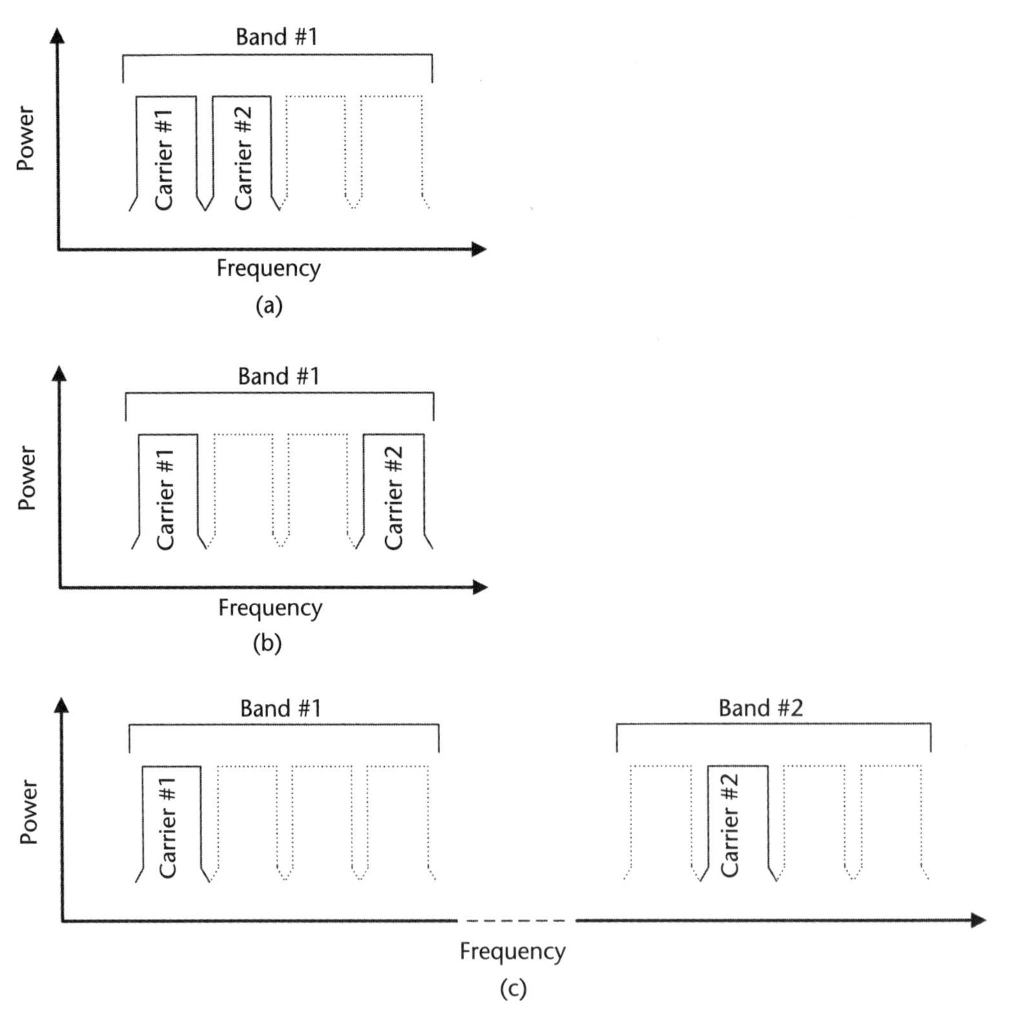

Figure 1.3 CA concept: (a) CA with contiguous carriers, (b) CA with intraband carriers, and (c) CA with interband carriers.

it is also dependent on the DSP part. Typically, the RF front-end of a transmitter includes frequency upconversion stage(s), filtering stage(s), amplification stage(s), and a transmitting antenna (or antennas in case of MIMO). It might also include IQ modulation in case the DSP block outputs the in-phase (I) and quadrature (Q) components separately. Similarly, the RF front-end of a receiver is made of a receiving antenna, low noise amplification stage(s), frequency downconversion stage(s), and filtering stage(s). When the DSP block is configured to receive I and Q components separately, analog IQ demodulation is also implemented within the receiver's RF front-end. Hence, RF transmitters and receivers are classified based on whether the IQ modulation is performed in the analog or digital domains. Another important factor in distinguishing RF transmitters and receivers architectures is the number of stages used for frequency upconversion and downconversion, respectively.

1.2.1 RF Transmitter Architectures

Figure 1.4 presents a block diagram of the direct conversion transmitter architecture also known as homodyne or zero intermediate frequency (IF) transmitter architecture. In this system, the I and Q components of the signal to be transmitted are output by the DSP unit and converted into analog domain through two separate digital-to-analog converters (DACs). The analog I and Q signals are first lowpass-filtered and then modulated and upconverted around the carrier frequency using a single-stage analog modulator. The resulting RF signal is then fed to the power amplifier (PA) stage, and next to the transmitting antenna. A clear advantage of the homodyne transmitter is its low number of components and the simplicity of the architecture, which makes it suitable for low cost and highly integrated systems. However, the main problem of this architecture resides in the frequency planning since the local oscillator (LO) signal falls within the transmit band. Therefore, the RF to LO isolation (LO leakage to the RF port) of the mixers is critical. Moreover, any impairments (such as gain and phase imbalance, DC offset) between the I and Q branches, which are unavoidable in analog implementations, would result in unwanted frequency components at the output of the upconversion stage, which will further degrade the transmitter performance. In such case, additional algorithms are needed in the DSP block to compensate for the impairments generated in the analog domain.

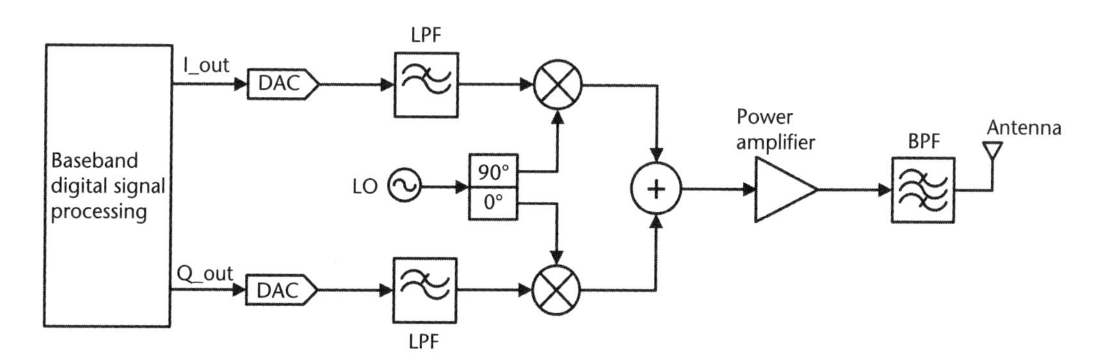

Figure 1.4 Homodyne (direct conversion or zero IF) transmitter architecture.

In order to alleviate the frequency-planning problem observed in homodyne transmitters, a two-step modulation and frequency upconversion is adopted in heterodyne transmitters. As depicted in Figure 1.5, each of the digital I and Q signals pass through a digital-to-analog converter (DAC) and a lowpass filter. The resulting analog I and Q signals are then modulated using an analog IQ modulator. However, unlike the homodyne architecture, the IQ modulator in the heterodyne architecture outputs a signal at an IF that is later frequency-upconverted around the desired RF carrier frequency. The resulting RF signal is then bandpass-filtered and fed to the amplification stage and then to the transmit antenna. A key aspect in the heterodyne transmitter is related to the selection of the IF. Selecting a lower IF will simplify the design of the IQ modulator. However, this will bring closer together the unwanted image and the wanted useful signal at the output of the second frequency upconversion stage. Typically, the IF should be high enough to allow for a feasible image rejection bandpass filtering at the output of the last frequency upconversion stage. An additional advantage of the heterodyne architecture is that it allows for the reuse of components between several designs. For example, in a multiband transmitter, the baseband signals corresponding to each frequency band can all be modulated around the same IF. This would enable the use of identical LOs, mixers, and filters for the IQ modulation stage of each band. Later, band-dependent frequency upconversion can be used to translate each IF around the desired final RF carrier frequency.

To reduce the system complexity in heterodyne transmitters, the two LO signals can be derived from the same RF synthesizer which generates a signal that is an integer multiple of each LO signal. Such constraint can be taken into account while performing the frequency planning of the heterodyne transmitter. Despite its numerous advantages over the homodyne architecture, the heterodyne architecture results in more bulky designs with higher power consumption. Also, it still suffers from the impairments due to the analog implementation of the IQ modulation.

Taking into consideration the continuous developments in terms of speed of the DACs as well as the processing capabilities of DSP and field programmable gate arrays (FPGA), it became possible in most applications to implement the IQ modulation in the digital domain. Hence, the impairments that were observed in the analog implementation of the IQ modulation can be circumvented. This leads to the digital-IF architecture reported in Figure 1.6. As shown, the IQ modulation is performed in the digital domain and the resulting signal is converted to the analog domain using a single DAC. The RF signal is obtained by frequency-upconverting of the IF signal and then bandpass filtering of the mixer's output signal. The RF signal is later fed to the amplification stage and the transmit antenna. The selection of the IF in the digital IF architecture is constrained by the speed of the DAC, and therefore aliasing effects should be taken into consideration. The same considerations mentioned for the heterodyne architecture about the bandpass filtering after the second frequency upconversion stage also apply in the digital IF architecture. Most importantly, the digital IF architecture pushes the DSP functionality closer to the amplification stage and antenna, and therefore reduces the number of functions implemented in the always imperfect analog domain. Hence, it is one step closer to the highly sought all digital transmitters. The only drawback of this architecture is the need for high-speed high dynamic range DAC.

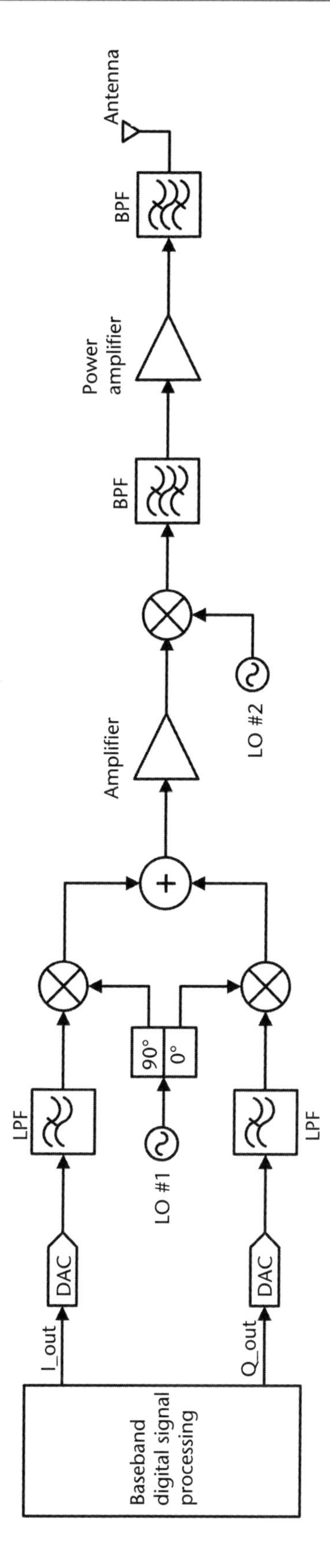

Figure 1.5 Heterodyne transmitter architecture.

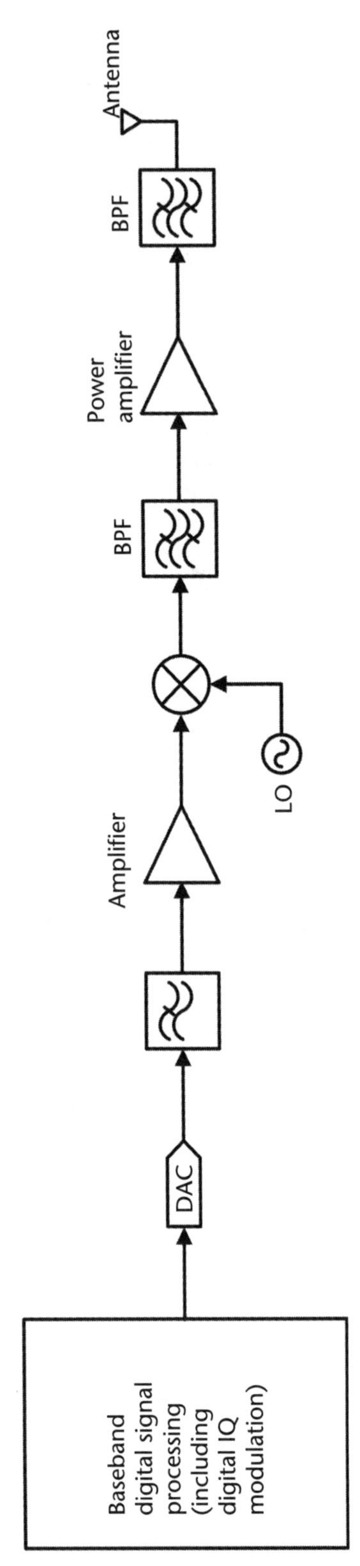

Figure 1.6 Digital IF transmitter architecture.

1.2.2 RF Receiver Architectures

Similar to the RF transmitter architectures discussed in the previous section, one can identify three major receiver architectures, namely, the homodyne (or direct conversion) receiver, the heterodyne receiver, and the digital IF receiver. The homodyne receiver, shown in Figure 1.7, is the simplest architecture in which the signal picked by the receiving antenna is bandpass-filtered and then fed to a low noise amplifier (LNA). The amplified signal is then frequency downconverted and IQ demodulated. The resulting analog I and Q signals are then lowpass-filtered before being digitized and processed in the DSP unit to extract the transmitted information.

In the homodyne architecture, a band-selection filter can be used between the antenna and the LNA. However, it is practically quasi-impossible to design a channel selection filter at the RF. Hence, the signal input to the LNA as well as the IQ demodulator will include potential in-band blocking signals and interferers. Moreover, homodyne receivers share many of the disadvantages of their transmitter counterparts such as frequency planning and the critical high-frequency implementation of analog IQ demodulators. Naturally, to address most of these limitations, one can consider the heterodyne receiver in which the frequency downconversion and analog IQ demodulation are performed in two separate stages as illustrated in Figure 1.8. The first band-selection filter is the same as the one used in the homodyne receiver. However, following the first frequency downconversion stage, a second bandpass filter can be used. A proper choice of the IF frequency will make it possible for this second bandpass filter to operate as a channel selection filter, therefore preventing in-band blockers and interferers from propagating further within the receiver lineup. Moreover, the implementation of analog IQ demodulation at lower frequency can result in better performances in terms of IQ imbalance. As mentioned in the case of heterodyne transmitter, bringing the RF signal to an IF can also be beneficial for the reuse, between different designs, of circuits and components operating at this frequency.

The digital IF receiver architecture illustrated in Figure 1.9 alleviates the problems due to the analog domain implementation of the IQ demodulation. By selecting a low enough IF at the output of the first frequency downconversion stage, and taking advantage of the capabilities of modern ADCs, it is possible to digitize the modulated IF signal and perform the IQ demodulation in the digital domain. Careful frequency planning needs to be performed when designing heterodyne and digital IF receivers to take into account the location of the unwanted signals and spurious with respect to the useful signal and to ensure that these can be filtered out.

1.2.3 Digital IF RF Transceivers

Even though conventional architectures of RF transmitters and receivers were discussed separately in the previous sections, these always coexist within the same RF transceiver system while often operating at different carrier frequencies. In fact, for most communication systems, uplink and downlink frequency bands are distinct. In Figure 1.10, the digital IF transmitter and receiver architectures are shown. Indeed, these are widely used in modern communication applications due to their superior performances. As illustrated in the block diagram of Figure 1.10, in a transceiver system, the same antenna is habitually used to transmit and receive since the uplink and downlink bands are fairly close. Hence, in full-duplex systems, a diplexer is

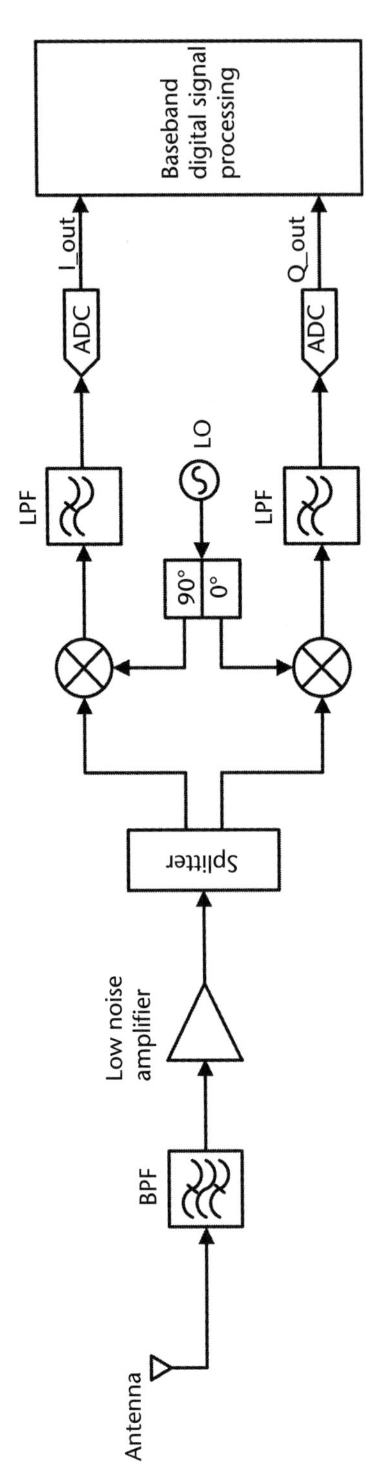

Figure 1.7 Homodyne (direct conversion, zero IF) receiver architecture.

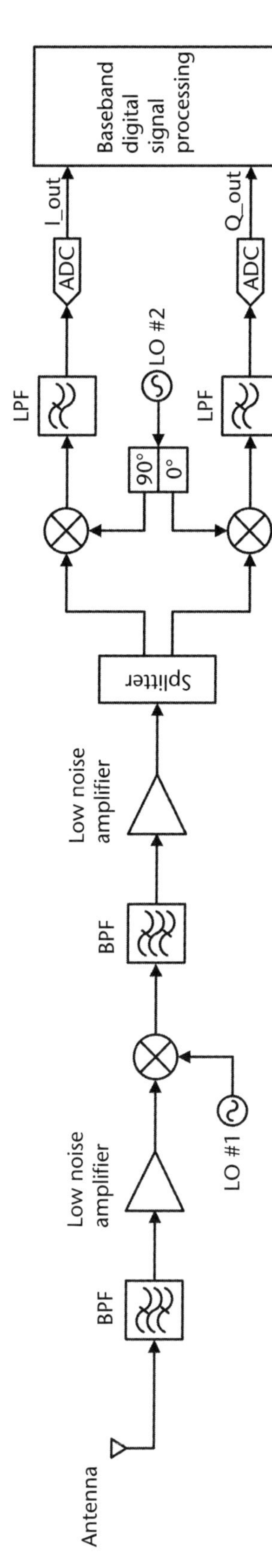

Figure 1.8 Heterodyne receiver architecture.

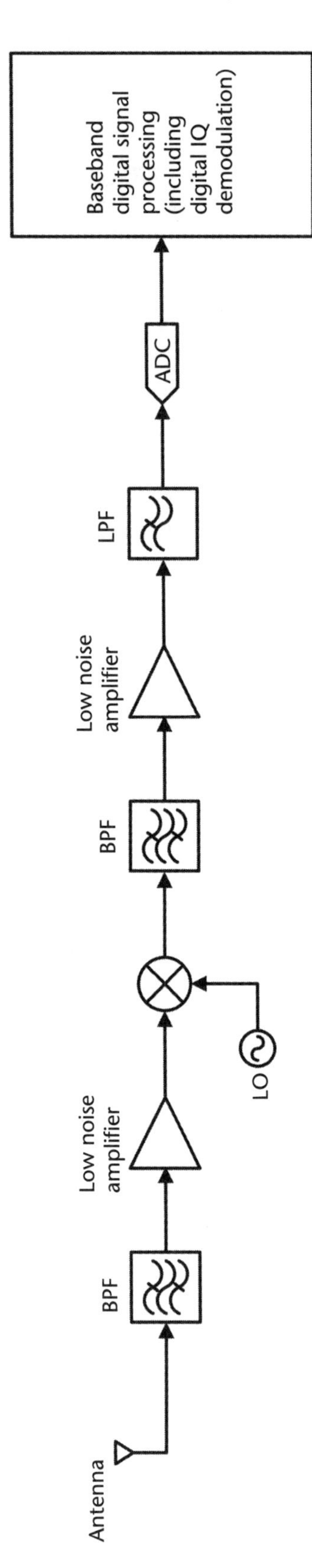

Figure 1.9 Digital IF receiver architecture.

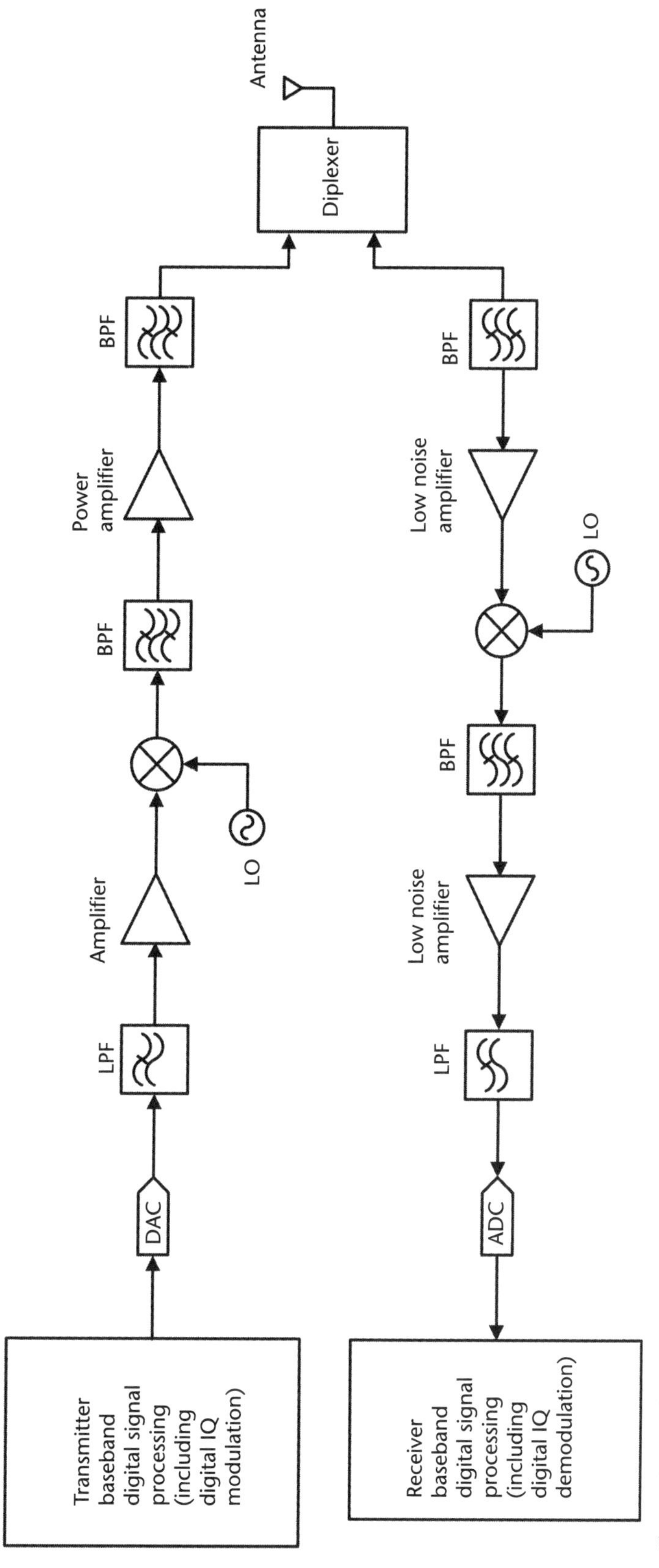

Figure 1.10 Block diagram of a digital IF transceiver.

needed in order to route the signal to be transmitted from the PA to the antenna and the received signal from the antenna to the receiver LNA based on their frequency bands. In half-duplex systems, the diplexer can be replaced by a switch since transmit and receive functions are not active concurrently.

Before concluding this section, it is important to mention that due to the unavoidable nonlinear distortions caused by the PA of the transmitter's front-end, additional circuitry is often used to compensate for these distortions in base stations mainly through digital predistortion (DPD) technique [13, 14]. This requires the addition of a feedback path that samples the signal at the output of the PA in order to provide a signal to be used in synthesizing the DPD function. Figure 1.11 illustrates the typical arrangement of additional circuits needed to implement baseband DPD based linearizers in digital IF transmitters.

1.3 Technology Trends

Classical RF transmitters and receivers architectures are unlikely to cope with the recent trends in wireless communication systems, which require high-efficiency, as well as MIMO, broadband/multiband capabilities. Communication signals are amplitude modulated with high peak-to-average power ratio (PAPR), which makes them sensitive to the distortions of the transmitter's PA. Moreover, the high PAPR of these signals forces the amplifier to operate in deep backoff. With the large proliferation in the number of base stations, power efficiency of the transmitter's amplifier is a critical issue when one aims to reduce the overall power consumption and the running costs of base stations, as well as their carbon footprint. Advanced amplification systems such as dual-input power amplification systems (such as Doherty, linear amplification using nonlinear components [LINC], and envelope-tracking PAs) are perceived as enabling technologies for high-efficiency transmitters [15]. However, the transmitter's architecture has to be built around these amplification systems. In LINC and dual-input Doherty amplifiers, two RF signals need to be fed to the power amplification stage as depicted in Figure 1.12(a). Therefore, two frequency upconversion stages are needed. Conversely, in the envelope-tracking architecture, a variable supply signal and an RF signal are input to the amplifier. Hence, the architecture is illustrated in Figure 1.12(b).

To enable multiband transmitters and receivers, the most straightforward approach is to duplicate the entire RF front-end for each operation band. A more efficient and optimized alternative is to design multiband RF amplifiers, filters, and antennas in order to minimize the number of components. Such architecture is depicted in Figure 1.13 for a multiband transmitter. Obviously, RF circuits designed to operate in multibands concurrently will unavoidably have less performance than circuits optimized for a single frequency band. However, reasonable performance degradation in multiband circuits can be tolerated given the size and cost reductions to which such circuits yield. The concept of AIAs comes to fill this requirement by providing an integrated solution that can provide close to single-frequency solutions with higher integration between the RF front-end components, namely, the antenna and the amplifier (PA or LNA) in both the transmitting and receiving paths as depicted in Figure 1.14(a, b), respectively. Such integration will provide a much

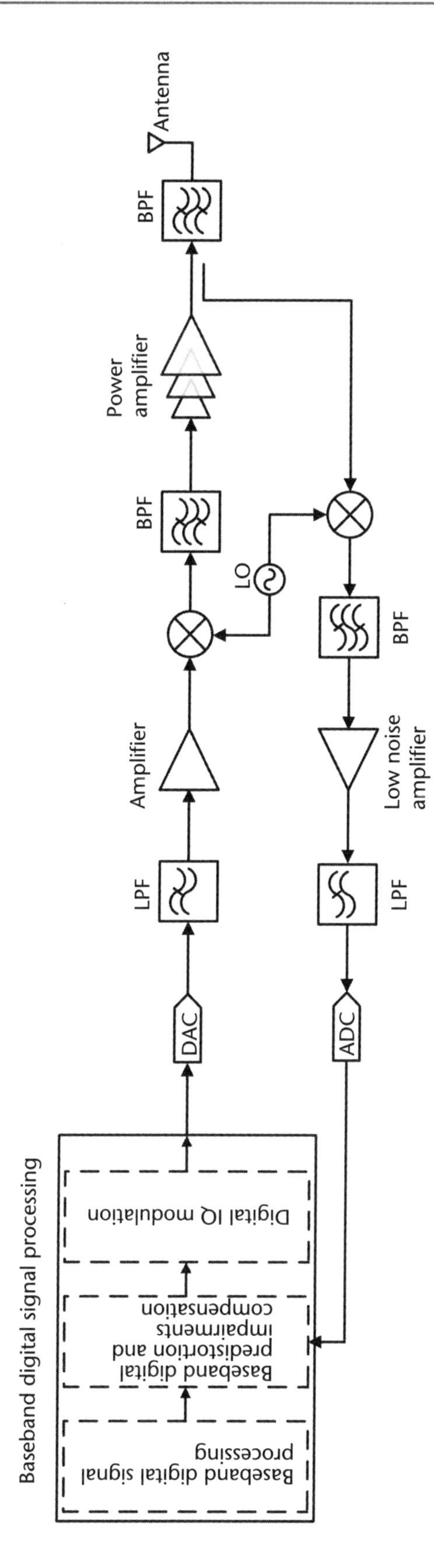

Figure 1.11 Digital IF transmitter with DPD.

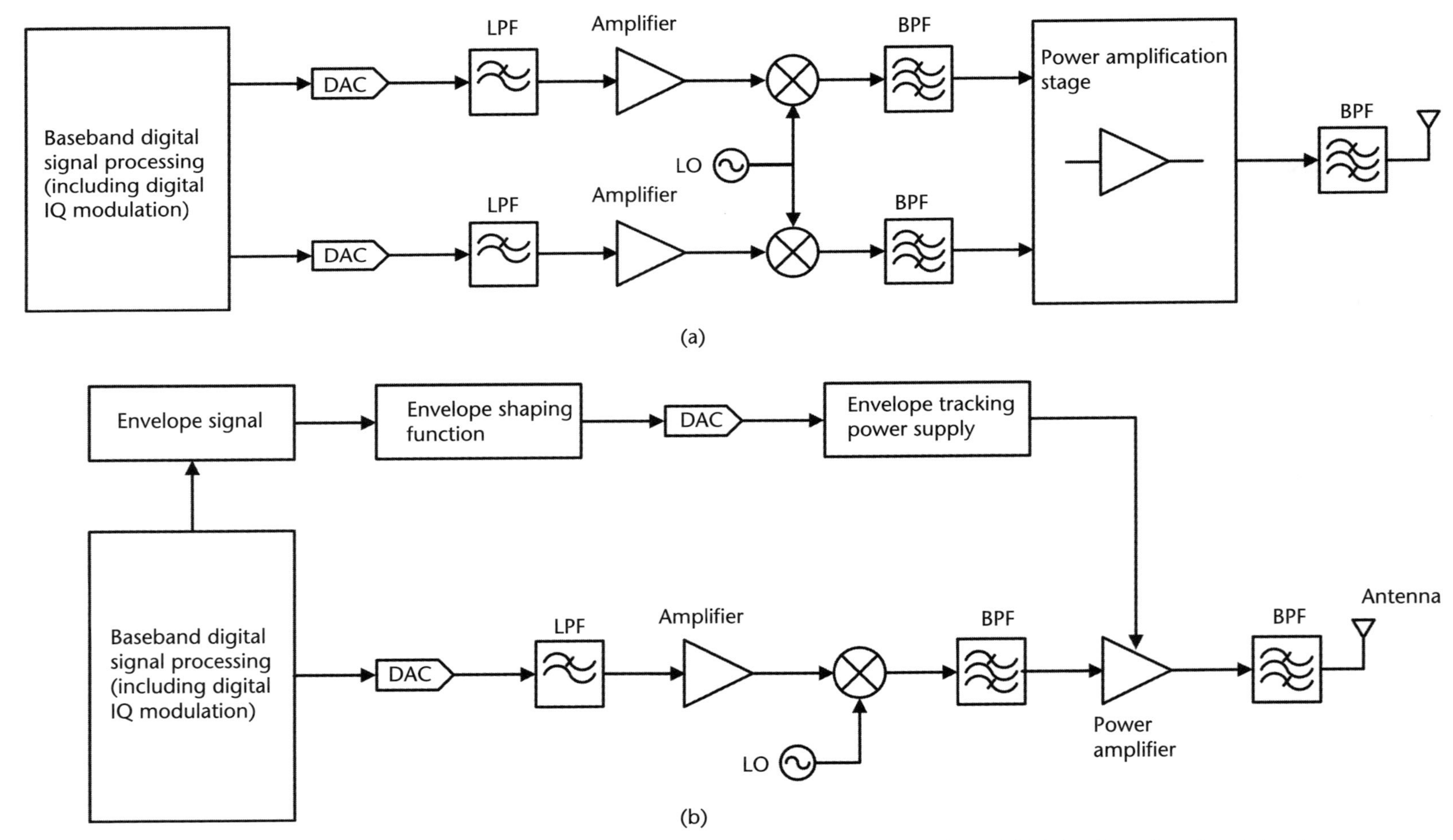

Figure 1.12 Digital IF transmitter for dual-input power amplification systems: (a) dual-RF input signals, and (b) RF and envelope input signals.

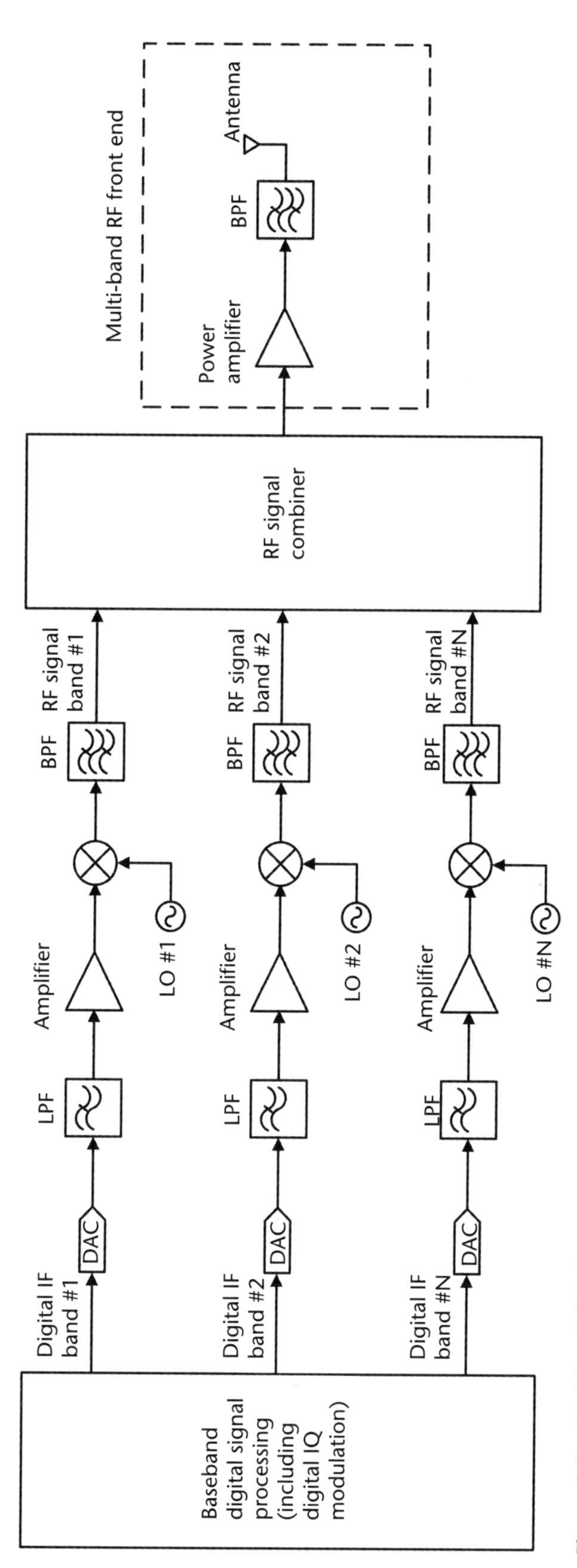

Figure 1.13 Multiband digital IF transmitter.

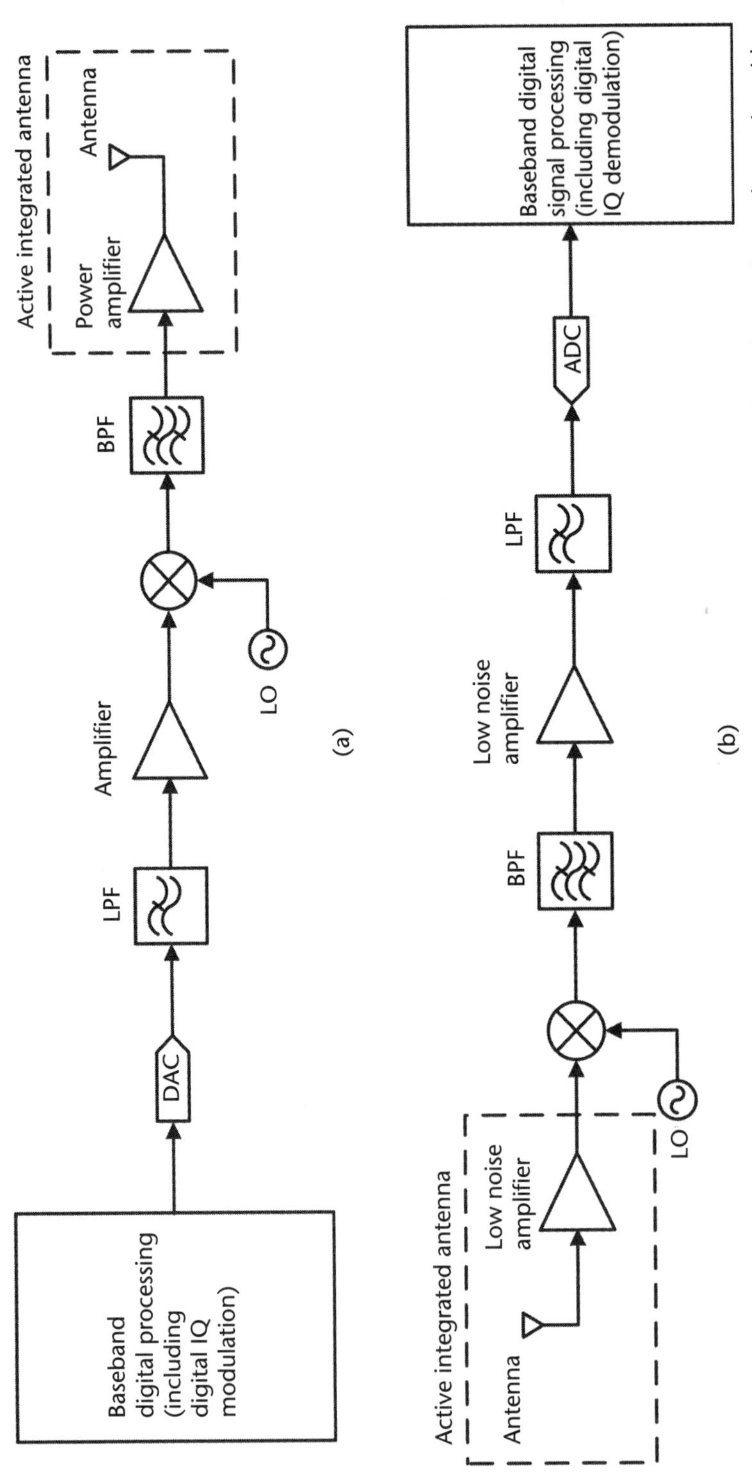

Figure 1.14 AIA block diagram. (a) PA and antenna integration at the transmitter side. (b) Antenna and LNA integration at the receiver side.

smaller design footprint and can relax the matching and bandwidth requirements when utilized properly. The concept of AIAs is currently being deployed in future wireless devices to provide more capabilities such as beam-switching and steering to provide better communication links for the users and allow for higher data rates for beyond 4G architectures.

With the recent interest in millimeter-wave frequencies, which are being considered for the upcoming 5G communication systems, the significant decrease in the wavelength enables MIMO arrays with large number of antenna elements. Consequently, highly directional antennas can be synthesized by implementing beamforming in either the analog or in the digital domain. Also, the concept of AIAs will play a major role in 5G as the antennas will be designed and placed closer to the active circuits on the chip or on the package, allowing for higher integration and more compact sizes.

1.4 Conclusions

In this chapter, the evolution of wireless communication systems was revisited to highlight the key challenges set on modern systems. Then typical RF transmitter and receiver architectures were introduced and discussed in terms of their advantages and drawbacks. Finally, technology trends and their impact on the transceiver architectures were discussed. Throughout this introductory chapter, it appears that despite of all the variations of transmitters and receiver architectures, the amplifier and the antenna are indissociable blocks that are closely placed. Therefore, it is important to codesign these two elements and integrate them to further enhance the performance of transmitters and receivers, thus highlighting the importance of the AIA concept, which is the focus of this book.

References

[1] Yang, S. C., *OFDM System Analysis and Design*, Norwood, MA: Artech House, 2010.

[2] Rohling, H., and D. Galda, "OFDM Transmission Technique: A Strong Candidate for Next Generation Mobile Communications," *The URSI Radio Science Bulletin*, Vol. 2004, No. 310, September 2004, pp. 47–58.

[3] Zaidi, A. A., et al., "Waveform and Numerology to Support 5G Services and Requirements," *IEEE Communications Magazine*, Vol. 54, No. 11, November 2016, pp. 90–98.

[4] Winters, J. H., "On the Capacity of Radio Communication Systems with Diversity in a Rayleigh Fading Environment," *IEEE Journal on Selected Areas in Communications*, Vol. SAC-5, No. 5, June 1987, pp. 871–878.

[5] Rusek, F., et al., "Scaling Up MIMO: Opportunities and Challenges with Very Large Arrays," *IEEE Signal Processing Magazine*, Vol. 30, No. 1, January 2013, pp. 40–60.

[6] Larsson, E. G., et al., "Massive MIMO for Next Generation Wireless Systems," *IEEE Communications Magazine*, Vol. 52, No. 2, February 2014, pp. 186–195.

[7] Sharawi, M. S., *Printed MIMO Antenna Engineering*, Norwood, MA: Artech House, 2014.

[8] Yuan, G., et al., "Carrier Aggregation for LTE-Advanced Mobile Communication Systems," *IEEE Communications Magazine*, Vol. 48, No. 2, February 2010, pp. 88–93.

[9] Pedersen, K. I., et al., "Carrier Aggregation for LTE-Advanced: Functionality and Performance Aspects," *IEEE Communications Magazine*, Vol. 49, No. 6, June 2011, pp. 89–95.

[10] Shen, Z., et al., "Overview of 3GPP LTE-Advanced Carrier Aggregation for 4G Wireless Communications," *IEEE Communications Magazine*, Vol. 50, No. 2, February 2012, pp. 122–130.

[11] Lee, S., et al., "The Useful Impact of Carrier Aggregation: A Measurement Study in South Korea for Commercial LTE-Advanced Networks," *IEEE Vehicular Technology Magazine*, Vol. 12, No. 1, March 2017, pp. 55–62.

[12] 3GPP TS 36.141 v14.1.0, Evolved Universal Terrestrial Radio Access (E-UTRA); Base Station (BS) Conformance Testing (Release 14), 3GPP, Technical Report, September 2016.

[13] Wood, J., *Behavioral Modeling and Linearization of RF Power Amplifiers*, Norwood, MA: Artech House, 2014.

[14] Ghannouchi, F. M., O. Hammi, and M. Helaoui, *Behavioral Modeling and Predistortion of Wideband Wireless Transmitters*, New York: John Wiley & Sons, 2015.

[15] Cripps, S. C., *RF Power Amplifiers for Wireless Communications*, Second Edition, Norwood, MA: Artech House, 2006.

CHAPTER 2

Impedance Matching Methods

Impedance matching is a very important concept by which engineers ensure that a generator is presenting the appropriate impedance value as seen by the loading circuit, or, equivalently, the load presents a specific impedance value that is required by the driving circuit in order to ensure that a certain set of performance metrics are met. Commonly, impedance matching is performed to guarantee maximum power transfer between two circuits. This is known as the conjugate matching condition. However, as will be discussed in the upcoming chapters, a conjugate match is not always needed or desired. For example, in Chapter 3, we will see that, in amplifier design, specific nonconjugate impedance matching must be performed at the input of a transistor to achieve certain noise performance.

In the field of active integrated antennas (AIAs), impedance matching is of great importance as well. Indeed, the rationale of AIA systems is to codesign the antenna and the amplifier (or other active element) as a single system without having to match each to the standard 50Ω impedance. Hence, matching networks are a core part of the AIA design process.

In this chapter, the basics of impedance matching will be reviewed. Narrowband matching using lumped and distributed elements will be discussed along with some standard matching network topologies. After that, wideband matching techniques employing constant Q-circles, real frequency, and non-Foster techniques will be introduced. Several theoretical aspects are supplemented by detailed examples. An example illustrating the use of a commercial software package (using Microwave Office [MWO]) in the design of a matching network is given at the end of this chapter.

2.1 Introduction to Impedance Matching

At radio and microwave frequencies, the wavelength of the signal, which corresponds to its spatial period, approaches the same order of magnitude as the physical size of the circuit. In free space, the signal wavelength (λ_0) is related to its frequency (f) according to

$$\lambda_0 = \frac{c}{f} \tag{2.1}$$

where $c = 3 \times 10^8$ m/s is the speed of light in free space. Within a dielectric material having a dielectric constant denoted by ε_r, (2.1) becomes

$$\lambda = \frac{c}{f\sqrt{\varepsilon_r}} = \frac{\lambda_0}{\sqrt{\varepsilon_r}} \tag{2.2}$$

where $\varepsilon_r \approx 1$ in air. However, for all other propagation media, $\varepsilon_r > 1$. Hence, the wavelength (λ) in any medium (referred to as the guided wavelength) will always be shorter than the free space wavelength (λ_0).

According to (2.1), at an operating frequency of 3 GHz, the free-space wavelength is 10 cm, which is in the same range as the physical dimensions of the circuit, especially in the widely adopted microwave integrated circuit (MIC) technology. Therefore, transmission line concepts are used for the analysis of such circuits where the signal is perceived as a traveling wave that can be incident and/or reflected. We assume here that the reader is familiar with basic concepts of transmission lines theory thoroughly covered in [1–3].

Some useful definitions are provided here for convenience. Standard notations will be used for the impedance (Z) and the admittance (Y). Their normalized values z and y, respectively, are given by

$$\begin{cases} z = \dfrac{Z}{Z_0} \\ y = Y \cdot Z_0 \end{cases} \tag{2.3}$$

Throughout this chapter, the characteristic impedance of the systems (Z_0) is considered to be 50Ω.

The reflection coefficient (Γ) associated with an impedance (Z) is defined as

$$\Gamma = \frac{Z - Z_0}{Z + Z_0} = \frac{z - 1}{z + 1} \tag{2.4}$$

Using (2.4), one can relate the impedance (Z) to the corresponding reflection coefficient (Γ) through

$$Z = Z_0 \cdot \frac{1 + \Gamma}{1 - \Gamma} \tag{2.5}$$

As illustrated in Figure 2.1, impedance matching consists of designing a matching network that will change the impedance value seen at reference plane A into a

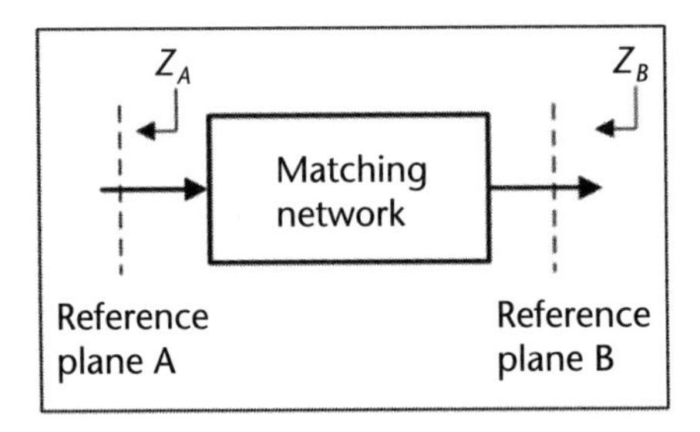

Figure 2.1 Illustration of the impedance matching concept.

desired impedance at reference plane B. These matching networks can be designed using either reactive lumped components (inductors or capacitors) or transmission line sections, mainly microstrip transmission lines for the applications considered in this book. Hence, such matching networks can be considered lossless.

In amplifier design, two matching networks are needed as shown in Figure 2.2. An input matching network, placed between the source and the active element, is needed to change the source impedance from 50Ω to the value Z_S. The output matching network, between the transistor and its load, is designed to change the load impedance from a typical 50Ω value into an impedance Z_L. The theory behind the selection of the values of Z_S and Z_L (or, equivalently, their corresponding reflection coefficients Γ_S and Γ_L) in amplifier design is thoroughly discussed in Chapter 3. In this chapter, the focus is on the internal design of the matching network in order to change an initial impedance Z_A, at reference plane A, into a final impedance Z_B, at reference plane B. The type of the elements used to build the matching network as well as their arrangements will be discussed.

2.2 Narrowband Matching

The use of reactive elements to build matching networks allows for the synthesis of lossless networks. However, this inevitably introduces a frequency response and hence limits the bandwidth of the matching network. In this section, narrowband matching networks will be covered. This implies that the frequency response of the synthesized network is not a prime concern, and the focus is only on the performance at a single frequency. Matching networks can be built using either lumped components or transmission lines. Lumped elements are easy to use, but their frequency of application is limited to several gigahertz. Moreover, at low frequencies, transmission line-based matching networks are usually larger in size when compared to lumped element-based networks. Indeed, as described by (2.1), the wavelength of a signal is inversely proportional to the frequency of operation. Therefore, the dimensions of transmission lines increase as the frequency of operation decreases. However, at higher frequencies, parasitic effects of lumped components become too cumbersome, and transmission line-based matching networks are more compact, and hence more suitable. Furthermore, the use of lumped elements is also limited by their power handling. Thus, for high-power applications, transmission lines are often preferred.

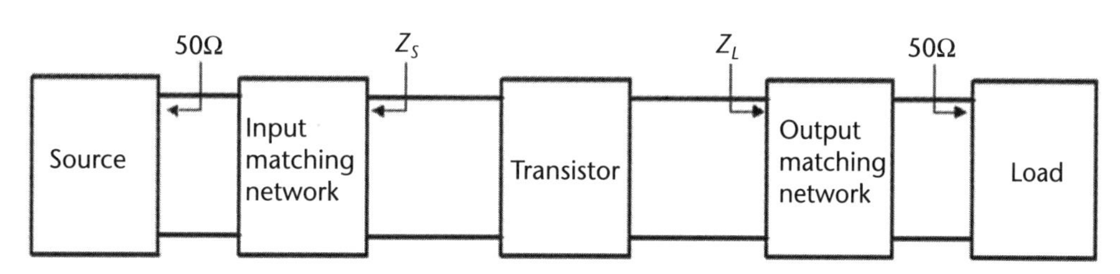

Figure 2.2 Impedance matching concept for amplifier design.

2.2.1 Lumped Element Matching Using the L-Network

By adding two elements (one connected in series and one in shunt), it is possible to change the value of any impedance into any desired impedance value at a single frequency of operation. Using reactive elements, one can distinguish eight possible configurations for two-element-based matching networks. The topologies illustrated in Figure 2.3 are also known as L-networks.

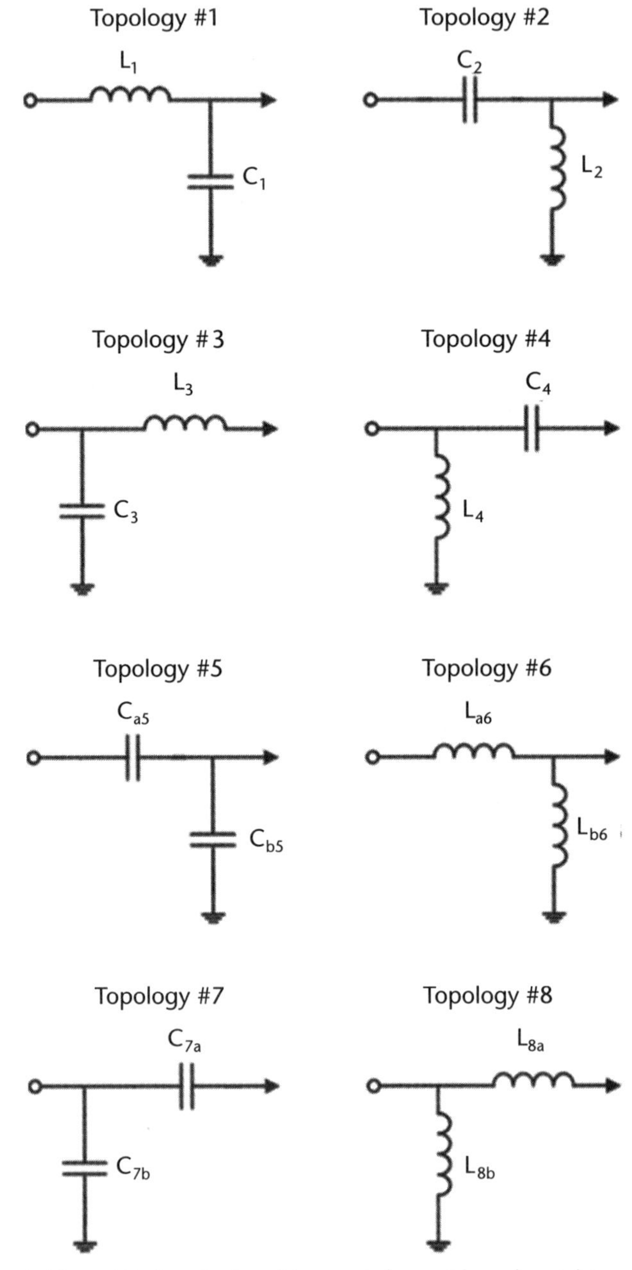

Figure 2.3 The eight possible topologies of lumped element-based matching using the L-network.

One major advantage of the L-network topologies is the possibility to analytically calculate the values of the lumped components that constitute the matching network [2–4]. Let's consider the standard schematic shown in Figure 2.4(a) in which an L-network is used to match an impedance Z_s to a load impedance Z_L. Unless otherwise specified, this schematic will be used throughout this section discussing L-network matching circuits. The eight topologies of Figure 2.3 can be classified into two configurations: the series-shunt configuration as illustrated in Figure 2.4(b) and the shunt-series configuration as shown in Figure 2.4(c).

For the circuit schematic of Figure 2.4, assuming that the matching network can have any of the eight topologies depicted in Figure 2.3, the values of the components

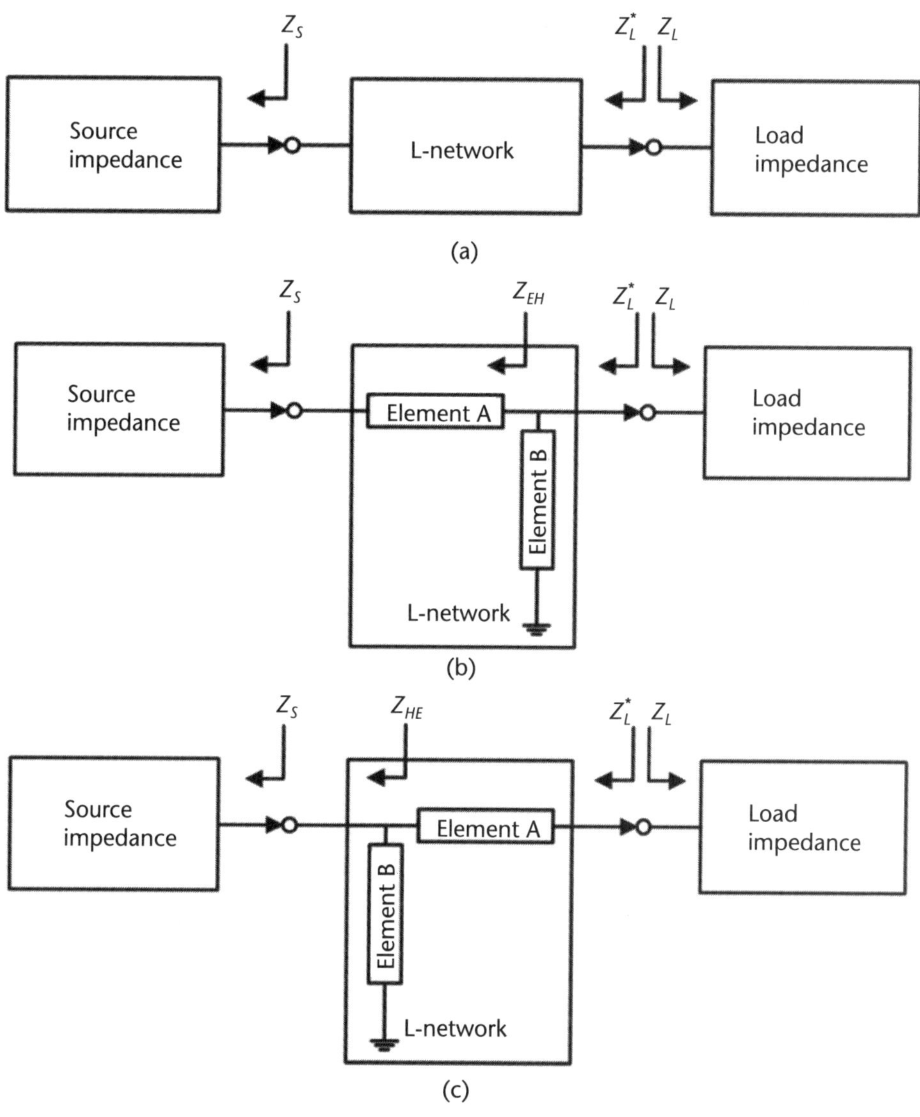

Figure 2.4 Circuit schematic used to calculate the values of lumped elements in L-network. (a) Generic schematic, (b) series-shunt configuration, and (c) shunt-series configuration.

needed in each of the eight L-networks can be calculated using (2.7) through (2.14). These analytical solutions are derived for Z_S and Z_L defined as [4]

$$\begin{aligned} Z_S &= R_S \\ Z_L &= R_L + jX_L \\ Y_L &= G_L + jB_L = \frac{R_L}{R_L^2 + X_L^2} - j\frac{X_L}{R_L^2 + X_L^2} \end{aligned} \tag{2.6}$$

$$\begin{cases} L_1 = \dfrac{A}{2\pi f\left(G_L^2 + A^2\right)} \\ C_1 = \dfrac{A - B_L}{2\pi f} \end{cases} \tag{2.7}$$

$$\begin{cases} L_2 = \dfrac{1}{2\pi f\left(A + B_L\right)} \\ C_2 = \dfrac{G_L^2 + A^2}{2\pi f A} \end{cases} \tag{2.8}$$

$$\begin{cases} L_3 = \dfrac{B}{2\pi f} \\ C_3 = \dfrac{B + X_L}{2\pi f\left[R_L^2 + \left(B + X_L\right)^2\right]} \end{cases} \tag{2.9}$$

$$\begin{cases} L_4 = -\dfrac{\left[R_L^2 + \left(C + X_L\right)^2\right]}{2\pi f\left(C + X_L\right)} \\ C_4 = \dfrac{-1}{2\pi f C} \end{cases} \tag{2.10}$$

$$\begin{cases} C_{5a} = \dfrac{G_L^2 + A^2}{2\pi f A} \\ C_{5b} = -\dfrac{A + B_L}{2\pi f} \end{cases} \tag{2.11}$$

$$\begin{cases} L_{6a} = \dfrac{A}{2\pi f\left(G_L^2 + A^2\right)} \\ L_{6b} = -\dfrac{1}{2\pi f\left(A - B_L\right)} \end{cases} \tag{2.12}$$

$$\begin{cases} C_{7a} = -\dfrac{1}{2\pi f B} \\ C_{7b} = \dfrac{B + X_L}{2\pi f\left[R_L^2 + \left(B + X_L\right)^2\right]} \end{cases} \tag{2.13}$$

$$\begin{cases} L_{8a} = \dfrac{C}{2\pi f} \\ L_{8b} = -\dfrac{\left[R_L^2 + \left(C + X_L\right)^2\right]}{2\pi f\left(C + X_L\right)} \end{cases} \tag{2.14}$$

where

$$A = \sqrt{\frac{G_L}{R_S} - G_L^2} \tag{2.15}$$

$$B = \sqrt{R_L\left(R_S - R_L\right)} - X_L \tag{2.16}$$

$$C = -\sqrt{R_L\left(R_S - R_L\right)} - X_L \tag{2.17}$$

At this point, two important facts need to be highlighted. First, some of the values predicted by the previous set of equations might not be realistic for implementation (i.e., negative values, complex values, or positive values, but too large or too small to be realized practically). Moreover, each of the eight L-network topologies has a limited range of applications in the sense that a given L-network cannot match any two random impedances in the Smith chart [4, 5]. Hence, for a given value of the impedances to be matched, only a limited set of L-network topologies can be adopted. One additional factor that can influence the selection of a specific L-network topology is its frequency response. Indeed, while some topologies have a lowpass frequency response, others behave as high-pass networks. Typically, software utilities such as the Smith chart and impedance matching in Keysight's Advanced Design System (ADS) software or the iMatch wizard in MWO from AWR software are handy tools that allow the user to automatically synthesize an L-network matching circuit and visualize its frequency response.

A graphical approach using the Smith chart can also be employed to synthesize an L-network matching circuit. Here, it is important to classify the eight L-network topologies according to the two configurations described in Figure 2.4. Thus, topologies 1, 2, 5, and 6 are all built using the series-shunt configuration. However, topologies 3, 4, 7, and 8 have an arrangement similar to the shunt-series configuration. For the series-shunt configuration, it is anticipated that the first reactive component, which is connected in series with the impedance Z_0, will result in a displacement along the constant resistance circle $R = \mathrm{Re}(Z_0)$, and that the second reactive component, which is connected in shunt, will result in a displacement along a constant conductance circle $G = \mathrm{Re}(1/Z^*_L)$ that passes through the final impedance Z^*_L. Hence, the impedance Z_{EH} seen after element A in Figure 2.4(b) is such that

$$\begin{cases} \mathrm{Re}(Z_{EH}) = \mathrm{Re}(Z_0) = R_0 \\ \text{and} \\ \mathrm{Re}\left(\dfrac{1}{Z_{EH}}\right) = \mathrm{Re}\left(\dfrac{1}{Z^*_L}\right) \end{cases} \tag{2.18}$$

where Re(Z) refers to the real part of the complex number Z.

Accordingly, the impedance Z_{EH} corresponds to any of the intersection points of the two circles defined by (2.18). Once the impedance Z_{EH} is selected, the values of the components can be determined by reading their normalized reactance and/or susceptance value from the Smith chart.

Similarly, in the shunt-series configuration, the shunt component will change the value of the impedance Z_0 into the impedance Z_{HE} by moving on a constant conductance circle, while the second component (connected in series) will change the impedance Z_{HE} into the impedance Z^*_L by moving along a constant resistance circle. Hence, the impedance Z_{HE} must satisfy

$$\begin{cases} \mathrm{Re}\left(\dfrac{1}{Z_{HE}}\right) = \mathrm{Re}\left(\dfrac{1}{Z_0}\right) = \dfrac{1}{R_0} \\ \text{and} \\ \mathrm{Re}(Z_{HE}) = \mathrm{Re}(Z^*_L) \end{cases} \tag{2.19}$$

Therefore, the impedance Z_{HE} lies on the intersection of the two circles defined by (2.19). Knowing the values of Z_{HE}, one can determine the values of the two reactive components using the Smith chart.

2.2.2 Example of the Lumped Element L-Matching Method

To illustrate the design of L-networks using the analytical and graphical approaches, let's consider the design of a matching network that matches a source impedance $Z_0 = 50\Omega$ to a load impedance $Z_L = 20 - j40$ at a frequency of operation $f = 2$ GHz.

- *Analytical approach:* Using (2.7) to (2.17), it appears that among the eight L-network topologies of Figure 2.3, only four topologies can be used to design this matching network. These are topologies 2, 3, 6, and 8. The values of the components needed for each topology are summarized in Table 2.1.
- *Graphical approach:* First, we locate the starting impedance $Z_0 = 50\Omega$, and the final impedance $Z_f = Z^*_L = (20 + j40)\Omega$. Their normalized values on the Smith chart are $z_0 = 1\Omega$ and $z_f = (0.4 + j0.8)\Omega$, respectively. These impedances are shown in Figure 2.5. The admittance y_f corresponding to z_f is $y_f = (0.5 - j1)$S.

Table 2.1 Theoretical Calculations of the Components' Values for the Possible L-Network Topologies Able to Match $Z_0 = 50\Omega$ to $Z_L = 20 - j40$

	Component Name	*Component Value*
Topology 2 (series C, shunt L)	L_2	2.65 nH
	C_2	1.59 pF
Topology 3 (shunt C, series L)	L_3	5.13 nH
	C_3	1.95 pF
Topology 6 (series L, shunt L)	L_{6a}	3.98 nH
	L_{6b}	7.96 nH
Topology 8 (shunt L, series L)	L_{8a}	1.23 nH
	L_{8b}	3.25 nH

As one can observe, it is possible to use either the series-shunt configuration or the shunt-series configuration to build the desired matching network. When considering the series-shunt configuration, z_{EH} can be any of the two points at which the circles $r = r_0 = 1\Omega$ and $g = g_f = 0.5S$ intersect. The normalized impedances at these two points are $z_{EH1} = (1 + j1)\Omega$ and $z_{EH2} = (1 - j1)\Omega$, which correspond to normalized admittances $y_{EH1} = (0.5 - j0.5)S$ and $y_{EH2} = (0.5 + j0.5)S$, respectively.

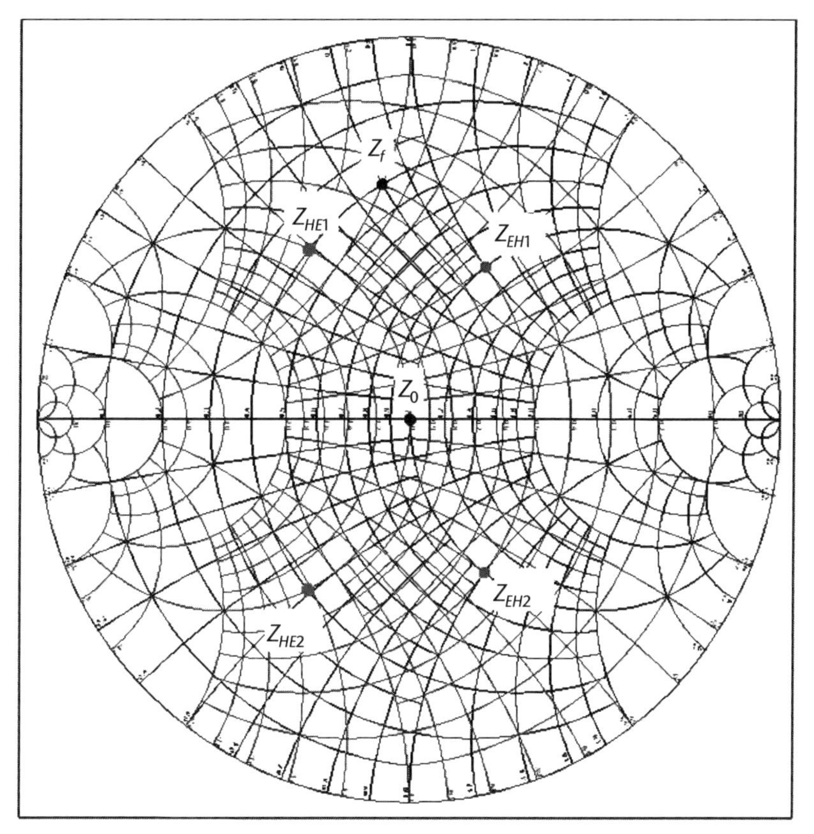

Figure 2.5 Smith chart-based approach for calculating the values of the L-network components.

Conversely, when the shunt-series configuration is adopted, z_{HE} can be any of the two points at which the circles $g = g_0 = 1S$ and $r = r_f = 0.4\Omega$ intersect. The coordinates of these two points are $z_{EH1} = (0.4 + j0.49)\Omega$, and $z_{EH2} = (0.4 - j0.49)\Omega$, which correspond to admittances $y_{HE1} = (1 - j1.22)S$ and $y_{HE2} = (1 + j1.22)S$, respectively. The details about the calculations of the components values based on the Smith chart approach are summarized in Table 2.2.

2.2.3 Quality Factor of a Matching Network

The quality factor of a matching network is an important metric that quantifies its bandwidth and frequency selectivity. While in some applications wideband matching networks are sought to accommodate wideband signals or multiband operation, highly selective narrowband matching networks are useful when filtering capabilities are needed within the matching network. The loaded quality factor (Q_L) of a network relates to its bandwidth through

$$Q_L = \frac{f_0}{BW_{3\text{dB}}} \tag{2.20}$$

where f_0 and $BW_{3\text{dB}}$ are the center frequency and the 3-dB bandwidth of the network, respectively [5].

In an L-matching network, the quality factor (Q) is related to the loaded quality factor (Q_L) through [5]

Table 2.2 Smith Chart-Based Estimation of the Components' Values for the Possible L-Network Topologies Able to Match $Z_0 = 50\Omega$ to $Z_L = (20 - j40)\Omega$

	Series-Shunt Configuration		*Shunt-Series Configuration*	
Intermediate point	Z_{EH1}	Z_{EH2}	Z_{HE1}	Z_{HE2}
Normalized impedance at intermediate point	$(1 + j1)\Omega$	$(1 - j1)\Omega$	$(0.4 + j0.49)\Omega$	$(0.4 - j0.49)\Omega$
Impedance at intermediate point	$(50 + j50)\Omega$	$(50 - j50)\Omega$	$(20 + j24.5)\Omega$	$(20 - j24.5)\Omega$
Normalized admittance at intermediate point	$(0.5 - j0.5)$S	$(0.5 + j0.5)$S	$(1 - j1.22)$S	$(1 + j1.22)$S
Admittance at intermediate point	$(10 - j10)$ mS	$(10 + j10)$ mS	$(20 - j24.4)$ mS	$(20 + j24.4)$ mS
Series element reactance	$j50\Omega$	$-j50\Omega$	$j15.5\Omega$	$j64.5\Omega$
Type and value of series element	Inductor $L = 3.98$ nH	Capacitor $C = 1.59$ pF	Inductor $L = 1.23$ nH	Inductor $L = 5.13$ nH
Parallel element susceptance	$-j10$ mS	$-j30$ mS	$-j24.4$ mS	$j24.4$ mS
Type and value of parallel element	Inductor $L = 7.96$ nH	Inductor $L = 2.65$ nH	Inductor $L = 3.26$ nH	Capacitor $C = 1.94$ pF
Corresponding topology (using theoretical approach)	Topology 6	Topology 2	Topology 8	Topology 3

$$Q = 2 \cdot Q_L \tag{2.21}$$

To assess the quality factor (Q) of an L-matching network, we will consider the case of a purely resistive source and load impedances. Hence assuming that in Figure 2.4, $Z_S = R_S$ and $Z_L = R_L$. From the analytical solutions of the L-network component values ([2.7] to [2.14]) as well as the Smith chart based graphical approach, we can conclude that the series-shunt configuration of Figure 2.4(b) can only be used if $R_S < R_L$ while the shunt-series configuration of Figure 2.4(c) can only be used for cases where $R_S > R_L$. More specifically, for purely resistive source and load impedances, only L-network topologies 1 and 2 can be used when $R_S < R_L$, and only topologies 3 and 4 can be selected when $R_S > R_L$.

The quality factor of a two-element matching network arranged according to the series-shunt configuration (Q_{EH}) is [3]

$$Q_{EH} = \sqrt{\frac{R_S}{R_L} - 1} \tag{2.22}$$

Similarly, the quality factor of a two-element matching network using the shunt-series configuration (Q_{HE}) is [3]

$$Q_{HE} = \sqrt{\frac{R_L}{R_S} - 1} \tag{2.23}$$

Equations (2.22) and (2.23) can be rewritten in a unified manner to express the quality factor the L-network as

$$Q = \sqrt{\frac{\max(R_S, R_L)}{\min(R_S, R_L)} - 1} \tag{2.24}$$

The values of the reactances of elements A (connected in series) and B (connected in shunt), as illustrated in Figure 2.4(a, b), are [3]

$$X_A = \pm\min(R_S, R_L) \times Q \tag{2.25}$$

and

$$X_B = \frac{\mp\max(R_S, R_L)}{Q} \tag{2.26}$$

The $\pm$ and $\mp$ signs in (2.25) and (2.26) are to reflect that the two elements A and B cannot be of the same type. One is an inductor and the other is a capacitor. Hence, using (2.25) and (2.26), one can calculate the values of the inductor and capacitor in the L-network needed to match a resistive source to a resistive load.

It is important to note that the constraint imposed earlier on the resistive nature of the impedances Z_S and Z_L is not restricting the validity of the analysis performed within this section. The case of complex source and load impedances can be dealt with by generalizing the case of purely resistive impedances using either the absorption or resonance approaches [3]. In the absorption-based approach, the reactive parts of the source and/or load impedance are used as part of the matching network. Conversely, in the resonance based approach, an additional reactive component is used to resonate with the reactive part of the source and/or the load impedance.

2.2.4 Lumped Element Matching Using the T-Networks and π-Networks

A closer look at the value of the quality factor in an L-network as defined by (2.24) shows that its value is independent of the matching network, and is only function of the source and load resistances. The use of two elements in the matching network allows for only two degrees of freedom, which are used to match the complex source impedance to the complex load impedance. In order to control the quality factor of the matching network, an additional degree of freedom is required. This is achieved by using three-element networks, which can be either of the T or π types. The T and π matching networks are illustrated in Figure 2.6. As shown, the T-network can be described as a combination of two series-shunt L-networks, whereas the π-network is a combination of two shunt-series L-networks. According to this figure, the shunt reactance X_{TB} in the T-network is the equivalent of the reactances X_{TB1} and X_{TB2} of its expanded version. Similarly, the series reactance $X_{\pi A}$ in the π-network is the equivalent of the reactances $X_{\pi A1}$ and $X_{\pi A2}$ shown in its expanded version.

To discuss the design procedure of the T-networks and the π-networks, we will consider without loss of generality that the source and load impedances are purely resistive ($Z_S = R_S$ and $Z_L = R_L$, respectively). As discussed earlier for the case of the L-networks, the analysis can be extended to the case of complex impedances using the absorption and or resonance concepts. Moreover, for the purpose of this analysis, the expanded versions of the T-networks and π-networks will be used. A straightforward design methodology consists of placing a virtual resistive load at the junction connecting the two L-networks as illustrated in Figure 2.7.

In the T-network, the L-networks connecting the virtual resistance R_T to R_S and R_L from the left and right sides, respectively, are of the series-shunt configuration. Hence, R_T must satisfy

$$R_T > \max\left(R_S, R_L\right) \tag{2.27}$$

Moreover, according to (2.24), the quality factor of the left-side L-network ($Q_{T\text{Left}}$) made of the reactances X_{TA1} and X_{TB1} is

$$Q_{T\text{Left}} = \sqrt{\frac{R_T}{R_S} - 1} \tag{2.28}$$

In the same way, the quality factor of the right-side L-network ($Q_{T\text{Right}}$) made of the reactances X_{TB2} and X_{TA2} is

$$Q_{T\text{Right}} = \sqrt{\frac{R_T}{R_L} - 1} \tag{2.29}$$

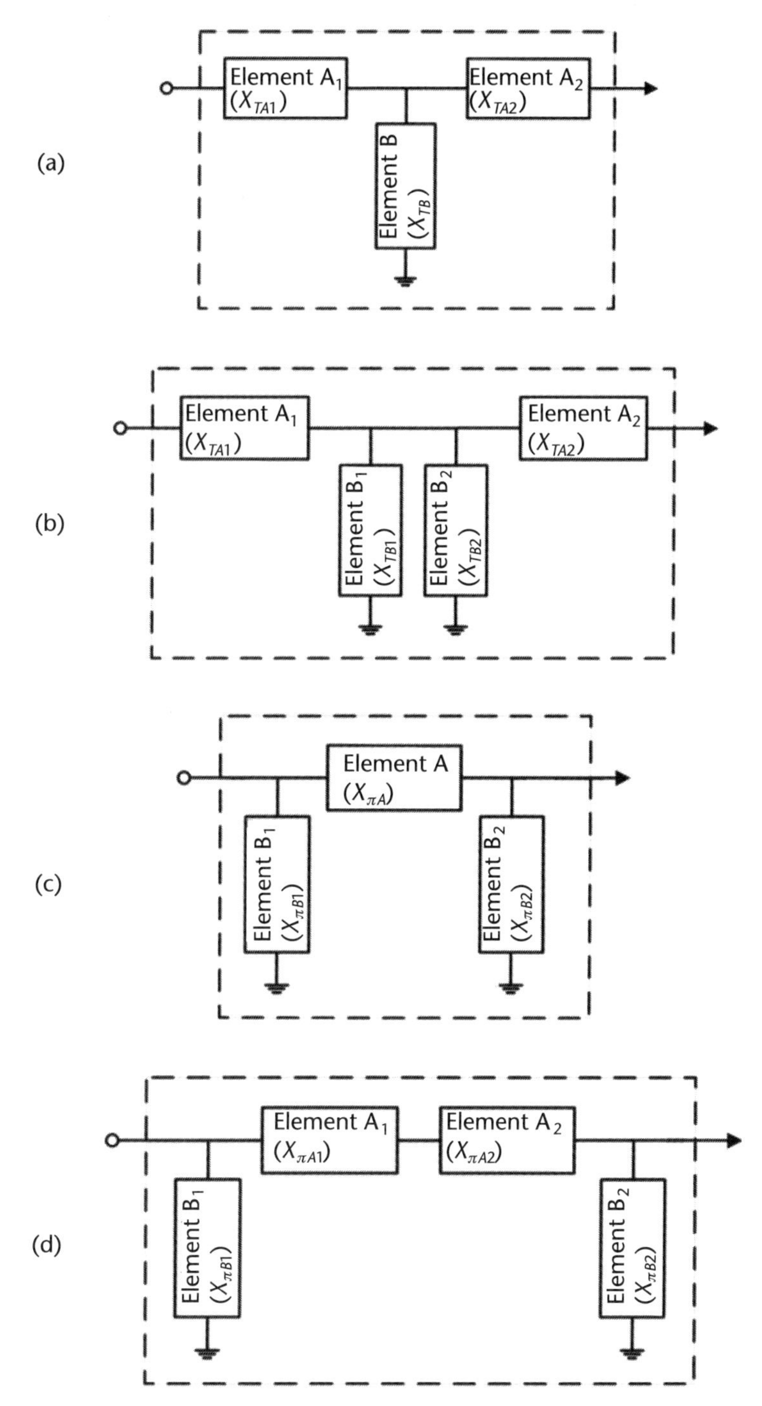

Figure 2.6 Schematic of the T-networks and π-networks. (a) T-network circuit, (b) T-network circuit expanded representation as two series-shunt L-networks, (c) π-network circuit, and (d) π-network circuit expanded representation as two shunt-series L-networks.

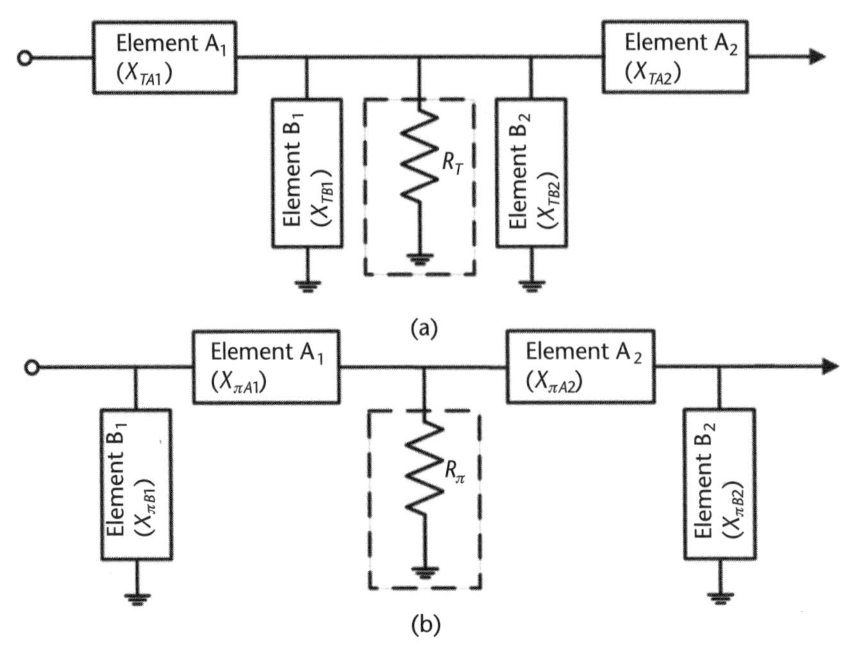

Figure 2.7 Schematic of the T-networks and π-networks including the virtual resistance needed for the design procedure. (a) T-network circuit, and (b) π-network circuit.

The quality factor of the T-network can be approximated according to

$$Q_T = \max\left(Q_{T\text{Left}}, Q_{T\text{Right}}\right) = \sqrt{\frac{R_T}{\min\left(R_S, R_L\right)} - 1} \tag{2.30}$$

Since the main motivation for using the three-element based T-network rather than the two-element L-network is the extra degree of freedom made possible with the third component, which allows for the control of the quality factor, the design of the T-network often starts by calculating the required value of the virtual resistance R_T for a given quality factor requirement, and source and load impedances. Once the value of R_T is selected, the two L-networks are designed using the procedure described in the previous section, and then the two reactances X_{TB1} and X_{TB2} are combined. Finally, the presence of reactive parts in the source and load impedances is included in the matching network using absorption and/or resonance concepts.

2.2.5 Example of T-Matching Using Lumped Elements

To better comprehend these design steps, let's consider the design of a T-matching network having a quality factor $Q_T = 2$ for

$$Z_S = (20 + j30)\ \Omega$$

and

$$Z_L = (60 + j10)\ \Omega$$

at f_0 = 1 GHz. In this example, we will consider that the source and load impedances are both made of a series connection of a resistor and an inductor with R_S = 20Ω, L_S = 4.775 nH, R_L = 60Ω, and L_L = 1.592 nH.

The first step is to calculate the value of the virtual resistance R_T. In this case, R_T = 100Ω. Consequently, Q_{TLeft} = 2 and Q_{TRight} = 0.816. In the second step, using the values of R_S, R_T, and Q_{TLeft}, we can find that X_{TA1} = ±40Ω and X_{TB1} = ∓50Ω. Hence, one possible solution is X_{TA1} = 40Ω and X_{TB1} = −50Ω, which correspond to L_{TA1} = 6.366 nH and C_{TB1} = 3.183 pF, respectively. The second step needs to be repeated to calculate the values of the components of the right-side L-network. In this case, given R_T, R_L, and Q_{TRight}, we can find that X_{TA2} = ±48.96Ω and X_{TB2} = ∓122.55Ω. One possible solution is X_{TA2} = 48.96Ω and X_{TB2} = −122.55Ω, which correspond to L_{TA2} = 7.792 nH and C_{TB2} = 1.299 pF, respectively. The two capacitors C_{TB1} and C_{TB2} can be combined into a single capacitor C_{TB} = 4.482 pF.

The initial matching network designed using the T-network to match R_S = 20Ω to R_L = 60Ω is reported in Figure 2.8. The source and load reactances (L_S and L_L) can be absorbed in the inductors L_{TA1} and L_{TA2}, respectively. Therefore, following the absorption process, the new values of the series inductors L_{TA1} and L_{TA2} are L'_{TA1} = 1.591 nH and L'_{TA2} = 6.2 nH, respectively. The final matching network is illustrated in Figure 2.9.

The design of the π-network is conceptually similar to that of the T-network. In the π-network, the L-networks connecting the virtual resistor R_π to R_S and R_L from the left and right sides, respectively, are of the shunt-series configuration. Hence, R_π must satisfy

$$R_\pi < \min(R_S, R_L) \tag{2.31}$$

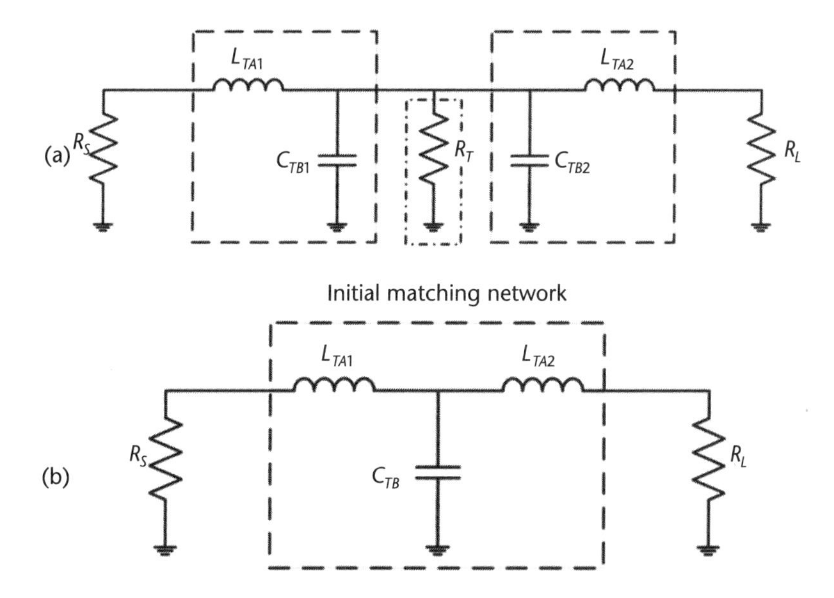

Figure 2.8 Schematic of the initial T-network needed to match R_S = 20Ω to R_L = 60Ω. (a) T-network expanded circuit (including the virtual resistor), and (b) T-network circuit.

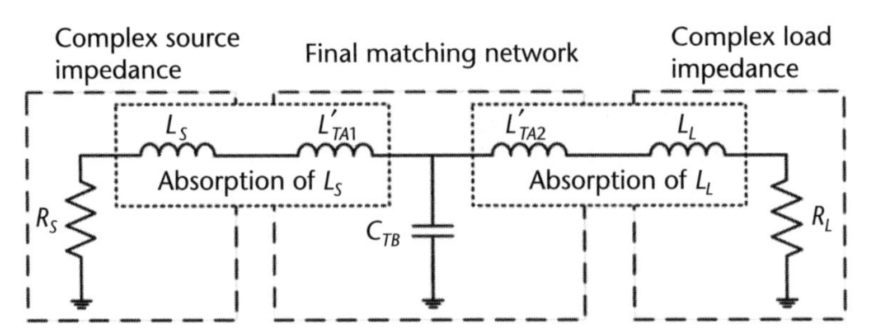

Figure 2.9 Schematic of the final T-network needed to match $Z_S = (20 + j30)\Omega$ to $Z_L = (60 + j10)\Omega$ showing the absorption of the source and load reactances.

In addition, according to (2.24), the quality factor of the left side L-network ($Q_{\pi\text{Left}}$) made of the reactances $X_{\pi A1}$ and $X_{\pi B1}$ is

$$Q_{\pi\text{Left}} = \sqrt{\frac{R_S}{R_\pi} - 1} \tag{2.32}$$

Similarly, the quality factor of the right side L-network ($Q_{\pi\text{Right}}$) made of the reactances $X_{\pi B2}$ and $X_{\pi A2}$ is

$$Q_{\pi\text{Right}} = \sqrt{\frac{R_L}{R_\pi} - 1} \tag{2.33}$$

Accordingly, the quality factor of the π-network can be approximated according by

$$Q_\pi = \max\left(Q_{\pi\text{Left}}, Q_{\pi\text{Right}}\right) = \sqrt{\frac{\max\left(R_S, R_L\right)}{R_\pi} - 1} \tag{2.34}$$

Here also the design of the π-network usually starts by calculating the required value of the virtual resistance (R_π) for a given quality factor requirement, and source and load impedances. Once the value of R_π is calculated, the two L-networks are designed using the procedure described previously, and then the two reactances $X_{\pi A1}$ and $X_{\pi A2}$ are combined. Finally, the presence of reactive parts in the source and load impedances is taken into account in the final matching network using absorption and/or resonance concepts.

Before concluding this section, it is important to note that the use of a three-element-based matching network (according to either the T-network or the π-network), will unavoidably increase the quality factor of the circuit and hence reduce its bandwidth. In fact, combining (2.27) and (2.30) leads to $Q_T > Q$. Likewise, (2.31) and (2.34) result in $Q_\pi > Q$ where Q is the quality factor of the L-network.

2.2.6 Distributed Element Matching

Another approach to design matching networks is based on the use of microstrip transmission lines. This technique is easily performed graphically using the Smith chart. A transmission line can be characterized by its characteristic impedance and electrical length. These electrical characteristics can be mapped to the physical dimensions of the transmission line since the width of the line will set its impedance and the physical length defines its electrical length. Tools like "LineCalc" in ADS or "TxLine" in MWO can be used to design (that is, determine the physical dimensions based on the electrical properties) or analyze (that is, calculate the electrical properties for given physical dimensions) a transmission line for a specific substrate. In this section, matching networks using transmission lines having an impedance equal to the system's characteristic impedance ($Z_0 = 50\Omega$) will be discussed. Other approaches based on non-50Ω transmission lines can be found in specialized references such as [2, 3].

Two types of transmission lines can be used when designing matching networks: series transmission line and stub transmission line. A series transmission line having the system's characteristic impedance will change the phase of the reflection coefficient and hence results in a rotation along the constant $|\Gamma|$ circle on the Smith chart as shown in Figure 2.10(a). However, a stub will introduce a susceptance that appears in shunt with the initial impedance. Hence, this will cause a rotation on the constant conductance circle as illustrated in Figure 2.10(b).

Figure 2.11 illustrates the layout of some possible topologies commonly referred to as the single-stub and double-stub topologies. Even though the stubs shown in this figure are open-circuited, the same topologies can be built using short-circuited stubs. An open-circuited stub having a length shorter than 0.25λ can be used to implement a capacitive susceptance (positive), while a short-circuited stub shorter than 0.25λ is equivalent to an inductive susceptance (negative). Even though the length of transmission lines is commonly taken into consideration, in some cases, open-circuited stubs are preferred since they are easier to tune on a circuit prototype and avoid using grounded vias. In such cases, an inductive susceptance can be implemented using an open-circuited stub with a length between 0.25λ and 0.5λ.

Given two arbitrary complex impedances Z_S and Z_L, one of the two single-stub matching networks shown in Figure 2.12 can be designed. Moving from Z_S toward Z_L, a stub can be followed by a series transmission line, or a series transmission line can be followed by a stub. If any of these two impedances (Z_S or Z_L) is equal to the system's characteristic impedance, then the stub must be directly connected to this impedance.

To graphically design the matching networks of Figure 2.12, it is essential to determine the location, on the Smith chart, of the intermediate impedances Z_1 and Z_2 shown in Figure 2.12(a, b), respectively.

In the first configuration (Figure 2.12[a]), the two conditions on the impedance Z_X are

$$\begin{cases} \mathrm{Re}(y_1) = \mathrm{Re}(y_S) \\ \text{and} \\ |\Gamma_1| = |\Gamma_L| \end{cases} \tag{2.35}$$

where y_1 and y_S are the normalized admittances associated with the impedances Z_1 and Z_S, respectively. Γ_1 and Γ_L are the reflection coefficients associated with the

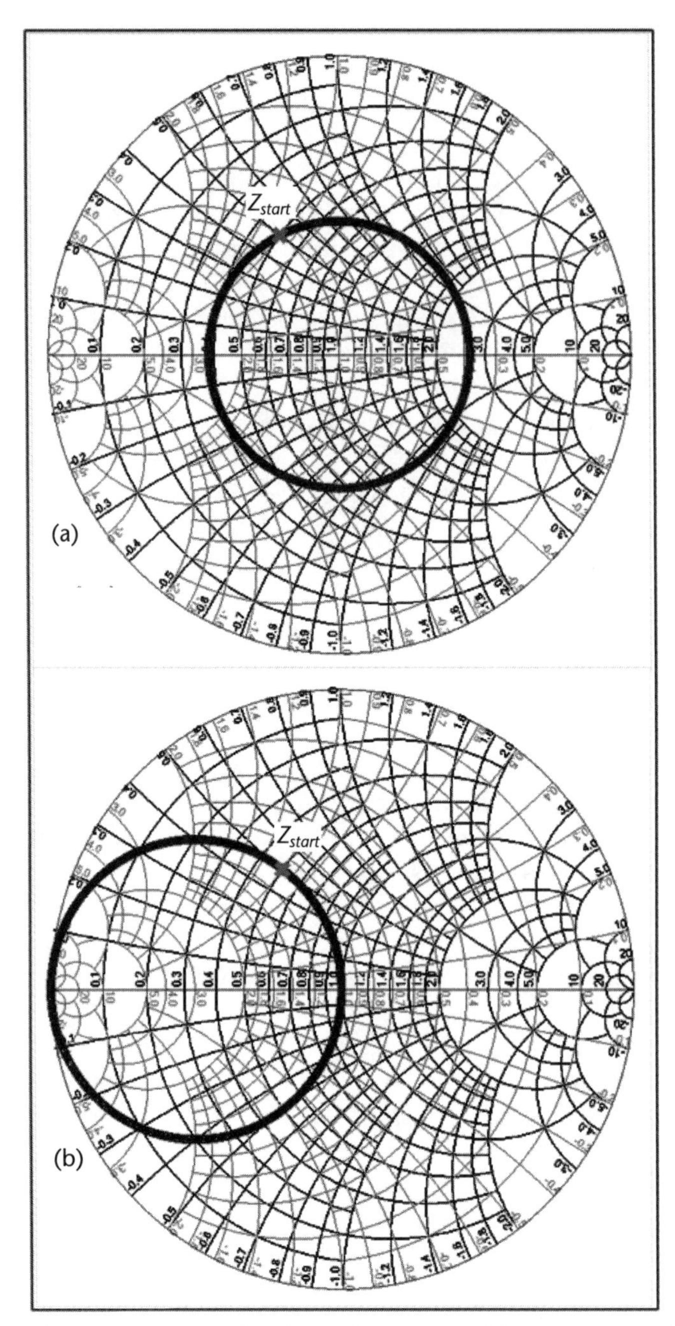

Figure 2.10 Displacement on the Smith chart following the addition of a transmission line to an impedance Z_{start}. (a) Effect of adding a series transmission line. (b) Effect of adding a stub.

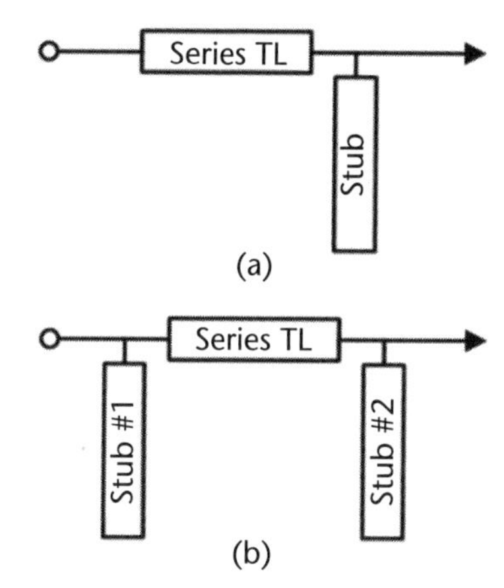

Figure 2.11 Matching networks using transmission line. (a) Single-stub matching network topology. (b) Double-stub matching network topology.

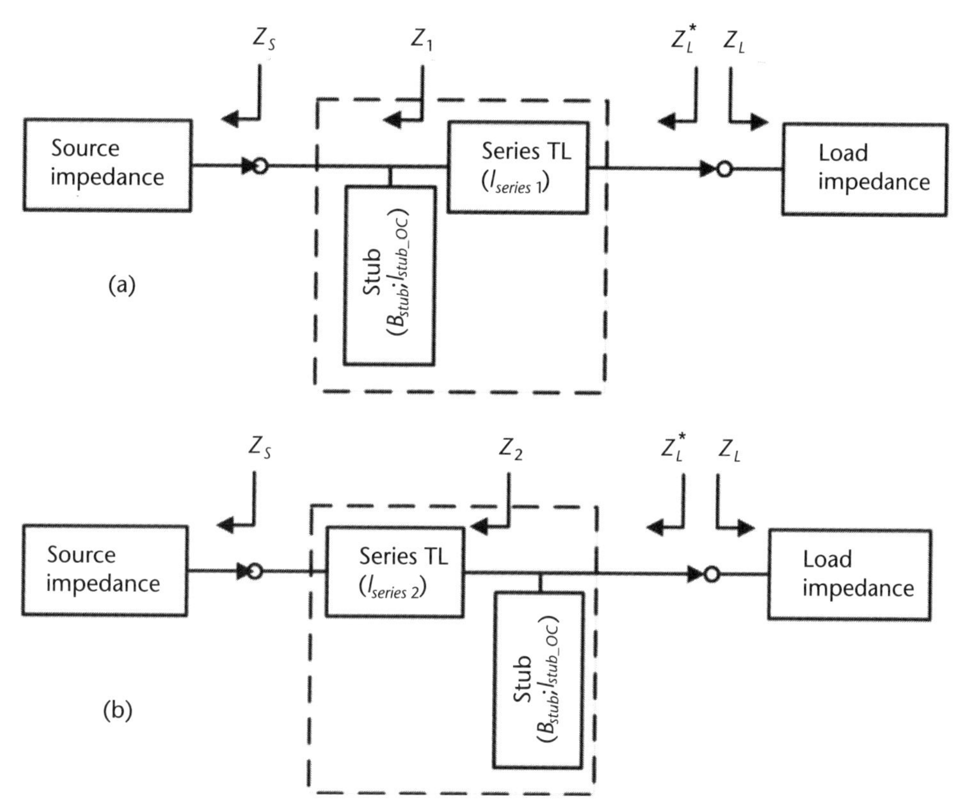

Figure 2.12 Single-stub matching network topologies. (a) Configuration 1 showing a stub followed by a series transmission line, (b) Configuration 2 showing a series transmission line followed by a stub.

impedances Z_1 and Z_L, respectively. Therefore, the impedance Z_1 will correspond to the intersection of the constant conductance circle $g = \text{Re}(y_S)$ and the constant reflection coefficient magnitude circle $|\Gamma| = |\Gamma_L|$. The length of the series transmission line can be read from the Smith chart in terms of the wavelength (λ) based on the angular rotation from Z_1 to Z_L^*. The length of the stub is the length needed to create a normalized susceptance B_{stub} such that

$$B_{\text{stub}} = \text{Im}(y_1) - \text{Im}(y_S) \tag{2.36}$$

where Im(X) refers to the imaginary part of the complex number X.

In configuration 2, illustrated in Figure 2.12(b), the constraints on the intermediate impedance Z_2 are

$$\begin{cases} |\Gamma_2| = |\Gamma_S| \\ \text{and} \\ \text{Re}(y_2) = \text{Re}(y_L) \end{cases} \tag{2.37}$$

where Γ_2 and Γ_S are the reflection coefficients associated with the impedances Z_2 and Z_S respectively. y_2 and y_L refer to the normalized admittances associated with the impedances Z_2 and Z_L, respectively. Accordingly, the impedance Z_2 is located at the intersection of the constant reflection coefficient magnitude circle $|\Gamma| = |\Gamma_S|$ and the constant conductance circle $g = \text{Re}(y_L)$. Similarly to the case of the matching network on Figure 2.12(a), the length of the series transmission line can be read from the Smith chart in terms of λ based on the angular rotation from Z_S to Z_2. The length of the stub is the length needed to create a normalized susceptance B'_{stub} given by

$$B'_{\text{stub}} = \text{Im}(y_L^*) - \text{Im}(y_2) \tag{2.38}$$

The single stub can be replaced by a double stub as illustrated in Figure 2.13. Since the two stubs appear as two shunt connected susceptances, one has to ensure that

$$B_{\text{stub1}} + B_{\text{stub2}} = B_{\text{stub}} \tag{2.39}$$

where B_{stub1} and B_{stub2} are the susceptances of stub 1 and stub 2, respectively. This also applies to the impedance matching network depicted in Figure 2.12(b).

2.2.7 Example of Distributed Element Matching

As an example, the circuit of configuration 1 is now applied to design a matching network for $Z_S = 50\Omega$ to $Z_L = (60 + j80)\Omega$. Based on the Smith chart, and using $Z_0 = 50\Omega$, $z_S = 1\Omega$ and $z_L = (1.2 + j1.6)\Omega$. The normalized admittances and reflection coefficients associated with these impedances are $y_S = 1\text{S}$ and $y_L = (0.3 - j0.4)\text{S}$, and

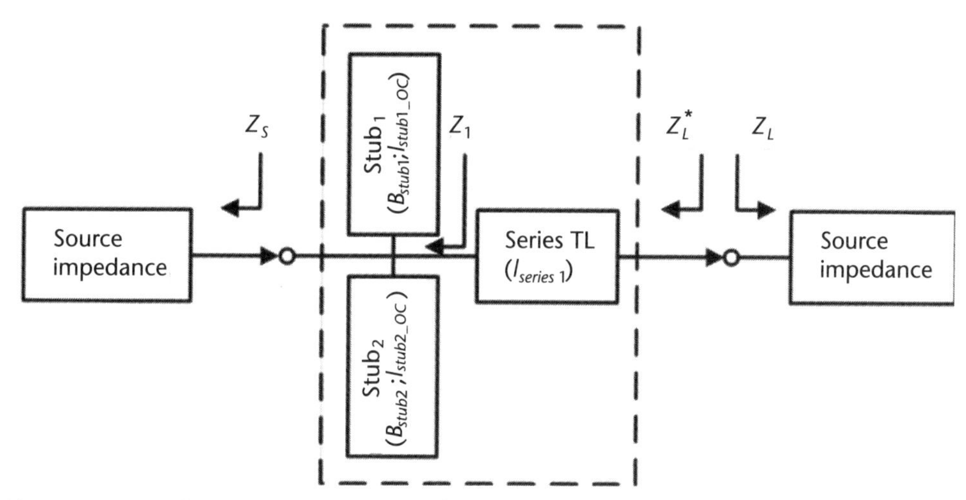

Figure 2.13 Single-stub matching network topology of Figure 2.12(a) implemented using double-stub.

$\Gamma_S = 0$ and $\Gamma_L = 0.593\angle 46.85^\circ$, respectively. As discussed earlier, the intermediate impedance Z_1 is located on the circles $\text{Re}(y) = \text{Re}(y_S) = 1$ and $|\Gamma| = |\Gamma_L| = 0.593$. As depicted in Figure 2.14, there are two possible choices for the intermediate point Z_1. The values of the series transmission line and the stub lengths as well as key intermediate calculations for this matching network are summarized in Table 2.3.

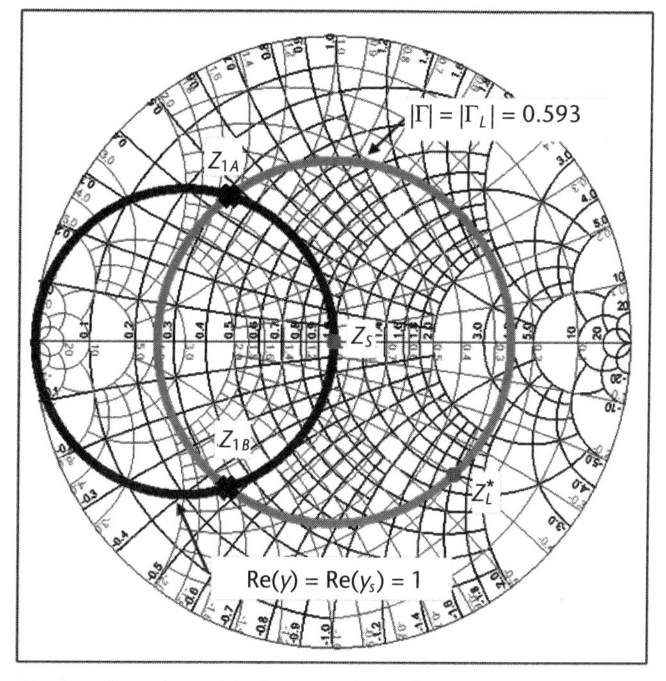

Figure 2.14 Smith chart-based graphical approach for the design of a single-stub matching network (following configuration 1) for $Z_S = (60 + j80)\Omega$ and $Z_L = 50\Omega$.

Table 2.3 Smith Chart-Based Design of the Single-Stub Matching Network for $Z_S = (60 + j80)\Omega$ and $Z_L = 50\Omega$ Using Circuit Configuration 1

	Configuration 1	
Intermediate point	Z_{1A}	Z_{1B}
Normalized impedance at intermediate point	$(0.316 + j0.465)\Omega$	$(0.316 - j0.465)\Omega$
Impedance at intermediate point	$(15.8 + j23.25)\Omega$	$(15.8 - j23.25)\Omega$
Normalized admittance at intermediate point	$(1 - j1.47)$S	$(1 + j1.47)$S
Admittance at intermediate point	$(20 - j29.42)$ mS	$(20 + j29.42)$ mS
Γ at the intermediate point	$0.593\angle 126.33^\circ$	$0.593\angle -126.33^\circ$
Series transmission line length	$L_{series1\_A} = 0.240\lambda$	$L_{series1\_B} = 0.390\lambda$
Normalized admittance of the stub	$B_{stub\_A} = -1.47 - 0 = -1.47$	$B_{stub\_B} = 1.47 - 0 = 1.47$
Stub length (if open-circuited)	$l_{stub\_OC\_A} = 0.345\lambda$	$l_{stub\_OC\_B} = 0.155\lambda$
Stub length (if short-circuited)	$l_{stub\_SC\_A} = 0.095\lambda$	$l_{stub\_SC\_B} = 0.405\lambda$

2.3 Wideband Matching

The impedance matching techniques discussed in the previous section mainly focus on the design of a matching network optimized at the frequency of operation while disregarding its performance over frequency. In this section, the focus is on the design of wideband matching networks. Three approaches will be discussed: the use of constant-Q circles, the real-frequency technique, and the non-Foster technique.

2.3.1 Constant-Q Circles Technique

In this technique, the bandwidth of the matching network is optimized by minimization of the value of the quality factor or at least maintaining it below a certain value. The bandwidth of the matching network is inversely proportional to the quality factor of the circuit which corresponds to the maximum of the quality factors at each of the circuits' nodes.

For a given complex impedance (Z), Q is defined as:

$$Q = \left| \frac{\mathrm{Im}(Z)}{\mathrm{Re}(Z)} \right| \tag{2.40}$$

Hence, contours corresponding to constant-Q can be drawn on the Smith chart as illustrated in Figure 2.15. To ensure wideband operation of the matching network, one has to maintain the circuit's quality factor to be as low as possible. Accordingly, for a specified quality factor, one must remain inside (or at worst on the edge of) a constant-Q contour.

Given a source and load impedances and their respective quality factors Q_{ZS} and Q_{ZL}, the minimum quality factor of the circuit is

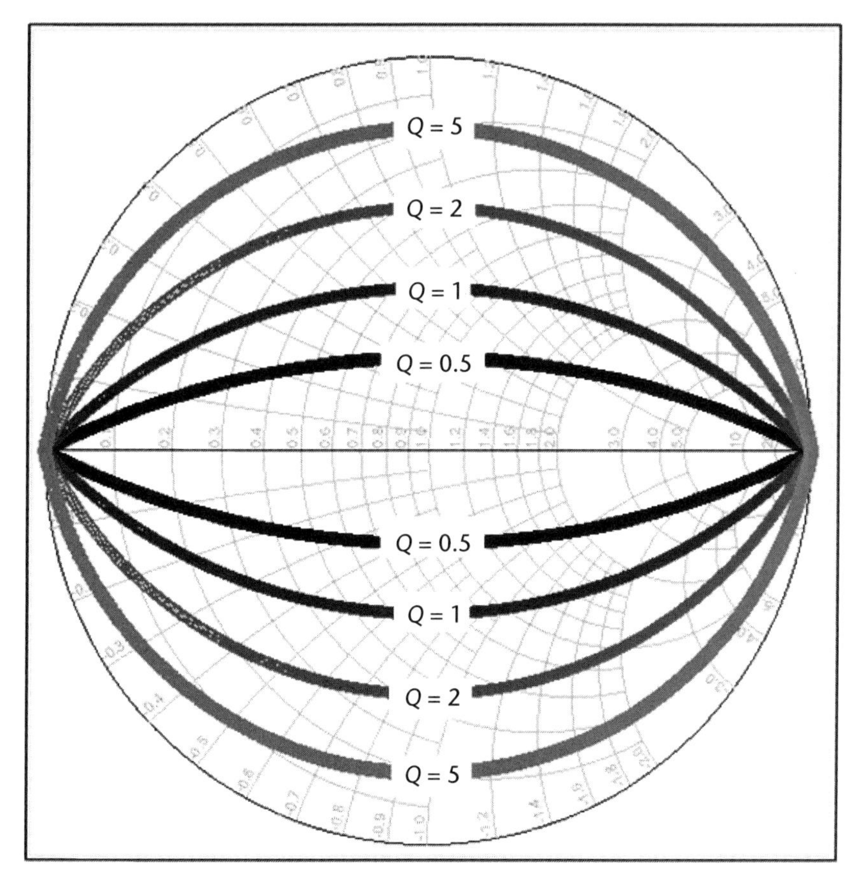

Figure 2.15 Constant-Q contours in the Smith chart.

$$Q_{\min} = \max\left(Q_{Z_S}, Q_{Z_L}\right) \tag{2.41}$$

When the source and load impedances are purely resistive, then $Q_{\min}$ is zero and hence, the designer has flexibility to choose any desired value for the quality factor. However, if at least one of these two impedances has a reactive part, then the value of $Q_{\min}$ will be fixed and the circuit cannot be designed for a quality factor better than $Q_{\min}$ defined by (2.41).

Designing a matching network for $Q = Q_{\min}$ can be achieved by ensuring that at all nodes of the circuit, the quality factor is lower than $Q_{\min}$. Graphically, this corresponds to forcing the impedances at all intermediate points of the circuit to remain within the constant-Q contour corresponding to $Q = Q_{\min}$. This graphical approach is particularly useful when designing transmission line-based matching networks.

When using lumped elements, it was shown in the previous section that the L-network has a quality factor that cannot be controlled since it is function of Z_S and Z_L. Moreover, three-element networks such as the T-networks and π-networks will unavoidably lead to higher quality factor and narrower bandwidth than the two-element L-network. To increase the bandwidth of the matching network, multiple L-networks can be cascaded. All these L-networks are of the same type (either series-shunt or shunt-series depending on the values of the source and load resistances). The number of networks is directly related to the desired quality factor. Figure 2.16 shows a block diagram of a multisection lumped elements based wideband matching network. In this figure, it is assumed that the source resistance (R_S) is lower than the load resistance (R_L). Thus, moving from the source to the load the L-sections are all of the series-shunt type. Conversely, if R_S was higher than R_L then shunt-series L-networks would have been used. Without loss of generality, the analysis of these networks will be carried out while assuming that the source and load impedances are purely resistive. When any of these impedances has a reactive part, the design approach can be generalized using the absorption and/or resonances techniques while keeping in mind the constraint on the minimum quality factor (described by [2.41]).

The design theory of multisection L-networks is based on the use of virtual resistances as described in the previous section for the T-networks and π-networks [3]. For the M-sections L-network of Figure 2.16, maximum bandwidth is obtained when the value of the virtual resistance at the output of the kth section (k between 1 and $M - 1$) is

$$R_k = \sqrt[M]{\left(R_S R_L\right)^k} \tag{2.42}$$

The number of stages needed depends on the desired quality factor. For a given quality factor and source and load resistances, the minimum number of sections is

$$M = ceil\left(\frac{\ln\left[\dfrac{\max\left(R_S, R_L\right)}{\min\left(R_S, R_L\right)}\right]}{\ln\left(1 + Q^2\right)}\right) \tag{2.43}$$

2.3.2 Real Frequency Technique

In several applications such as antenna and amplifiers design, the broadband impedance matching is delicate since the antenna's input impedance, as well as the amplifier's input and output impedances are frequency dependent. In such cases, the broadband matching concept translates into designing a single matching network that changes a starting impedance into a specific set of frequency-dependent impedances. A generic block diagram describing this problem is shown in Figure 2.17. The wideband matching network is ideally expected to change the source impedance Z_S into Z_M such that $Z_M(f) = Z^*_L(f)$ over the considered frequency band. This would results in maximum power transfer. However, practically the perfect complex conjugate match over a wide frequency range is not practically realizable. Hence, the design objective is to minimize, over the frequency band of operation, the mismatch losses or equivalently maximize the transducer power gain of the circuit, $T(\omega)$, which is defined as [6]

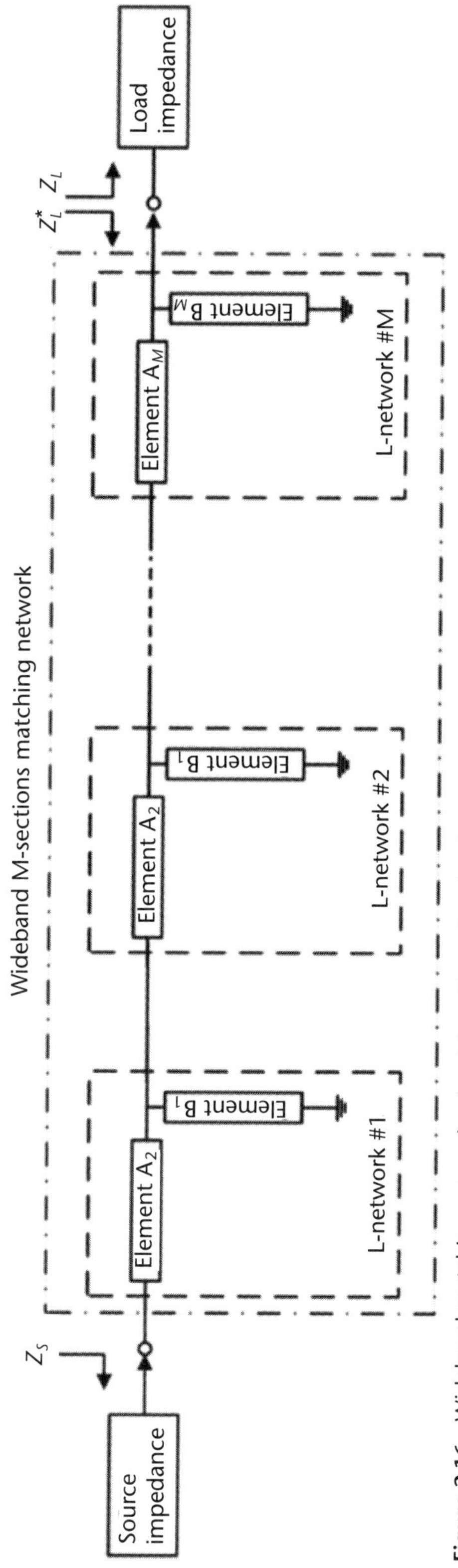

Figure 2.16 Wideband matching network using M-sections of series-shunt L-networks.

$$
\begin{aligned}
T\left(\omega^2\right) &= 1 - \left| \frac{Z_{MN\_S}(j\omega) - Z_S(-j\omega)}{Z_{MN\_S}(j\omega) + Z_S(j\omega)} \right|^2 \\
&= 1 - \left| \frac{Z_{MN\_L}(j\omega) - Z_L(-j\omega)}{Z_{MN\_L}(j\omega) + Z_S(j\omega)} \right|^2
\end{aligned}
\tag{2.44}
$$

where Z_{MN_S} and Z_{MN_L} are the impedances seen when looking into the matching network from the source and load sides, respectively. Z_S and Z_L are the source and load impedances, respectively.

The fundamental theoretical work developed by Fano and Youla in [7–9] can be used to design the broadband matching network. However, this requires a tedious process involving, among others, the development of an analytical approximation that fits the load impedance in the complex frequency plane. The real frequency technique was initially developed in [10] for the case of a resistive source impedance. The technique was later generalized for complex source impedances [6]. The real frequency technique is based on the use of available (ideally measured, otherwise simulated) impedances values of the load over a finite number of frequencies that cover the band over which the matching network is to be optimized. Then, numerical optimization techniques are used to estimate the best values of the resistance of the matching network at these discrete frequencies while maximizing the transducer power gain described by (2.44). These values will later be used to calculate the desired reactances of the matching network using the Hilbert transform. Once these values are estimated, a rationale function is formulated to approximate the impedance of the matching network as a function of frequency. Finally, a ladder LC-based network is synthesized to implement the desired impedance function. A detailed description of the equations as well as the step by step procedure for synthesizing matching networks using the real frequency technique is available in [6].

In [11], it was shown that alleviating the minimum impedance constraint used in [6], adds an additional degree of freedom that makes it possible to better optimize the design of the matching network. It was demonstrated in [11] that non-minimum reactance matching networks can lead to better performances than the minimum reactance matching networks and achieve superior performances than those obtained with the minimum reactance constraint of [6]. An enhanced fast simplified real frequency technique approach has recently been proposed in [12]. This technique improves the convergence time of conventional real frequency technique-based algorithm through the use of selective target data and constraint optimization approaches. The real frequency technique has been widely used to design

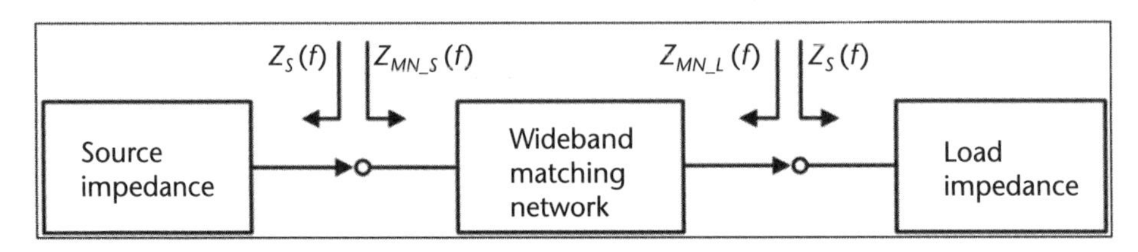

Figure 2.17 Generic block diagram for wideband matching problem.

wideband matching networks for antenna and amplifier applications. Sample results and examples of such applications are presented in [11–14].

2.3.3 Non-Foster Based Technique

The non-Foster based wideband matching technique is applied to wideband matching of antennas and to lesser extent amplifier systems [15–18]. It is particularly useful for electrically small antennas (ESA) for which it allows for enhancing the gain-bandwidth limitations of passive matching networks. Foster's theorem states that the input impedance of a passive network has a reactance that increases monotonically with respect to the angular frequency [19]. Hence, using conventional lossless matching networks based on inductors and/or capacitors will result in narrowband matching since it is only possible for the matching network to present a conjugate reactive part to the system to be matched over a single frequency point, and to maintain a quasi-conjugate relationship over a limited bandwidth. Figure 2.18(a) presents a conceptual block diagram of a one-port system (usually an antenna) and its input matching network. Assuming that the one-port system has a capacitive impedance, the matching networks needs to present an inductive impedance that would resonate with that of the system to be matched at the frequency of operation. In Figure 2.18(b), typical inductive and capacitive reactances are reported versus frequency for a system designed to resonate at 1 GHz. As illustrated, the bandwidth over which both reactances cancel each other is limited. This bandwidth problem is especially critical when the system to be matched presents reactive part that is a

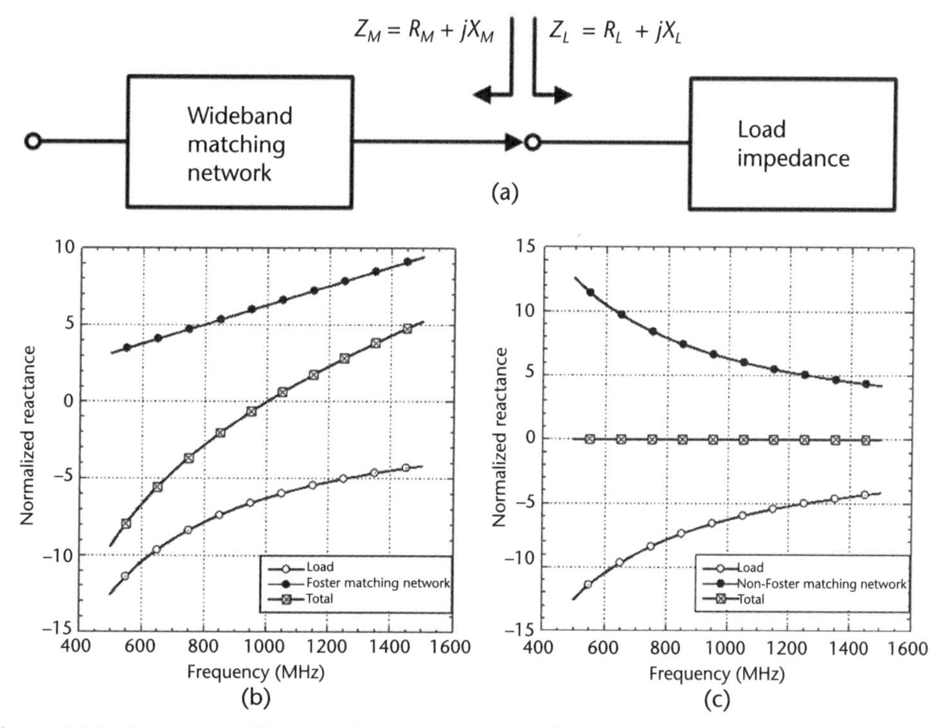

Figure 2.18 Reactance of Foster and non-Foster networks versus frequency. (a) Conceptual block diagram. (b) Case of a Foster network. (c) Case of a non-Foster network.

strong function of the frequency. In order to extend the reactance cancellation bandwidth, it is crucial to synthesize a component whose reactive part is monotonically decreasing versus frequency as depicted in Figure 2.18(c). Such an ideal component would result in a perfect conjugate match and thus broadband matching.

The matching networks that can implement reactive impedances that have a negative slope with respect to frequency are known as non-Foster networks. These correspond to negative capacitors and/or negative inductors and are physically realizable mainly by using negative impedance converters (NICs) that are based on active elements. An NIC is a circuit designed to change the value of the impedance that loads its output port into its negative counterpart as illustrated in Figure 2.19. The impedances seen at the input and output of a NIC are related through [17]

$$Z_1 = -k \cdot Z_2 \tag{2.45}$$

where k ($k > 0$) is the transformation ratio, Z_1 and Z_2 are the impedances shown in Figure 2.19 at reference planes 1 and 2, respectively.

Several NIC topologies have been reported in the literature [17, 20–23], and we will discuss them again in Chapter 5. However, it is important to recall that due to the presence of active components, NICs often result in signal to noise ratio degradation and power-dependent distortion, which are critical for antenna applications [15, 17, 23].

2.4 Use of CAD for Matching Network Design

Now that we have discussed the most used methods for matching network designs and highlighted their features and design procedures, we can now move forward and rely on some readymade solutions within some of the software packages mentioned in this book. More specifically we will consider an example using the automated matching network feature in MWO (available from v12 onward).

The iMatch matching tool in MWO is under the iFilter design wizard. The wizard panel can be expanded upon opening a new project, from where the desired matching option can be selected after selecting the iFilter wizard. Let us assume that we want to match a 50Ω source to a $20 + j\,62.83\Omega$ load at the center frequency of 1 GHz. We can put the values of source side (Z1) and load side (Z2) as shown in Figure 2.20. Note that we converted the value of the reactance to its corresponding inductance value at 1 GHz. Furthermore, we can specify the range of frequencies

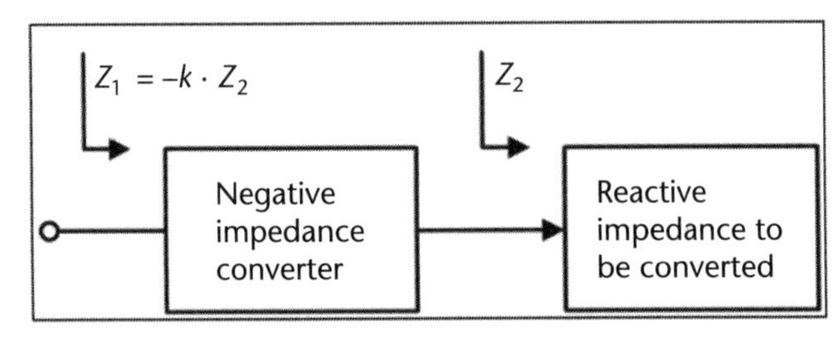

Figure 2.19 Simplified block diagram of a negative impedance converter.

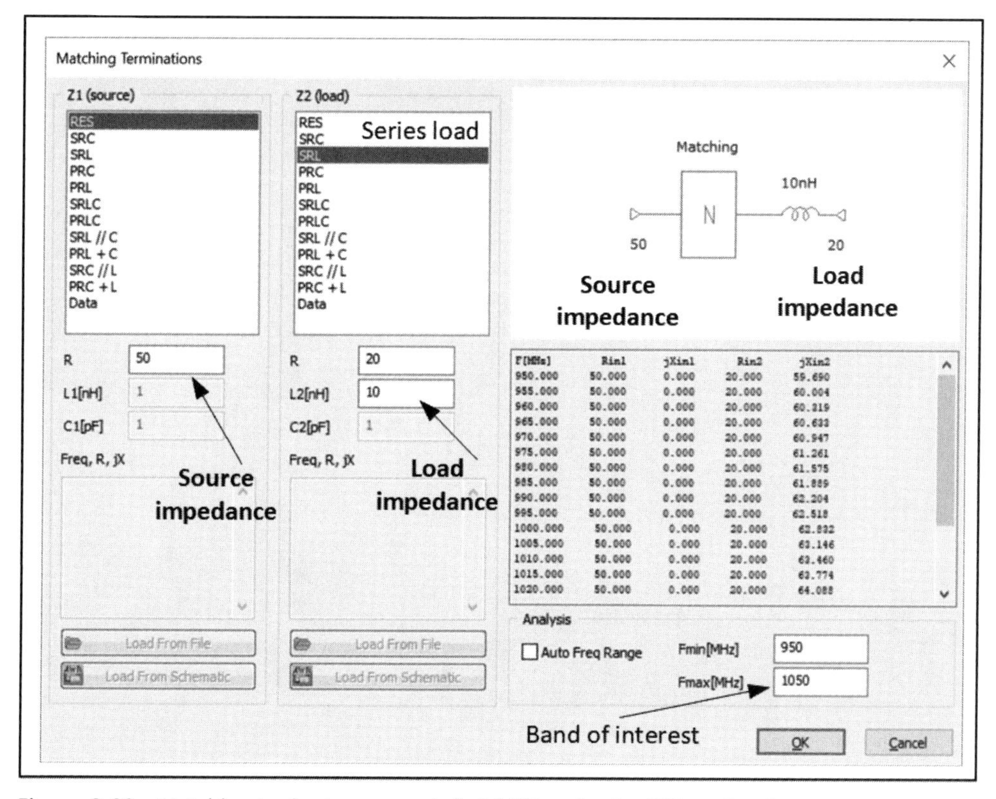

Figure 2.20 Matching tool setup example in MWO under the iFilter wizard.

that we want to check the matching over (i.e., 950–1,050 MHz in this example with 1 GHz as the center frequency). The source (Z1) and load (Z2) networks can be of any RLC combination (i.e., parallel, series, two, or three elements as shown under each category). We will choose a "Res" for Z1 and "SRL" for series RL Z2 with $R = 20\Omega$ and $L = 10$ nH as illustrated in the figure. We click "OK," and the wizard brings up the results window in Figure 2.21.

The results window has several parts to give the user flexibility in choosing the type of the matching network (1), the schematic and positions of source and load (3), its behavior in terms of return loss and insertion loss (2) within the frequency range requested, as well as the impedance locus (behavior) on the Smith chart at the center frequency (4). If we select a highpass-based L-type matching network from (1) as shown in Figure 2.21, the solution is directly calculated in (3) with actual inductor and capacitor values at the center frequency of 1 GHz. The inductor values are in nanohenries while capacitor values are in picofarads. The return loss and insertion loss curves are shown in (2) for the band of interest (the band can be expanded by pressing a button below window (2) with "WS" meaning "wide Frequency Span." The smith chart showing the impedance going from the load to the source is shown in window (4). Other views can be obtained by changing the various setting below the window as the user wishes (i.e., tracing the impedance behavior versus frequency).

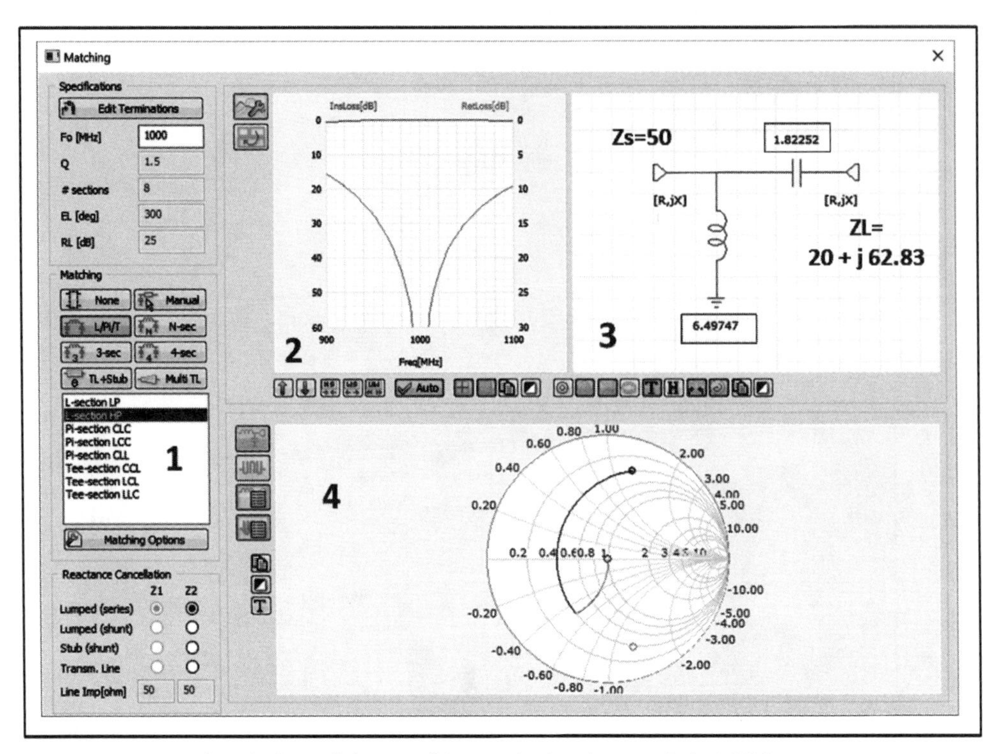

Figure 2.21 Results window of the matching tool wizard example in MWO.

This tool provides different options, with all previously mentioned matching network types supported and automatically calculated. Different matching network configurations can be selected (i.e., π or T) for various L and C combinations by choosing the options in window (1) and getting the solution in (3) instantly along with the response curves in (2) and (4). This will provide the designer the ability to check the various what-if solutions and effects of various matching networks on the design behavior.

2.5 Conclusions

In this chapter, impedance matching methods that are crucial for antenna, amplifier, and AIA designs were discussed. The basics of impedance matching were first reviewed. Then several narrowband matching techniques were discussed and their design equations presented. This included the design of L, T, and π matching networks. All of these matching networks can be synthesized using the analytical approach presented. They are suitable for narrowband applications where the matching network is also used for filtering unwanted signals. The use of the graphical tool was also utilized which is widely used. Broadband matching techniques using constant-Q circles, real frequency technique, and non-Foster matching concepts were then discussed. Although these methods provide wider matching bandwidth, some of their drawbacks are hardware complexity, use of active components (i.e., amplifiers) and sometimes end up with nonrealizable network values. The chapter is closed with a matching example using the matching wizard in MWO.

References

[1] Maas, S. A., *Practical Microwave Circuits*, Norwood, MA: Artech House, 2014.

[2] Pozar, D. M., *Microwave Engineering*, Fourth Edition, New York: John Wiley & Sons, 2011.

[3] Steer, M., *Microwave and RF Design: A Systems Approach*, 2nd Edition, SciTech Publishing, 2013.

[4] Rhea, T., "The Yin-Yang of Matching: Part 1: Basic Matching Concepts," *High Frequency Electronics*, Vol. 5, No. 3, March 2006, pp. 16–25.

[5] Gonzalez, G., *Microwave Transistor Amplifiers Analysis and Design*, Second Edition, Upper Saddle River, NJ: Prentice Hall, 1997.

[6] Carlin, H. J., and B. S. Yarman, "The Double Matching Problem: Analytic and Real Frequency Solutions," *IEEE Transactions on Circuits and Systems*, Vol. CAS-30, January 1983, pp. 15–28.

[7] Fano, F. M., "Theoretical Limitations on the Broadband Matching of Arbitrary Impedances," *Journal of Franklin Institute*, Vol. 249, January 1960, pp. 57–83.

[8] Fano, F. M., "Theoretical Limitations on the Broadband Matching of Arbitrary Impedances," *Journal of Franklin Institute*, Vol. 249, February 1960, pp. 139–155.

[9] Youla, D. C., "A New Theory of Broadband Matching," *IEEE Transactions on Circuit Theory*, Vol. CT-11, March 1964, pp. 30–50.

[10] Carlin, H. J., "A New Approach to Gain-Bandwidth Problems," *IEEE Transactions on Circuits and Systems*, Vol. CAS-24, April 1977, pp. 170–175.

[11] Newman, E. H., "Real Frequency Wide-Band Impedance Matching with Nonminimum Reactance Equalizers," *IEEE Transactions on Antennas and Propagation*, Vol. 53, No. 11, November 2005, pp. 3597–3063.

[12] Kopru, R., "FSRFT-Fast Simplified Real Frequency Technique Via Selective Target Data Approach for Broadband Double Matching," *IEEE Transactions on Circuits and Systems II: Express Briefs*, Vol. 64, No. 2, February 2017, pp. 141–145.

[13] Sun, G., and R. H. Jansen, "Broadband Doherty Power Amplifier Via Real Frequency Technique," *IEEE Transactions on Microwave Theory and Techniques*, Vol. 60, No. 1, January 2012, pp. 99–111.

[14] An, H., B. K. J. C. Nauwelaers, and A. R. Van de Capelle, "Broadband Microstrip Antenna Design with the Simplified Real Frequency Technique," *IEEE Transactions on Antennas and Propagation*, Vol. 42, No. 2, February 1994, pp. 129–136.

[15] Jacob, M. M., and D. F. Sievenpiper, "Non-Foster Matched Antennas for High Power Applications," *IEEE Transactions on Antennas and Propagation*, Vol. 65, No. 9, September 2017, pp. 4461–4469.

[16] Albarracin-Vargas, F., et al., "Design Method for Actively Matched Antennas with Non-Foster Elements," *IEEE Transactions on Antennas and Propagation*, Vol. 64, No. 9, September 2016, pp. 4118–4123.

[17] Susan-Fort, S. E., and R. M. Rudish, "Non-Foster Impedance Matching of Electrically-Small Antennas," *IEEE Transactions on Antennas and Propagation*, Vol. 57, No. 8, August 2009, pp. 2230–2241.

[18] Lee, S., et al., "A Broadband GaN pHEMT Power Amplifier Using Non-Foster Matching," *IEEE Transactions on Microwave Theory and Techniques*, Vol. 63, No. 12, December 2015, pp. 4406–4414.

[19] Foster, R. M., “A Reactance Theorem,” *The Bell System Technical Journal*, Vol. 3, No. 2, April 1924, pp. 259–267.

[20] Horowitz, I., “Negative-Impedance Converters,” *IRE Transactions on Component Parts*, Vol. 9, No. 1, March 1962, pp. 33–38.

[21] Kuo, C. K., and K. L. Su, “Some New Four Terminal NIC Circuits,” *IEEE Proceedings on Circuit Theory*, August 1969, pp. 379–381.

[22] Myers, B. R., “Some Negative Impedance Converters,” *The Proceedings of IEEE*, Vol. 61, No. 5, May 1973, pp. 669–670.

[23] Jacob, M. M., and D. F. Sievenpiper, “Gain and Noise Analysis of Non-Foster Matched Antennas,” *IEEE Transactions on Antennas and Propagation*, Vol. 64, No. 12, December 2016, pp. 4993–5004.

CHAPTER 3

Amplifier Design

Amplifiers are essential building blocks in any wireless system at both transmitter and receiver sides. Disregarding the fact that all amplifiers are designed for increasing the power of the signal passing through them, their design approaches differ depending on the specific requirement of the application. For example, in receivers, an amplifier needs to be placed as close as possible to the receiving antenna. This amplifier needs to be optimized for low noise performance since it will be the main contributor to the overall noise performance of the receiver chain. Conversely, in transmitters, a power amplifier (PA) is needed right before the antenna to boost the power of the output signal. For such amplifiers, noise performance is not the main concern; however, power efficiency is of paramount importance. Moreover, linearity is also critical in most modern applications when PAs are driven by amplitude modulated signals. Therefore, the design of amplifiers can be perceived as application dependent. From such a perspective, one can distinguish four types of amplifiers: low noise amplifiers (LNAs), maximum gain amplifiers, gain-noise trade-off amplifiers, and PAs. While the first three types are considered small-signal amplifiers and their design is based on small-signal *S*-parameters techniques, PAs are often referred to as large-signal amplifiers and their design calls for load-pull techniques [1–3].

It is a common misconception to refer to PAs as amplifiers having high-power handling capabilities, and small-signal amplifiers as amplifiers with low-power handling capabilities. A small-signal amplifier is an amplifier that is designed to operate with fairly low-power levels compared to the capabilities of the transistor used. Conversely, a PA is an amplifier that is designed to operate at power levels very close to its maximum power handling irrespective of the absolute value of such power. PAs come in very different sizes and packages with dimensions in the range of few millimeters and powers around 0.5 to 1W for cell phone applications and tens of inches, with up to a few hundred watts of peak output power for base station applications.

3.1 Generic Approach for Amplifiers Design

The design of an amplifier starts by setting the operating bias condition using the current and voltage (I-V) characteristics of the transistor. The DC bias point will affect the AC performance of the amplifier such as stability, gain, and noise characteristics. Often, for small-signal applications, linear operation is desired and therefore class A biasing conditions is commonly adopted. For large-signal operation, as it is the case in PAs, power-efficient classes of operation are sought. Consequently, the transistor can be biased at lower quiescent current in class AB, B, or even C conditions. Following the selection of the bias point, the stability of the transistor can be evaluated. A transistor can be either unconditionally stable or potentially unstable. A transistor is said to be unconditionally stable if for any value of passive source and load terminations (i.e., $|\Gamma_S| < 1$ and $|\Gamma_L| < 1$), the input and output ports present a positive resistance value (i.e., $|\Gamma_{in}| < 1$ and $|\Gamma_{out}| < 1$). A transistor is said to be potentially unstable if there exist some passive source and load terminations, which might cause the input and output ports to present a negative resistance value and therefore cause oscillations [3–5].

In the case of a unilateral transistor (for which $|S_{12}| \approx 0$), the input and output reflection coefficients are independent of the source and load reflection coefficients, and therefore a necessary and sufficient stability condition is:

$$\begin{cases} |S_{11}| \left(= |\Gamma_{in}|\right) < 1 \\ \text{and} \\ |S_{22}| \left(= |\Gamma_{out}|\right) < 1 \end{cases} \tag{3.1}$$

When a transistor is considered to be bilateral, the input and output reflections coefficients (Γ_{in} and Γ_{out}, respectively) depend on the source and load reflection coefficients (Γ_S and Γ_L, respectively). The equations relating the various reflection coefficients in such two-port networks, which can be derived from signal flow graphs theory (thoroughly discussed in [3]), are

$$\Gamma_{in} = S_{11} + \frac{S_{12}S_{21}\Gamma_L}{1 - S_{22}\Gamma_L} \tag{3.2}$$

$$\Gamma_{out} = S_{22} + \frac{S_{12}S_{21}\Gamma_S}{1 - S_{11}\Gamma_S} \tag{3.3}$$

where S_{11}, S_{12}, S_{21}, and S_{22} are the S-parameters of the transistor.

In such network, the stability can be assessed using Rollet's stability factor or K-factor defined as [2–6]:

$$K = \frac{1 - |S_{11}|^2 - |S_{22}|^2 + |\Delta|^2}{2|S_{12} \cdot S_{21}|} \tag{3.4}$$

where Δ is given by

$$\Delta = S_{11}S_{22} - S_{12} \cdot S_{21} \tag{3.5}$$

To ensure the unconditional stability of a bilateral transistor, two necessary and sufficient conditions are needed:

$$K > 1 \tag{3.6}$$

and

$$|\Delta| < 1 \tag{3.7}$$

The condition expressed in (3.7) can be substituted by

$$B_1 > 0 \tag{3.8}$$

with

$$B_1 = 1 + |S_{11}|^2 - |S_{22}|^2 - |\Delta|^2 \tag{3.9}$$

Here, it is important to note that for most of the devices intended for amplifier applications, the conditions stated by (3.7) and (3.8) are normally satisfied.

When a transistor is potentially unstable, it is highly recommended to stabilize it. However, transistor stabilization often degrades its gain and/or noise performance. Therefore, the stabilization needs to be performed while carefully considering the resulting minimum noise figure of the transistor and its maximum gain. The designer can also choose not to stabilize the transistor but to ensure that it will always see source and load reflection coefficients that will guarantee stable operation.

For a potentially unstable transistor, input and output stability circles can be derived to delimit the regions of the Smith chart that correspond to values of source reflection coefficient (Γ_S) and load reflection coefficient (Γ_L), respectively, that will result in stable operation. At a given frequency of operation, the input stability circle is drawn in the Γ_S plane and it corresponds to Γ_S values that will result in $|\Gamma_{\text{out}}| = 1$, which represent the edge of stability. Similarly, the output stability circle is drawn in the Γ_L plane and corresponds to Γ_L values that will lead to $|\Gamma_{\text{in}}| = 1$.

Equation (3.3) can be used to derive the equations of the input stability circle's radius (r_S) and center (c_S) as [2–5]

$$r_S = \left| \frac{S_{12} \cdot S_{21}}{|S_{11}|^2 - |\Delta|^2} \right| \tag{3.10}$$

$$c_S = \frac{\left(S_{11} - \Delta \cdot S_{22}^*\right)^*}{|S_{11}|^2 - |\Delta|^2} \tag{3.11}$$

According to (3.3), for $\Gamma_S = 0$ (which corresponds to the center of the Smith chart in the Γ_S plane), $\Gamma_{\text{out}} = S_{22}$. Therefore, if $|S_{22}| < 1$, then the center of the Smith chart is in the stable region, conversely if $|S_{22}| > 1$, the center of the Smith chart is in the unstable region. This can be used to determine whether the inside of the input stability circle or its outside represents the stable region of the Γ_S plane.

The equations of the output stability circle's radius (r_L) and center (c_L) can be obtained from (3.2) as [2–5]

$$r_L = \left| \frac{S_{12} \cdot S_{21}}{\left|S_{22}\right|^2 - \left|\Delta\right|^2} \right| \quad (3.12)$$

$$c_L = \frac{\left(S_{22} - \Delta \cdot S_{11}^*\right)^*}{\left|S_{22}\right|^2 - \left|\Delta\right|^2} \quad (3.13)$$

An approach similar to that described for the input stability circle can be used to distinguish between the stable and unstable regions delimited by the output stability circle. The center of the Smith chart in the Γ_L plane corresponds to $\Gamma_{in} = S_{11}$ (using [3.2]). Thus, the center of the Smith chart is in the stable region if $|S_{11}| < 1$, and in the unstable region if $|S_{11}| > 1$. This enables the designer to distinguish if the stable region is inside or outside the output stability circle.

Following the study of the transistor's stability, the next design step is to select the source and load reflection coefficients to be presented to the transistor. This selection is often based on defining in the Γ_S and/or Γ_L planes the contours of equal performances, for example, constant gain contours, constant noise contours, and constant efficiency contours. This step differs from one type of amplifiers to the other and will be discussed in details in the subsequent sections.

3.2 LNA Design

Noise is a prime concern in receiver design since the signals picked by the antenna have very low-power levels. Accordingly, the signal-to-noise ratio (SNR) at the antenna output is expected to be barely enough to allow for information recovery with acceptable bit error rate (BER). The noise at the output of a receiver radio frequency (RF) front-end is made of two main contributions: the noise picked by the antenna, and the noise generated internally by the receiver. Therefore, the noise contribution of the receiver's RF front-end should be minimized.

Noise analysis in cascaded systems will be presented next. This will enable us to relate the noise contribution of each element to the overall noise figure of the receiver and hence better appreciate the need for LNAs and their critical role in receivers.

3.2.1 Noise Analysis in Cascaded Systems

The main sources of noise generated within RF receiver circuitry are thermal noise, shot noise, and flicker noise. Thermal noise is the predominant contributor to the overall noise generated by receivers, while shot and flicker noise contributions are negligible. Thermal noise is cause by thermal vibrations of bound charges, and is also known as Johnson noise or Nyquist noise [7, 8].

The thermal noise power in a circuit is given by

$$P_n = k \cdot T \cdot B \quad (3.14)$$

where P_n is the noise power in watts, k is Boltzmann's constant ($k = 1.38 \times 10^{-23}$ J/K), and T and B are the temperature in Kelvin and bandwidth in hertz, respectively.

The SNR, in decibels, at any point of a system is defined as:

$$SNR = 10\log_{10}\left(\frac{S}{N}\right) \tag{3.15}$$

where S and N are the signal and noise powers, respectively.

First, let's consider a noisy circuit as depicted in Figure 3.1. In this figure, S_i and N_i are the signal and noise powers at the circuit's input, respectively; S_o and N_o are the signal and noise powers at the circuit's output, respectively; G is the circuit's gain; and N_n is the noise added by the circuit. The noise added by the circuit can be either referenced to its input or to its output. To illustrate both cases, N_{ni} refers here to the noise added by the network with respect to its input, while N_{no} refers here to the noise added by the network with respect to its output.

Figure 3.1(a) depicts a generic representation of the noisy network. In Figure 3.1(b), the noisy network is represented while considering that the noise it adds is referenced to its input, while in Figure 3.1(c) the noise internally added by the network is considered to be referenced with respect to its output. Figure 3.1(b, c) shows the signal and noise paths from the input to the output of the noisy network. In both cases, the output signal S_o is obtained by applying the system's gain onto the input signal S_i. When the internal noise is referenced to the input of the noisy network, as illustrated in Figure 3.1(b), the total output noise is obtained by applying the system's gain (G) to the additive combination of the input noise (N_i) and the internal noise (N_{ni}). Conversely, when the internally generated noise is defined at

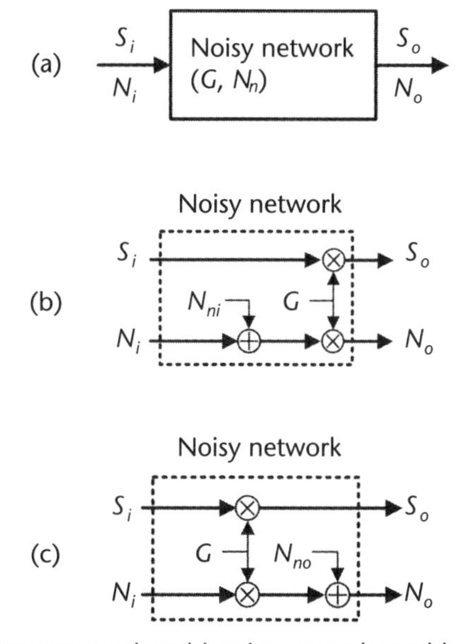

Figure 3.1 Noisy network representation: (a) noisy network as a black-box, (b) detailed signal and noise paths within a noisy network having internal noise referenced to its input, and (c) detailed signal and noise paths within a noisy network having internal noise referenced to its output.

the output of the noisy network, the output noise is obtained by passing the input noise (N_i) through the system's gain (G) and combining the resulting noise additively with the internally generated noise referenced with respect to the system's output (N_{no}) as shown in Figure 3.1(c).

The noise contribution of the circuit shown in Figure 3.1 can be quantified by its noise factor (F) defined as

$$F = \frac{SNR_{\text{in}}}{SNR_{\text{out}}} = \frac{S_i/N_i}{S_o/N_o} \tag{3.16}$$

The noise figure of a circuit is often used instead of its noise factor. The noise figure (NF) is the decibel value of the noise factor (F):

$$NF = 10\log_{10}(F) \tag{3.17}$$

The noise figure of a receiver or, equivalently, its noise factor quantifies the total amount of noise added by the various circuits of the receiver. As it can be understood from (3.16), this noise contribution will degrade the SNR of the signal, and therefore it has to be minimized. For each communication standard, minimum SNR levels are specified in order to ensure a given transmission performance. A sample of minimum SNR values required to ensure a probability of symbol error better than 10^{-3} is reported in Table 3.1 for various modulation schemes [9]. As it can be concluded, high-order constellations, which are more compact, are more sensitive to noise and therefore require higher SNR since their decision regions are smaller.

The noise (N_{out}) at the output of a noisy circuit is due to the circuit's input noise (N_0) when loaded with a matched resistor maintained at the reference temperature (T_0) and the noise generated by the circuit referenced at its input (N_n)

$$N_{\text{out}} = G \cdot \left(N_0 + N_n\right) \tag{3.18}$$

Thus, (3.16) becomes

$$F = 1 + \frac{N_n}{N_0} \tag{3.19}$$

Since the thermal noise power can be related to a temperature, one can define an equivalent noise temperature of a noisy circuit, which is the temperature at which a

Table 3.1 Required SNR for a 10^{-3} Probability of Symbol Error

Modulation Scheme	*Minimum SNR*
4 QAM	7 dB
16 QAM	12 dB
64 QAM	16.5 dB

matched resistor needs to be maintained in order to generate the same noise power as the circuit being considered [7]. Hence, N_n and N_0 can be expressed in terms of equivalent noise temperatures through

$$N_0 = k \cdot T_0 \cdot B \tag{3.20}$$

$$N_n = k \cdot T_e \cdot B \tag{3.21}$$

where T_0 is the reference temperature (T_0 = 290 K), and T_e is the equivalent noise temperature of the circuit.

Equations (3.19) to (3.21) lead to:

$$F = 1 + \frac{T_e}{T_0} \tag{3.22}$$

and

$$T_e = T_0 \cdot (F - 1) \tag{3.23}$$

Let's now consider the cascade of two circuits as depicted in Figure 3.2. The first circuit has a gain G_1, a noise factor F_1, and an equivalent noise temperature T_{e1}. Similarly, the second circuit is characterized by its gain G_2, its noise factor F_2, and its equivalent noise temperature T_{e2}. The equivalent circuit of the cascade made of circuits 1 and 2 has a gain G_e, a noise factor F_e, and an equivalent noise temperature T_e. All circuits are maintained at the reference temperature T_0.

Using (3.19), it is possible to calculate the noise generated by each circuit

$$\begin{cases} N_1 = (F_1 - 1) \cdot N_0 \\ N_2 = (F_2 - 1) \cdot N_0 \\ N_e = (F_e - 1) \cdot N_0 \end{cases} \tag{3.24}$$

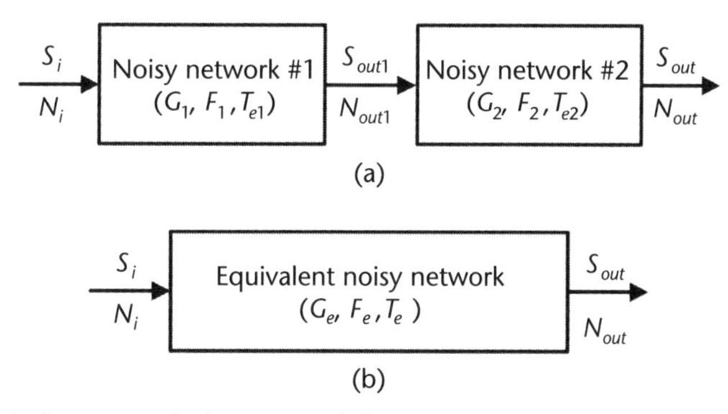

Figure 3.2 Equivalent network of two cascaded noisy networks: (a) two cascaded noisy networks, and (b) the equivalent noisy network of the cascade.

From Figure 3.2(a), the noise at the output of the cascade will be

$$N_{\text{out}} = G_2 \cdot \left(N_{\text{out1}} + N_2\right) \tag{3.25}$$

where N_{out1} is the noise at the output of the circuit 1, and input of circuit 2. N_{out1} is given by

$$N_{\text{out1}} = G_1 \cdot \left(N_i + N_1\right) \tag{3.26}$$

where N_i is the noise at the input of circuit 1. $N_i = N_0$ since the circuits are maintained at the reference temperature and matched.

By combining (3.24) to (3.25), we can rewrite the noise at the output of the cascade as

$$N_{\text{out}} = G_1 \cdot G_2 \cdot N_0 \cdot \left(F_1 + \frac{F_2 - 1}{G_1}\right) \tag{3.27}$$

Now, if we consider Figure 3.2(b), the noise at the output of the cascade is

$$\begin{aligned} N_{\text{out}} &= G_1 \cdot G_2 \cdot \left(N_0 + N_e\right) \\ &= G_1 \cdot G_2 \cdot N_0 \cdot F_e \end{aligned} \tag{3.28}$$

Equating (3.27) and (3.28), one can express the equivalent noise factor of the cascade as function of that of each circuit:

$$F_e = F_1 + \frac{F_2 - 1}{G_1} \tag{3.29}$$

Similarly, one can relate the equivalent noise temperature of the cascade as function of that of each circuit:

$$T_e = T_{e1} + \frac{T_{e2}}{G_1} \tag{3.30}$$

Equations (3.29) and (3.30) can be generalized to the case of K cascaded circuits similar to those illustrated in Figure 3.3 [7, 8]:

$$F_e = F_1 + \sum_{k=2}^{K} \left(\frac{F_k - 1}{\prod_{i=1}^{k-1} G_i} \right) \tag{3.31}$$

$$T_e = T_{e1} + \sum_{k=2}^{K} \left(\frac{T_{ek}}{\prod_{i=1}^{k-1} G_i} \right) \tag{3.32}$$

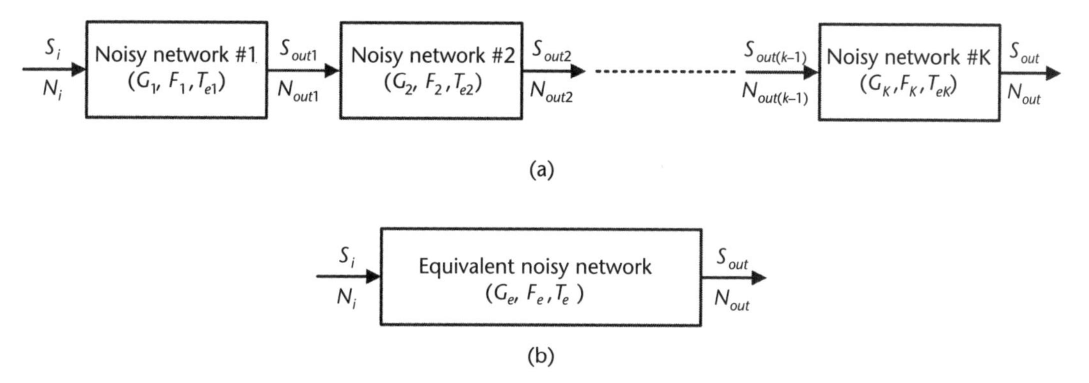

Figure 3.3 Equivalent network of *K* cascaded noisy networks: (a) *K* cascaded noisy networks, and (b) the equivalent noisy network of the cascade.

For a lossy element such as a filter, a cable, or an attenuator, having a characteristic impedance Z_0 and a loss L ($L > 1$), and maintained at a temperature T, the noise factor and the equivalent noise temperature are given by

$$T_e = (L - 1) \cdot T \tag{3.33}$$

and

$$F = 1 + (L - 1)\frac{T}{T_0} \tag{3.34}$$

Obviously, if the system is maintained at an ambient temperature equal to the reference temperature $T = T_0$, the noise factor of a lossy element is equal to its loss ($F = L$).

A closer look at the equivalent noise factor of a cascade ([3.31]) brings to light the importance of having a high gain and low noise factor at the first stage of a cascaded system to reduce its overall noise factor.

To better illustrate this, let's consider the two simplified RF front-end structures presented in Figure 3.4, which are made of the cascade of the same components but arranged in different orders. Table 3.2 summarizes the characteristics of each component.

The equivalent noise factor of the arrangement 1 shown in Figure 3.4(a) is

Table 3.2 Characteristics of the Receiver Components

Components	*Amplifier*	*Filter*	*Mixer*
Gain (dB)	20	−2	−4
Gain	100	0.63	0.40
Noise figure (dB)	1.5	2	4
Noise factor	1.41	1.58	2.51

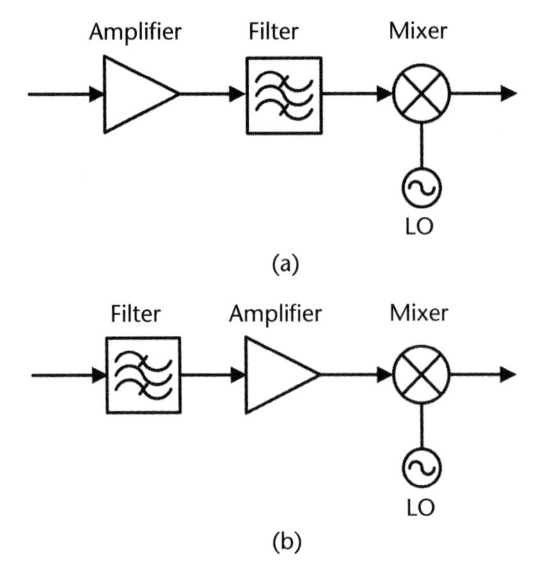

Figure 3.4 Simplified RF front-end architecture: (a) arrangement 1, and (b) arrangement 2.

$$F_a = 10^{0.15} + \frac{10^{0.2} - 1}{10^2} + \frac{10^{0.4} - 1}{10^2 \times 10^{-0.2}} = 1.44 \quad (3.35)$$

Conversely, the equivalent noise factor of the arrangement 2 reported in Figure 3.4(b) is

$$F_b = 10^{0.2} + \frac{10^{0.15} - 1}{10^{-0.2}} + \frac{10^{0.4} - 1}{10^{-0.2} \times 10^2} = 2.26 \quad (3.36)$$

The noise factor of the architecture in Figure 3.4(a) is almost equal to that of the amplifier alone. This is due to the high gain of the amplifier, which considerably reduces the contribution of the subsequent stages to the overall noise factor of the system. However, the overall noise factor of the architecture in Figure 3.4(b) is much higher than that of the filter alone. This is due to the filter's loss which makes the contribution of the amplifier's noise factor to the overall noise factor of the system significant. This clearly highlights the importance of LNAs in receiver chains and the need for low noise and high gain to significantly reduce the contribution of subsequent stages to the noise performance of the receiver.

3.2.2 Noise Analysis in Amplifiers

For an amplifier, the noise performance depends on Γ_S as well as other parameters inherent to the transistor. The noise parameters of a transistor are:

- $F_{\min}$ represents the minimum noise factor that can be obtained once an amplifier is built using this transistor. Equivalently, $NF_{\min}$ (the minimum noise figure) can be used.

- Γ_{opt} is the optimal source reflection coefficient that needs to be presented to the transistor in order to ensure that the noise factor of the designed amplifier is equal to the minimum noise factor of the transistor.
- R_n is the equivalent noise resistance.

F_{min} depends on the transistor's bias conditions as well as the frequency of operation. Moreover, Γ_{opt} is also bias and frequency-dependent. The noise parameters of a transistor are often available in its datasheet.

The noise factor of an amplifier can be expressed as a function of its noise parameters according to [3]

$$F = F_{min} + \frac{r_n}{g_s} \cdot \left| y_s - y_{opt} \right|^2 \tag{3.37}$$

where

- r_n is the normalized noise resistance ($r_n = R_n/Z_0$), and Z_0 is the characteristic impedance ($Z_0 = 50\Omega$).
- y_s and y_{opt} are the normalized admittances associated with the actual Γ_s and Γ_{opt}, respectively. The normalized admittance is related to the reflection coefficient through

$$y = \frac{1-\Gamma}{1+\Gamma} \tag{3.38}$$

- g_s is the normalized conductance associated with $y_s (g_s = \mathrm{Re}(y_s))$.

It is useful to rewrite (3.37) as a function of Γ_s so that one can easily determine on the Smith chart the locus of the source reflection coefficients that will lead to equal noise factors and equivalently noise figures.

$$F = F_{min} + \frac{4r_n \cdot \left| \Gamma_S - \Gamma_{opt} \right|^2}{\left(1 - \left|\Gamma_S\right|^2\right) \cdot \left|1 + \Gamma_{opt}\right|^2} \tag{3.39}$$

3.2.3 Design Procedure

This section discusses the design of LNAs in order to achieve a given noise factor or equivalently a given noise figure. Equation (3.39) can be used to locate on the Smith chart the values of Γ_s that result in a constant noise factor (F). This contour is a circle defined by its center (c_F) and radius (r_F) given by (3.40) and (3.41), respectively.

$$c_F = \frac{\Gamma_{opt}}{1+N} \tag{3.40}$$

$$r_F = \frac{1}{1+N}\sqrt{N^2 + N \cdot \left(1 - \left|\Gamma_{opt}\right|^2\right)} \tag{3.41}$$

where N is the noise figure parameter defined as

$$N = \frac{F - F_{\min}}{4r_n}\left|1 + \Gamma_{\text{opt}}\right|^2 \tag{3.42}$$

It is important to note that the centers of all noise circles are located on the same line connecting the center of the Smith chart and the point $\Gamma_S = \Gamma_{\text{opt}}$. According to (3.40) for any value of N (or equivalently F), the center of the corresponding noise circle will have the same phase as Γ_{opt}. Moreover, as the noise factor (F) increases, the noise figure parameter (N) increases, and therefore the radius of the noise circle increases. A sample plot of noise circles that illustrates these properties is reported in Figure 3.5.

Once the noise circles are drawn, the appropriate source reflection coefficient (Γ_S) can be selected. A Γ_S that lies on a noise circle will lead to a noise factor equal to the corresponding value on that circle, while a Γ_S that is located inside a specific noise circle will lead to a noise factor that is better (i.e., lower) than that associated with this noise circle. It is important to recall that as demonstrated by (3.39), the noise factor of an amplifier solely depends on the source reflection coefficient and not the load reflection coefficients. Therefore, once Γ_S is selected using the noise circles, the Γ_L will be chosen to ensure conjugate match at the output ($\Gamma_L = \Gamma^*_{\text{out}}$)

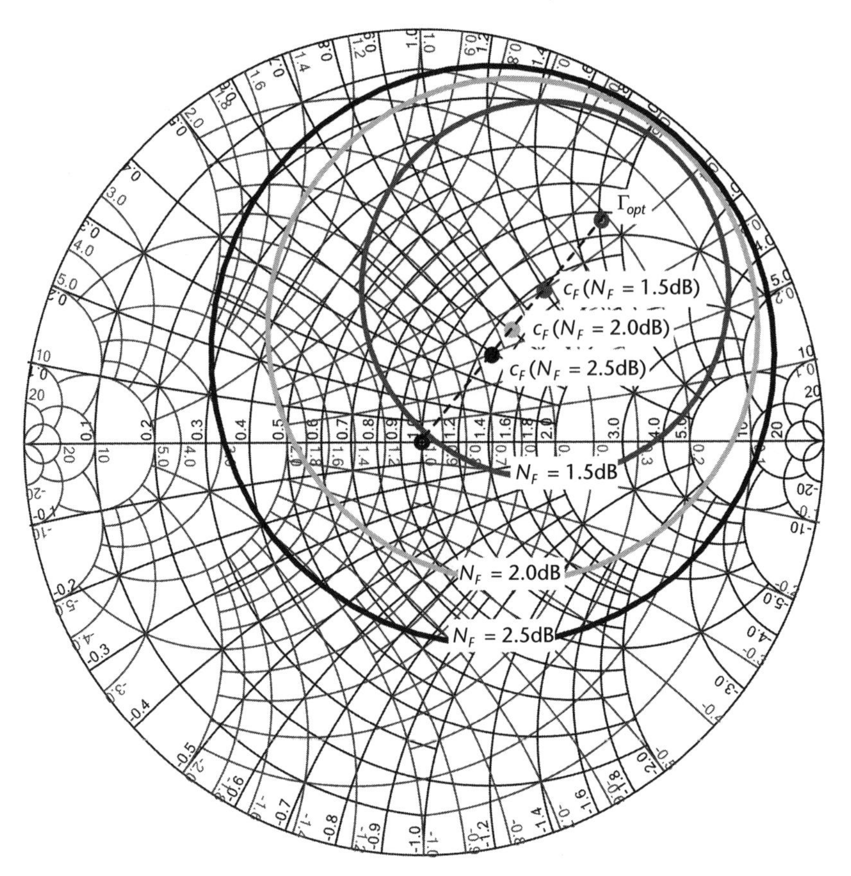

Figure 3.5 Sample plot of noise circles showing that all centers are aligned.

and hence achieve the maximum possible gain for the selected Γ_S. In some cases, a tradeoff between noise and gain is needed, and in such cases, noise and gain circles are simultaneously considered to select appropriate source reflection coefficient as will be discussed in Section 3.3.

3.2.4 Design Example

Let's consider a low noise transistor biased at $V_{DS} = 2\text{V}$ and $I_{DS} = 5$ mA and operated at 3 GHz. The transistor's noise parameters are $NF_{\min} = 0.48$ dB, $\Gamma_{\text{opt}} = 0.59\angle 98^\circ$, and $r_n = 0.11$. The S-parameters of the transistor are:

$$S = \begin{bmatrix} 0.71\angle -139^\circ & 0.05\angle 5^\circ \\ 3.34\angle 76^\circ & 0.40\angle -110^\circ \end{bmatrix}$$

The objective is to:

- Draw the noise circles for $NF = 1$ dB, $NF = 2$ dB, and $NF = 3$ dB.
- Determine the source and load reflection coefficients to build:
  - An LNA having the minimum NF with the highest possible gain.
  - An LNA having an $NF = 3$ dB.

Let's first assess the stability of this transistor. Equations (3.4) and (3.5) can be used to calculate the values of K and $|\Delta|$, respectively. These are found to be $K = 1.084$ and $|\Delta| = 0.162$. These values satisfy the unconditional stability conditions as expressed in (3.6) and (3.7). Therefore, the transistor is unconditionally stable at these operating conditions.

Next, let's verify whether the transistor can be considered as a unilateral device or as a bilateral device. The unilaterality figure of merit is given by [3, 10]:

$$U = \frac{|S_{12}||S_{21}||S_{11}||S_{22}|}{\left(1 - |S_{11}|^2\right)\left(1 - |S_{22}|^2\right)} \tag{3.43}$$

For this transistor, $U = 0.114$. Typically to consider a transistor unilateral, the unilaterality figure of merit has to satisfy [3]

$$U \leq 0.03 \tag{3.44}$$

Table 3.3 Coordinates of Noise Circle Centers and Their Radii

	$NF = 1\ dB$	$NF = 2\ dB$	$NF = 3\ dB$
Noise figure parameter (N)	0.3822	1.2593	2.3634
Noise circle center (c_F)	$0.4268\angle 98^\circ$	$0.2611\angle 98^\circ$	$0.1754\angle 98^\circ$
Noise circle radius (r_F)	0.4549	0.6867	0.7937

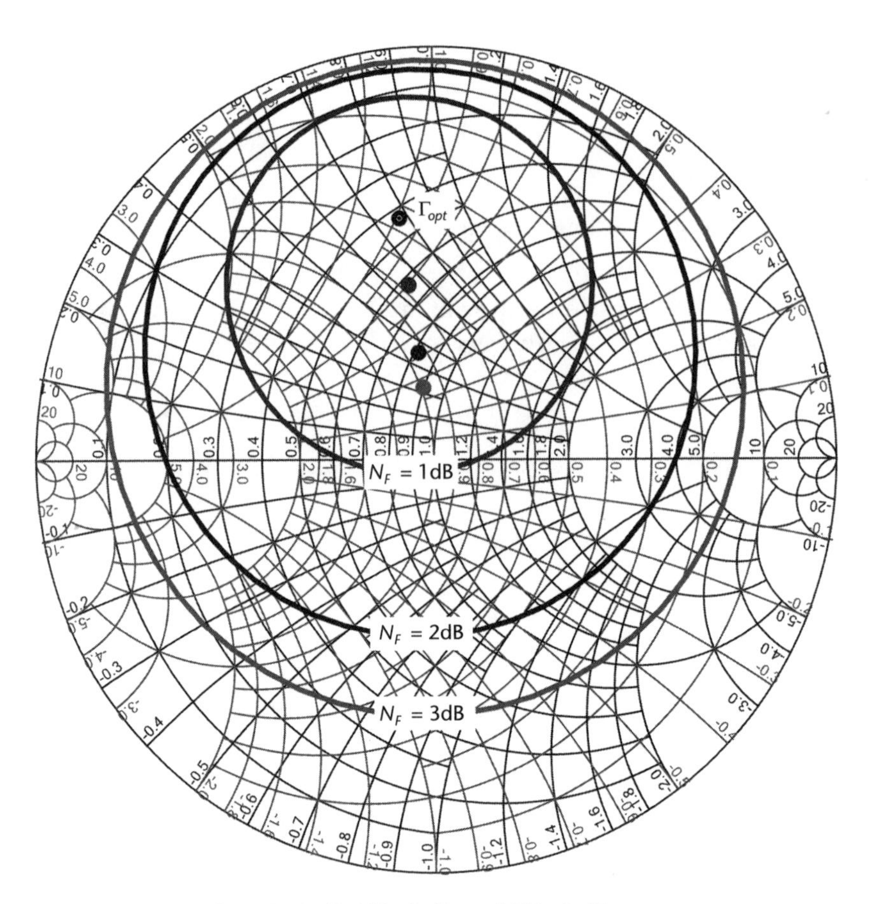

Figure 3.6 Noise circles for $NF = 1$ dB, $NF = 2$ dB, and $NF = 3$ dB.

Hence, this transistor is considered bilateral. Using (3.40) through (3.42), it is possible to calculate the coordinates of the noise circles centers as well as their radii. These are summarized in Table 3.3. The $NF = 1$ dB, $NF = 2$ dB, and $NF = 3$ dB noise circles are depicted in Figure 3.6.

To build an amplifier having minimum noise figure, Γ_S should be equal to Γ_{opt}. Thus, $\Gamma_S = 0.59\angle 98^\circ$. For the selected Γ_S, the value of the resulting transistor's Γ_{out} needs to be calculated using (3.3), giving $\Gamma_{out} = 0.415\angle -128.76^\circ$. In order to ensure maximum gain of this amplifier, Γ_L presented to the transistor needs to satisfy the complex conjugate match condition, and therefore $\Gamma_L = \Gamma^*_{out} = 0.415\angle +128.76^\circ$.

To design an amplifier such that $NF = 3$ dB, Γ_S has to be chosen on the 3-dB noise circle. Even though any of Γ_S points that lie on the 3-dB noise circle will lead to the same *NF*, they will result in different gain values. To optimize both the noise and gain of the designed amplifier, gain circles need to be drawn on the Γ_S plane as discussed in Section 3.4. For the time being, let's select an arbitrary Γ_S on the 3-dB noise circle since no specific gain requirements are defined. One possible choice is $\Gamma_S = 0.668\angle -32.71^\circ$. To maximize the gain of the amplifier for this Γ_S selection, conjugate match is needed at the output of the transistor. Consequently, $\Gamma_L = \Gamma^*_{out} = 0.332\angle +104.60^\circ$ where $\Gamma_{out} = 0.332\angle -104.60^\circ$ is the output reflection coefficient of the transistor calculated for the selected value of Γ_S using (3.3). The input and output matching network can be synthesized based on the techniques presented in Chapter 2.

3.3 Maximum Gain Amplifier Design

In most applications, a single-stage amplifier is unable to meet the gain requirements. Therefore, amplification systems are often built by cascading several stages. One or more of these stages is optimized for maximum gain. In a receiver setting, the first amplifier right after the antenna is designed for low noise, while the subsequent stages are often designed for maximum gain. A similar trend is observed in transmitters where the last amplifier (a power amplifier optimized for efficiency, output power, and/or linearity) is driven by preceding stages where gain is a more important concern.

This section discusses the design of maximum gain amplifiers for both unilateral and bilateral transistors.

3.3.1 Matching Requirements

In order to analyze the gain of a microwave amplifier, it is essential to define the various power quantities encountered in such systems. In a very general case, one can consider that an amplifier is driven by a source at the input and delivers its output power to a load as depicted in Figure 3.7. This generic representation replaces the circuit preceding the amplifier by its Thevenin equivalent and the circuit connected at the output of the amplifier by a load equivalent. The amplifier consists of the cascade of an input matching network (IMN), the transistor, and the output matching network (OMN).

As illustrated in Figure 3.7, one can define at least four power quantities: the power available from the source (P_S), the power at the input of the transistor (P_{in}), the power at the output of the transistor (P_{out}), and the power delivered to the load (P_L). Using these various powers, three gains can be distinguished for an amplifier circuit. These are namely: the available gain (G_A), the operating gain (G_P), and the transducer gain (G_T). These three gains are defined as [2–4]

$$G_A = \frac{P_{out}}{P_S} \tag{3.45}$$

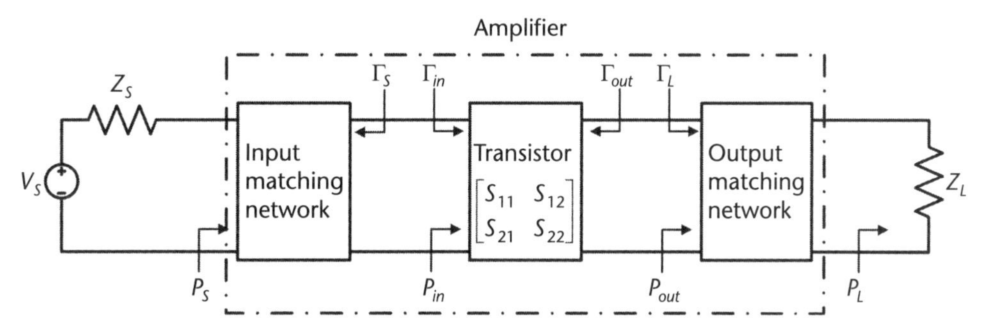

Figure 3.7 Power and reflection coefficients definitions in an amplifier circuit.

$$G_P = \frac{P_L}{P_{\text{in}}} \tag{3.46}$$

$$G_T = \frac{P_L}{P_S} \tag{3.47}$$

For a two-port bilateral network, the available, operating, and transducer gain can be expressed as functions of the device S-parameters and the source and load reflection coefficients according to (3.48)–(3.50), respectively [2–4].

$$G_A = \frac{1 - |\Gamma_S|^2}{|1 - S_{11}\Gamma_S|^2} |S_{21}|^2 \frac{1}{1 - |\Gamma_{\text{out}}|^2} \tag{3.48}$$

$$G_P = \frac{1}{1 - |\Gamma_{\text{in}}|^2} |S_{21}|^2 \frac{1 - |\Gamma_L|^2}{|1 - S_{22}\Gamma_L|^2} \tag{3.49}$$

$$\begin{aligned} G_T &= \frac{1 - |\Gamma_S|^2}{|1 - \Gamma_{\text{in}}\Gamma_S|^2} |S_{21}|^2 \frac{1 - |\Gamma_L|^2}{|1 - S_{22}\Gamma_L|^2} \\ G_T &= \frac{1 - |\Gamma_S|^2}{|1 - S_{11}\Gamma_S|^2} |S_{21}|^2 \frac{1 - |\Gamma_L|^2}{|1 - \Gamma_{\text{out}}\Gamma_L|^2} \end{aligned} \tag{3.50}$$

Based on the above definitions, it appears that G_A includes the IMN and the transistor, but is independent of the output matching. This is reflected in (3.48), which shows that G_A is function of the transistor's S-parameters and Γ_S. Equivalently, G_P takes into consideration the transistor and the OMN but not the IMN. This appears in (3.49) since G_P is only function of S-parameters of the transistor and Γ_L. Conversely, G_T provides a more comprehensive insight about the amplifier gain as it considers the transistor and both IMN and OMN. This is confirmed by (3.50), which reflects that G_T is function of the S-parameters, Γ_S and Γ_L.

In a maximum gain amplifier, Γ_S and Γ_L are selected to maximize G_T. This can be achieved by ensuring that there is a maximum power transfer from the source to the transistor's input ($P_{\text{in}} = P_S$), and from the transistor's output to the load ($P_L = P_{\text{out}}$). This can be achieved through simultaneous complex conjugate match at the transistor's input and output:

$$\begin{cases} \Gamma_S = \Gamma_{\text{in}}^* \\ \text{and} \\ \Gamma_L = \Gamma_{\text{out}}^* \end{cases} \tag{3.51}$$

3.3.2 Design Procedure

Implementing the simultaneous conjugate match condition of (3.51) is straightforward in the case of unilateral transistors. Indeed, as a consequence of the unilaterality property, one can consider that $S_{12} \approx 0$, and therefore, based on (3.2) and (3.3), the transistor's input reflection coefficient (Γ_{in}) will be constant ($\Gamma_{in} = S_{11}$) and independent of Γ_L. For the same reason, the transistor's output reflection coefficient (Γ_{out}) will also be constant ($\Gamma_{out} = S_{22}$) and independent of Γ_S. Consequently, for unilateral transistors, the simultaneous complex conjugate matching condition of (3.51) becomes

$$\begin{cases} \Gamma_S = S_{11}^* \\ \text{and} \\ \Gamma_L = S_{22}^* \end{cases} \tag{3.52}$$

For a unilateral transistor, simultaneous conjugate match according to the condition of (3.52) leads to a maximum transducer gain given by

$$G_{T\max Uni} = \frac{1}{1-|S_{11}|^2}|S_{21}|^2\frac{1}{1-|S_{22}|^2} \tag{3.53}$$

The design of maximum gain amplifiers in the case of a bilateral transistor requires solving simultaneously (3.2), (3.3), and (3.51). When the transistor is unconditionally stable, the simultaneous conjugate match condition leads to [3, 4, 10]

$$\begin{cases} \Gamma_{MS} = \dfrac{B_1 - \sqrt{B_1^2 - 4|C_1|^2}}{2C_1} \\ \text{and} \\ \Gamma_{ML} = \dfrac{B_2 - \sqrt{B_2^2 - 4|C_2|^2}}{2C_2} \end{cases} \tag{3.54}$$

where

$$B_1 = 1 + |S_{11}|^2 - |S_{22}|^2 - |\Delta|^2 \tag{3.55}$$

$$B_2 = 1 + |S_{22}|^2 - |S_{11}|^2 - |\Delta|^2 \tag{3.56}$$

$$C_1 = S_{11} - \Delta S_{22}^* \tag{3.57}$$

$$C_2 = S_{22} - \Delta S_{11}^* \tag{3.58}$$

$$\Delta = S_{11}S_{22} - S_{12}S_{21} \tag{3.59}$$

The resulting maximum transducer gain for a bilateral transistor is

$$G_{T\max} = \frac{1}{1 - |\Gamma_{MS}|^2}|S_{21}|^2 \frac{1 - |\Gamma_{ML}|^2}{|1 - S_{22}\Gamma_{ML}|^2} \tag{3.60}$$

For a potentially unstable transistor, the maximum transducer gain should not exceed the maximum stable gain. This latter can be calculated using

$$G_{T\max\text{Stable}} = \frac{|S_{21}|}{|S_{12}|} \tag{3.61}$$

In such case, to design an amplifier having the maximum stable gain, it is more appropriate to use the available and/or operating gain circles approach. The gain circles based approach will be thoroughly detailed when discussing the design for gain-noise trade-off in Section 3.4.

3.3.3 Design Example

Let's consider the transistor used in the design example of Section 3.2.4, and use it to design an amplifier with a maximum transducer gain.

Based on the stability and unilaterality analysis performed in Section 3.2.4, this transistor is bilateral and unconditionally stable. The maximum transducer gain that it can deliver, calculated using (3.60), is $G_{T\max} = 16.47$ dB. The simultaneous complex conjugate match can be obtained, using (3.54), by selecting $\Gamma_S = 0.90\angle 141.92°$ and $\Gamma_L = 0.80\angle 121.14°$. Figure 3.8 depicts the locus of Γ_S and Γ_L for maximum transducer gain.

Comparing the Γ_S for minimum noise (depicted in Figure 3.6) and that for maximum gain (shown in Figure 3.8), it appears that optimizing noise and gain cannot be achieved simultaneously since the values of Γ_{opt} and Γ_{MS} are away from each other. Hence, noise and gain circles are to be used when designing an amplifier to meet specific noise and gain performances.

3.4 Amplifier Design for Gain-Noise Trade-Off

When designing amplifiers for specific gain and noise performances, the design methodology consists of deriving the gain and noise circles, and using them to select the source reflection coefficient to be presented at the input of the transistor. The load reflection coefficient is later calculated once the source reflection coefficient is chosen. Based on the gain definitions introduced earlier in this chapter, the available gain of a bilateral amplifier is function of $\Gamma_S$while the operating gain is function of Γ_L. Accordingly, since the noise circles are drawn in the Γ_S plane, available gain

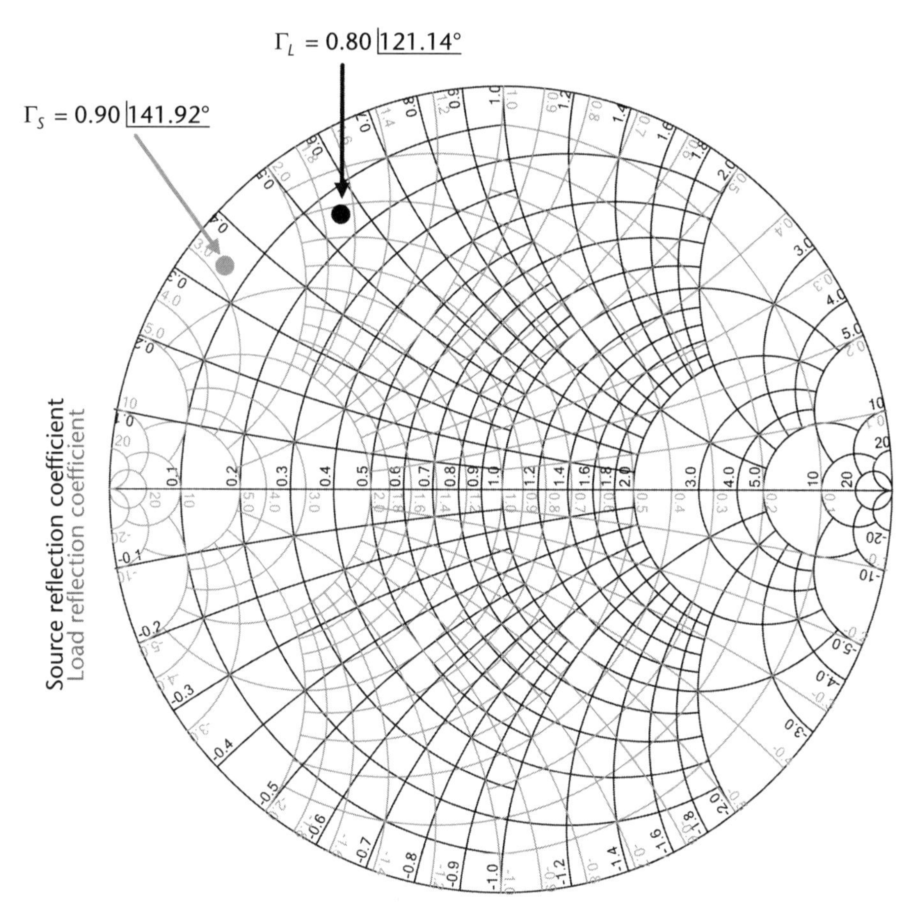

Figure 3.8 Γ_S and Γ_L for maximum transducer gain (simultaneous conjugate match).

circles are to be used when designing amplifiers for a gain-noise trade-off. This approach is generic, although there are some specificities to be distinguished based on whether a transistor is considered unilateral or bilateral. Since noise circles have already been introduced in Section 3.2.3, only gain circles will be discussed in the next section.

3.4.1 Gain Circles

3.4.1.1 Case of Unilateral Transistors

For a unilateral transistor, the transducer gain can be derived from (3.50) by setting $\Gamma_{in} = S_{11}$ and $\Gamma_{out} = S_{22}$ as a direct consequence of the unilaterality property. Hence, [3, 4, 10]

$$G_{TUni} = \frac{1-|\Gamma_S|^2}{|1-S_{11}\Gamma_S|^2}|S_{21}|^2\frac{1-|\Gamma_L|^2}{|1-S_{22}\Gamma_L|^2} \tag{3.62}$$

The independence between the input and output matching conditions for a unilateral device makes it possible to separate the gain of IMN and that of the OMN. The gain of the IMN (G_S) and that of the OMN (G_L) are defined as

$$G_S = \frac{1-|\Gamma_S|^2}{|1-S_{11}\Gamma_S|^2} \tag{3.63}$$

$$G_L = \frac{1-|\Gamma_L|^2}{|1-S_{22}\Gamma_L|^2} \tag{3.64}$$

It can be demonstrated that the values of Γ_S that will result in a constant G_S gain lie on a circle with radius and center are identified by (3.65) and (3.66), respectively [3, 4, 10].

$$r_{G_S} = \frac{\sqrt{1-g_S}\left(1-|S_{11}|^2\right)}{1-|S_{11}|^2(1-g_S)} \tag{3.65}$$

$$c_{G_S} = \frac{g_S S_{11}^*}{1-|S_{11}|^2(1-g_S)} \tag{3.66}$$

where

$$g_S = G_S\left(1-|S_{11}|^2\right) \tag{3.67}$$

It is worthy to mention that in (3.67), G_S is a linear gain (not in dB scale).

Similarly, one can define constant G_L circles which correspond to the values of Γ_L that will lead to a gain of the OMN equal to G_L. The radius (r_{GL}) and center (c_{GL}) of a constant G_L circle are [3, 4, 10]:

$$r_{G_L} = \frac{\sqrt{1-g_L}\left(1-|S_{22}|^2\right)}{1-|S_{22}|^2(1-g_L)} \tag{3.68}$$

$$c_{G_L} = \frac{g_L S_{22}^*}{1-|S_{22}|^2(1-g_L)} \tag{3.69}$$

where g_L is function of the linear gain G_L defined as

$$g_L = G_L\left(1-|S_{22}|^2\right) \tag{3.70}$$

From (3.66) and (3.69), it appears that the centers of the constant G_S and G_L circles are located on the line joining the center of the Smith chart and S^*_{11} for the constant G_S circles, and the center of the Smith chart and S^*_{22} for the constant G_L circles. Moreover, as the value of G_S (or equivalently G_L) increases, the radius of the corresponding gain circle decreases.

Sample constant G_S and constant G_L circles are shown in Figures 3.9 and 3.10, respectively. These circles are derived for a device whose S-parameters are:

$$S = \begin{bmatrix} 0.6\angle -170^\circ & 0 \\ 2.9\angle 46^\circ & 0.4\angle -80^\circ \end{bmatrix}$$

Constant G_S and G_L circles centers and radii can be calculated using (3.65) to (3.70). These values are summarized in Tables 3.4 and 3.5 for the constant G_S and the constant G_L circles, respectively. The maximum values of G_S and G_L, obtained for when $\Gamma_S = S^*_{11}$ and $\Gamma_L = S^*_{22}$, respectively; are $G_{S\max} = 1.94$ dB and $G_{L\max} = 0.76$ dB.

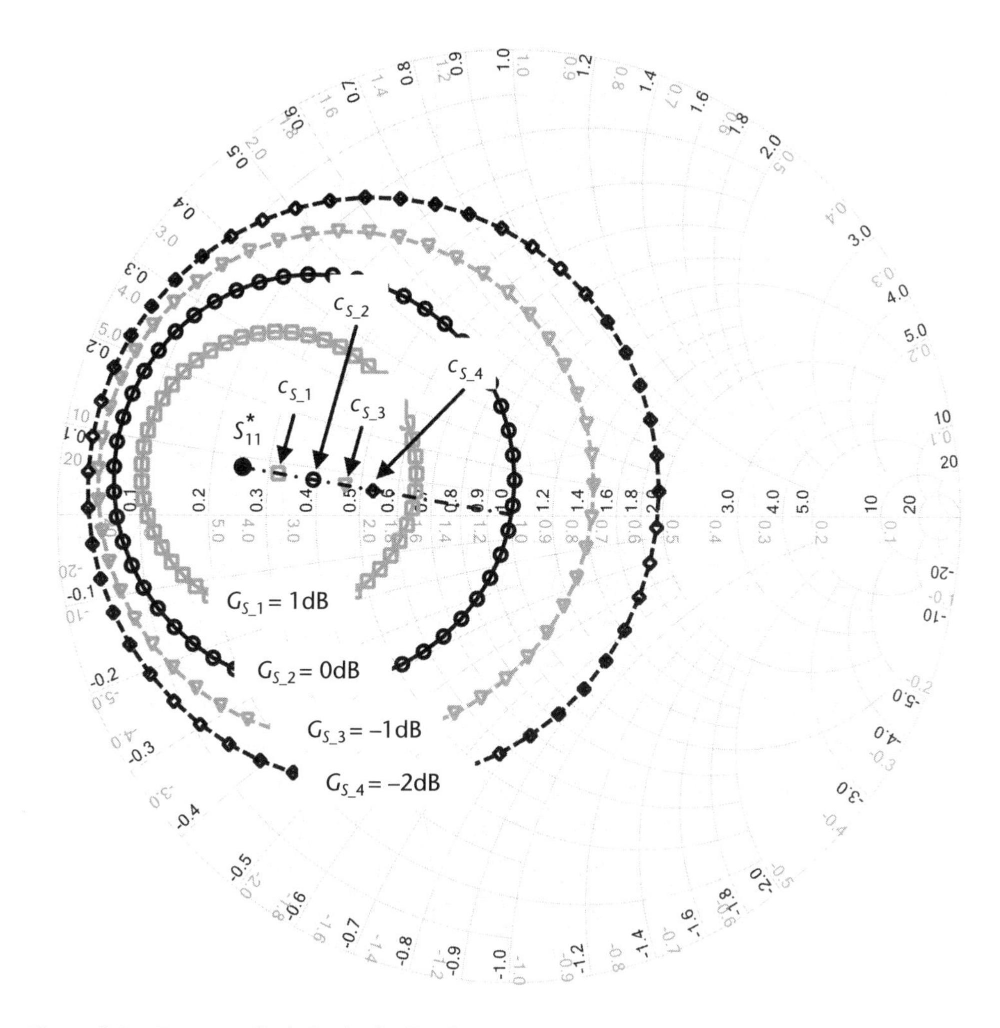

Figure 3.9 Constant G_S circles in the Γ_S plane.

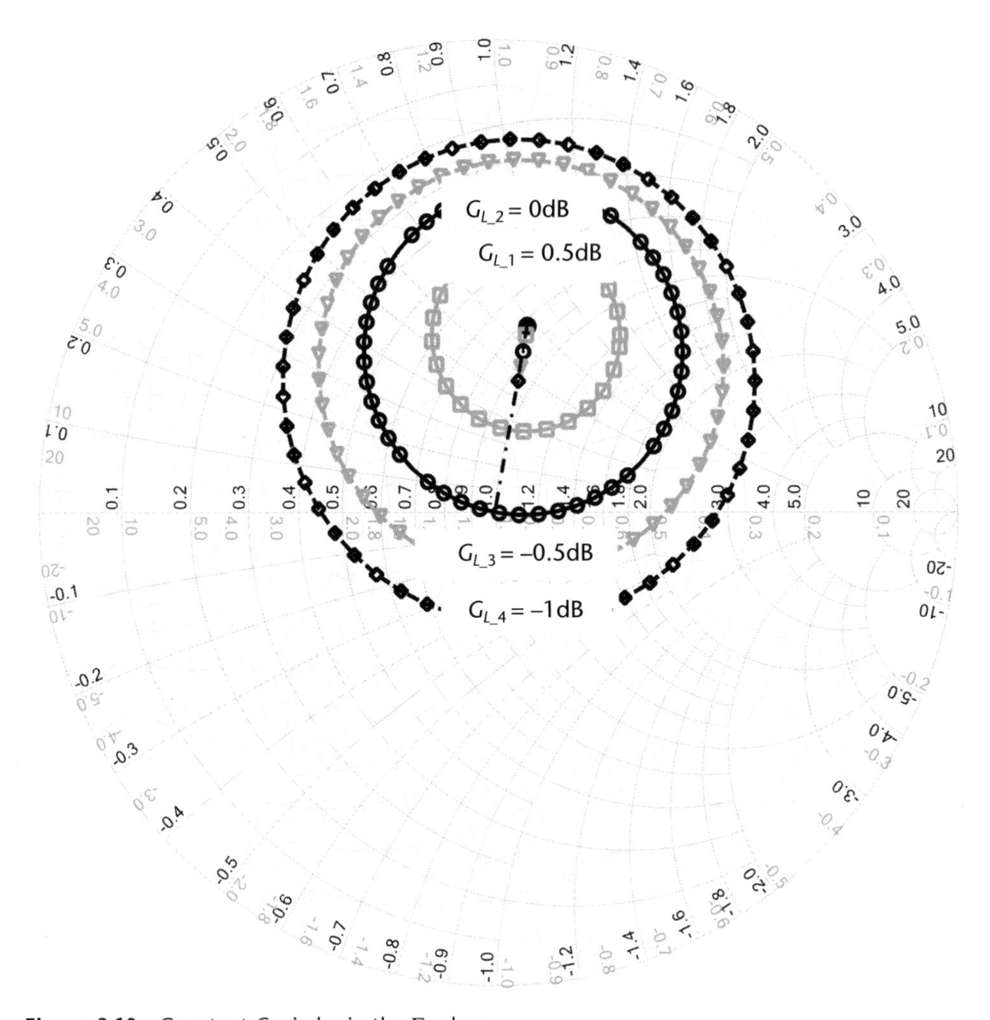

Figure 3.10 Constant G_L circles in the Γ_L plane.

Table 3.4 Coordinates of Constant G_S Circles Centers and Their Radii

	$G_S = 1\ dB$	$G_S = 0\ dB$	$G_S = -1\ dB$	$G_S = -2\ dB$
g_S	0.806	0.640	0.508	0.404
Gain circle center (c_{GS})	$0.520\angle 170°$	$0.441\angle 170°$	$0.371\angle 170°$	$0.309\angle 170°$
Gain circle radius (r_{GS})	0.303	0.441	0.545	0.629

Table 3.5 Coordinates of Constant G_L Circle Centers and Their Radii

	$G_L = 0.5\ dB$	$G_L = 0\ dB$	$G_L = -0.5\ dB$	$G_L = -1\ dB$
g_L	0.942	0.840	0.749	0.667
Gain circle center (c_{GL})	$0.380\angle 80°$	$0.345\angle 80°$	$0.312\angle 80°$	$0.282\angle 80°$
Gain circle radius (r_{GL})	0.203	0.345	0.439	0.512

3.4.1.2 Case of Bilateral Transistors

For a bilateral transistor, the design for a specific gain is carried out using the available gain and operating gain circles. As defined in (3.48), the available gain G_A is function of the transistor's S-parameters and Γ_S. The values of Γ_S that result in a specific value of the available gain G_A lie on a circle whose radius (r_{GA}) and center (c_{GA}) are given by

$$r_{G_A} = \frac{\sqrt{1 - 2K|S_{12} \cdot S_{21}|g_A + |S_{12} \cdot S_{21}|^2 \cdot g_A^2}}{\left|1 + g_A\left(|S_{11}|^2 - |\Delta|^2\right)\right|} \tag{3.71}$$

$$c_{G_A} = \frac{g_A C_1^*}{1 + g_A\left(|S_{11}|^2 - |\Delta|^2\right)} \tag{3.72}$$

where K, Δ, and C_1 are defined by (3.4), (3.5), and (3.57), respectively. g_A is defined as

$$g_A = \frac{G_A}{|S_{21}|^2} \tag{3.73}$$

In a similar way, it can be shown that the values of Γ_L for which a constant operating gain G_P is obtained are located on a circle whose radius (r_{GP}) and center (c_{GP}) are given by [3, 4, 10]

$$r_{G_P} = \frac{\sqrt{1 - 2K|S_{12} \cdot S_{21}|g_P + |S_{12} \cdot S_{21}|^2 \cdot g_P^2}}{\left|1 + g_P\left(|S_{22}|^2 - |\Delta|^2\right)\right|} \tag{3.74}$$

$$c_{G_P} = \frac{g_P C_2^*}{1 + g_P\left(|S_{22}|^2 - |\Delta|^2\right)} \tag{3.75}$$

where K, Δ, and C_2 are defined by (3.4), (3.5), and (3.58), respectively. g_P is defined as

$$g_P = \frac{G_P}{|S_{21}|^2} \tag{3.76}$$

Similarly to the case of the G_S and G_L circles, the centers of the G_A circles are located on the same line since all centers will have the same angle as C_1^* and the centers of the G_P circles are located on the same line since all centers will have the same angle as C_2^*. Moreover, as the value of the gain (G_A or G_P) increases, the radius of the corresponding gain circle decreases.

Sample constant G_A and constant G_P circles are presented in Figures 3.11 and 3.12, respectively. These circles are derived for an unconditionally stable bilateral device whose S-parameters are:

$$S = \begin{bmatrix} 0.6\angle{-120^\circ} & 0.05\angle{-132^\circ} \\ 3.6\angle{60^\circ} & 0.2\angle{-20^\circ} \end{bmatrix}$$

The radii and the center coordinates of the constant G_A and constant G_P circles are reported in Tables 3.6 and 3.7, respectively. As it was observed for the G_S and G_L circles, for all values of G_A, the centers of the corresponding circles are aligned. Furthermore, as the gain value increases, the radius of the constant gain circles decreases. The same also applies to the constant G_P circles.

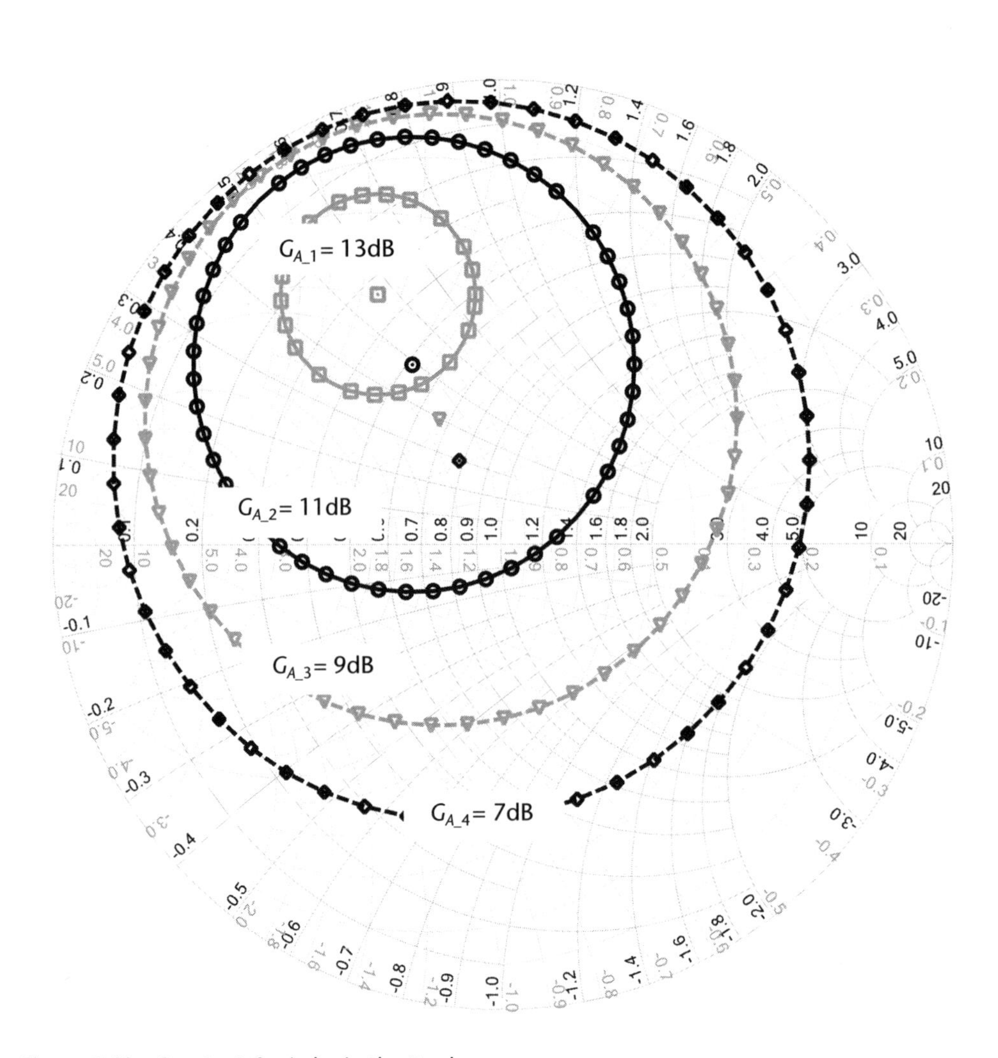

Figure 3.11 Constant G_A circles in the Γ_S plane.

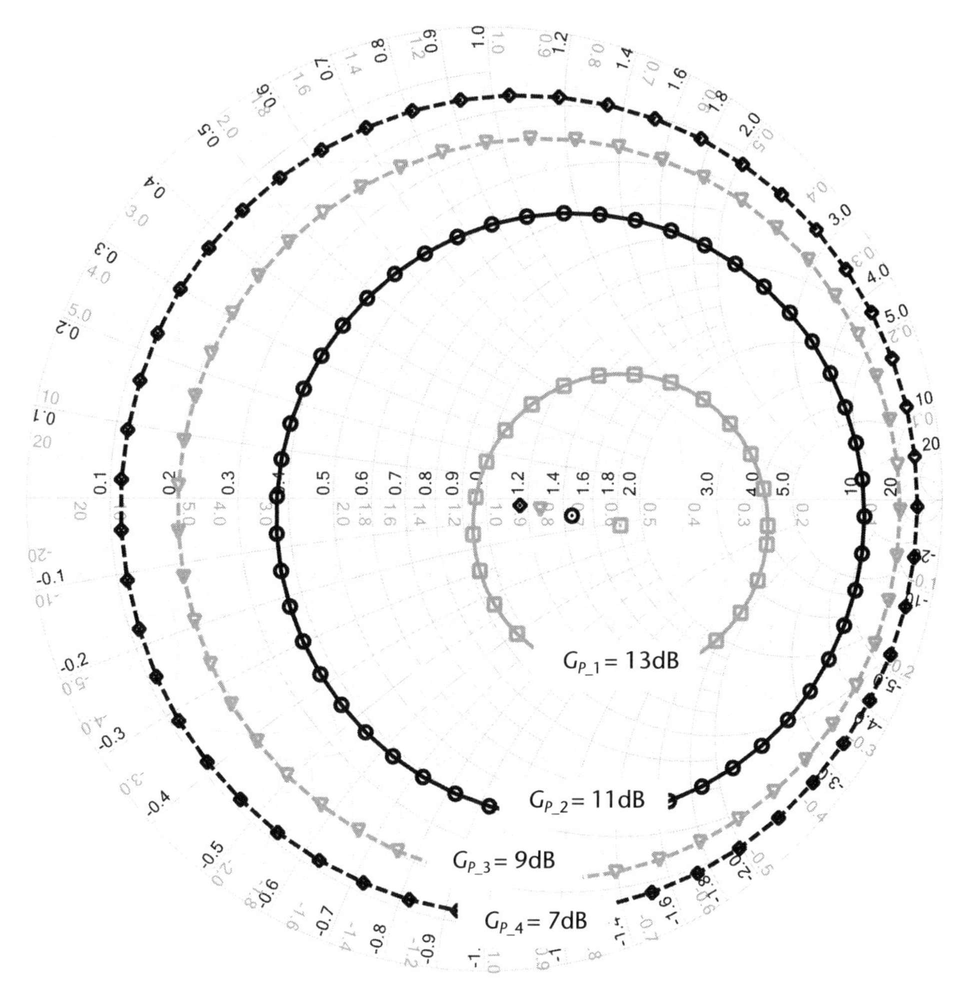

Figure 3.12 Constant G_P circles in the Γ_L plane.

Table 3.6 Coordinates of Constant G_A Circle Centers and Their Radii

	$g_A = 13\ dB$	$g_A = 11\ dB$	$g_A = 9\ dB$	$g_A = 7\ dB$
g_A	1.540	0.971	0.613	0.387
Gain circle center (r_{G_A})	$0.60\angle 116.76^\circ$	$0.44\angle 116.76^\circ$	$0.30\angle 116.76^\circ$	$0.20\angle 116.76^\circ$
Gain circle radius (r_{G_A})	0.216	0.490	0.658	0.774

Table 3.7 Coordinates of Constant G_P Circle Centers and Their Radii

	$g_P = 13\ dB$	$g_P = 11\ dB$	$g_P = 9\ dB$	$g_P = 7\ dB$
g_P	1.540	0.971	0.613	0.387
Gain circle center (c_{G_P})	$0.30\angle -10.73^\circ$	$0.19\angle -10.73^\circ$	$0.12\angle -10.73^\circ$	$0.08\angle -10.73^\circ$
Gain circle radius (r_{G_P})	0.321	0.641	0.786	0.869

3.4.2 Design Procedure

The design of any amplifier starts by selecting the DC bias point which will later set the AC characteristics of the amplifier (including S-parameters and noise parameters). The choice of the bias point for small-signal amplifiers is often similar to biasing a class A amplifier for maximum linearity.

The gain of the amplifier can be controlled by Γ_S and Γ_L presented to the transistor. However, its noise performance is only function of Γ_S. Therefore, when gain and noise requirements are defined for an amplifier, the design procedure always starts by determining the Γ_S. Hence, for a unilateral transistor, the design for specific gain and noise implies the use of constant G_S circles and constant noise circles. In the same way, for a bilateral transistor, constant G_A circles and constant noise circles are used.

It is important to recall here, that a constant gain circle represents the set of values of reflection coefficients to be presented to the transistor in order to achieve that specific gain. Likewise, a constant noise circle represents the set of values of Γ_S to be presented to the transistor to achieve that specific noise. In both cases, selecting a reflection coefficient inside the gain (or noise) circle results in better performance (i.e., higher gain for gain circles and lower noise for noise circles). Consequently, to meet predefined gain and noise performances, one has to select a reflection coefficient that is inside the corresponding gain and noise circles.

The design for gain and noise trade-off in the unilateral case is flexible in the sense that the designer can choose how to achieve the desired transducer gain by specifying values of the gains G_S and G_L. However, one should make sure that the selected G_S gain can be achieved along with the specified noise performance. This requires that the gain circle associated with the selected G_S and the noise circle associated with the desired noise performance have at least partial overlap in the Γ_S plane. If there is no overlap between the selected G_S circle and the desired noise performance circle, one can lower the selected value of the gain G_S while increasing the value of the gain G_L in a way that would maintain the transducer gain within the desired range. Reducing the value of the gain G_S will result in a larger gain circle and therefore lead to an overlap between the gain and noise circle to allow for the proper selection of the source reflection coefficient. The load reflection coefficient can be selected later based on the desired value of G_L. The design procedure in the case of a unilateral device is depicted in the flowchart of Figure 3.13.

In the case of a bilateral device, the available gain circle and the noise circle are considered. Let's assume that it is required to design an amplifier having a transducer gain of at least G_{T0}, and a noise figure better than NF_0. To design such amplifier, the constant available gain circle is drawn for $G_A = G_{T0}$ along with the constant noise circle corresponding to $NF = NF_0$. A source reflection coefficient that is located within the intersection of these two circles is then selected. To ensure that the transducer gain of the amplifier is equal to the desired value (G_{T0}), a complex conjugate match needs to be implemented at the transistor's output. This is performed by calculating the transistor's output reflection coefficient (Γ_{out}) for the selected source reflection coefficient (Γ_S) using (3.3), and synthesizing an output matching network such that $\Gamma_L = \Gamma^*_{out}$. Such output matching will result in $G_T = G_A$, and hence $G_T = G_{T0}$. This process is summarized in the flowchart of Figure 3.14.

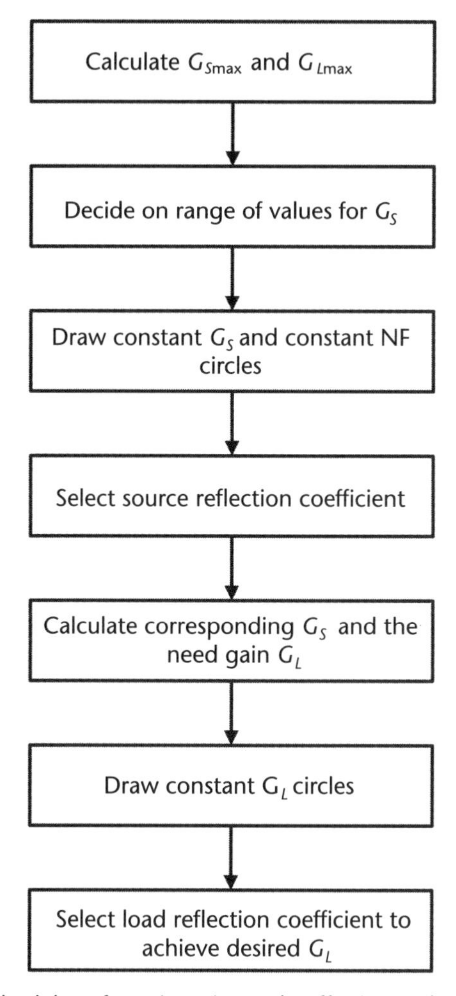

Figure 3.13 Design methodology for gain-noise trade-off using unilateral devices.

3.4.3 Design Examples

3.4.3.1 Case of Unilateral Transistors

Let's consider the example of a unilateral device having the S-parameters shown below, and for which the noise characteristics are $NF_{\min} = 0.4$ dB, $\Gamma_{\text{opt}} = 0.5\angle 30°$, and $R_n = 4\Omega$.

$$S = \begin{bmatrix} 0.6\angle -160° & 0 \\ 3.2\angle 46° & 0.4\angle -80° \end{bmatrix}$$

It is required to design, using this transistor, an amplifier that has a noise figure better than 0.6 dB and a transducer gain of at least 11.50 dB.

First, let's consider the maximum transducer gain that can be obtained using this transistor ($G_{T\max Uni}$). Using (3.53), it appears that $G_{T\max Uni} = 12.80$ dB. Hence, the designer has some flexibility while selecting Γ_S and Γ_L to achieve the desired

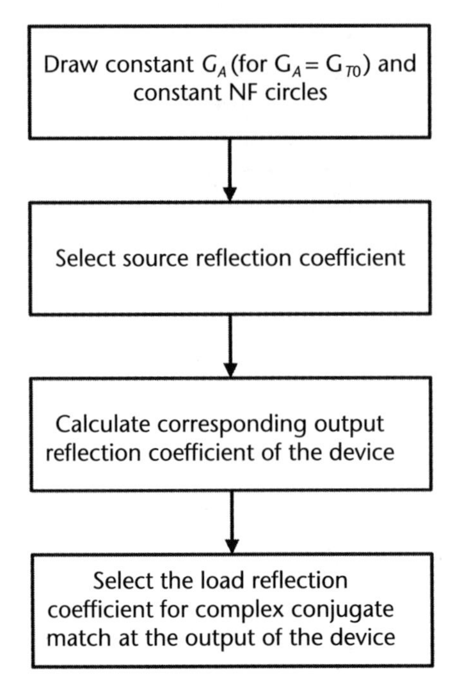

Figure 3.14 Design methodology for gain-noise trade-off using bilateral devices.

gain $G_{TUni} = 11.50$ dB. Indeed, considering the value of $|S_{21}| = 3.2$, Γ_S and Γ_L that need to be presented to the transistor must satisfy:

$$G_S + G_L = 11.50 - 10\log_{10}\left(|S_{21}|^2\right) = 1.4 \text{ dB} \tag{3.77}$$

Using (3.63) and (3.64), it is possible to estimate the maximum values of G_S and G_L, respectively. By setting $\Gamma_S = S^*_{11}$, one can find that $G_{S\max} = 1.94$ dB. Similarly, by setting $\Gamma_L = S^*_{22}$, it appears that $G_{L\max} = 0.76$ dB.

By taking advantage of the freedom (absence of constraint) to select any value for Γ_L, inherited by the unilaterality of the transistor and the fact that the noise performance solely depends on the Γ_S presented to the transistor, the value of G_L can be maximized. Therefore, one can select the load impedance such that $\Gamma_L = S^*_{22}$. This will lead to $G_L = G_{L\max} = 0.76$ dB. Hence, using (3.77), one can deduce that to obtain a transducer gain of at least 11.50 dB, it is needed to have $G_S \geq 0.64$ dB.

To meet the noise requirement, the noise circle corresponding to $NF = 0.6$ dB is plotted in the Γ_S plane. Furthermore, the constant G_S circle is also plotted for $G_S = 0.64$ dB in the same plane to select Γ_S that will meet both the noise and gain requirements. Figure 3.15 depicts these noise and gain circles. Any point located inside the gain circle will lead to a transducer gain that meets the requirements (subject to conjugate match at the output). Moreover, any point inside the noise circle will result in a noise performance that satisfies the specifications. Accordingly, any Γ_S that lie in the intersection area of the G_S and NF circles can be selected. It is important to note here that increasing the requirements on the G_S gain will make the corresponding circle smaller, which is likely to prevent the gain and noise circles from overlapping. In such case, there will be no Γ_S that will allow the designer to meet both gain and noise specifications. This illustrates the importance of maximizing the gain G_L through proper output matching in order to relax the requirements on the gain G_S and hence increase the overlap between the constant G_S and the constant NF circles.

3.4.3.2 Case of Bilateral Transistors

In this design example, an unconditionally stable bilateral transistor is used. The transistor's S-parameters are:

$$S = \begin{bmatrix} 0.6\underline{|-120^\circ} & 0.05\underline{|-132^\circ} \\ 3.6\underline{|60^\circ} & 0.2\underline{|-20^\circ} \end{bmatrix}$$

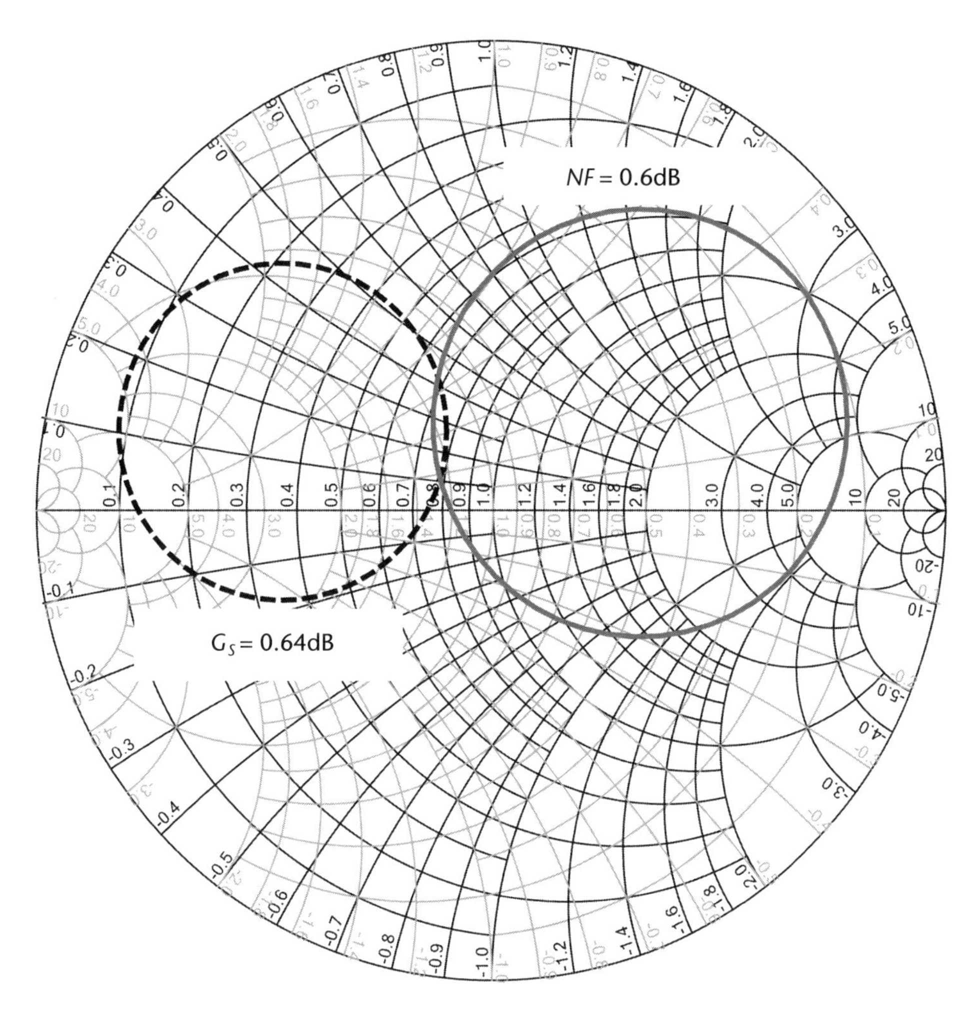

Figure 3.15 Constant G_S and NF circles for illustrating the values of Γ_S that can be selected to meet the design specifications.

Its noise parameters are $NF_{\min} = 0.6$ dB, $\Gamma_{\text{opt}} = 0.4\underline{|78^\circ}$, and $R_n = 5\Omega$. The objective is to design an amplifier that will have a transducer gain that is no more than 0.5 dB less than the maximum transducer gain, and a noise figure that does not exceed 0.75 dB.

Since the transistor being used is bilateral and unconditionally stable, one should focus on the selection of Γ_S to be presented to the device. Later, a conjugate match at the output will be ensured to get a transducer gain equal to the available gain. To meet the gain design requirement, the maximum available gain ($G_{A\max}$) is calculated and the constant available gain circle for $G_A = G_{A\max} - 0.5$ dB is drawn in the Γ_S plane. The constant noise circle for $NF = 0.75$ dB is also plotted in the same plane. Figure 3.16 depicts the constant gain and noise circles and confirms that the design specifications can be achieved since the two circles intersect. Figure 3.16 also shows that the value of Γ_S that will result in $G_A = G_{A\max}$. Once Γ_S is selected from the intersection region of the available gain and the noise circle, the corresponding Γ_{out} is calculated using (3.3), and Γ_L is set to $\Gamma_L = \Gamma^*_{\text{out}}$.

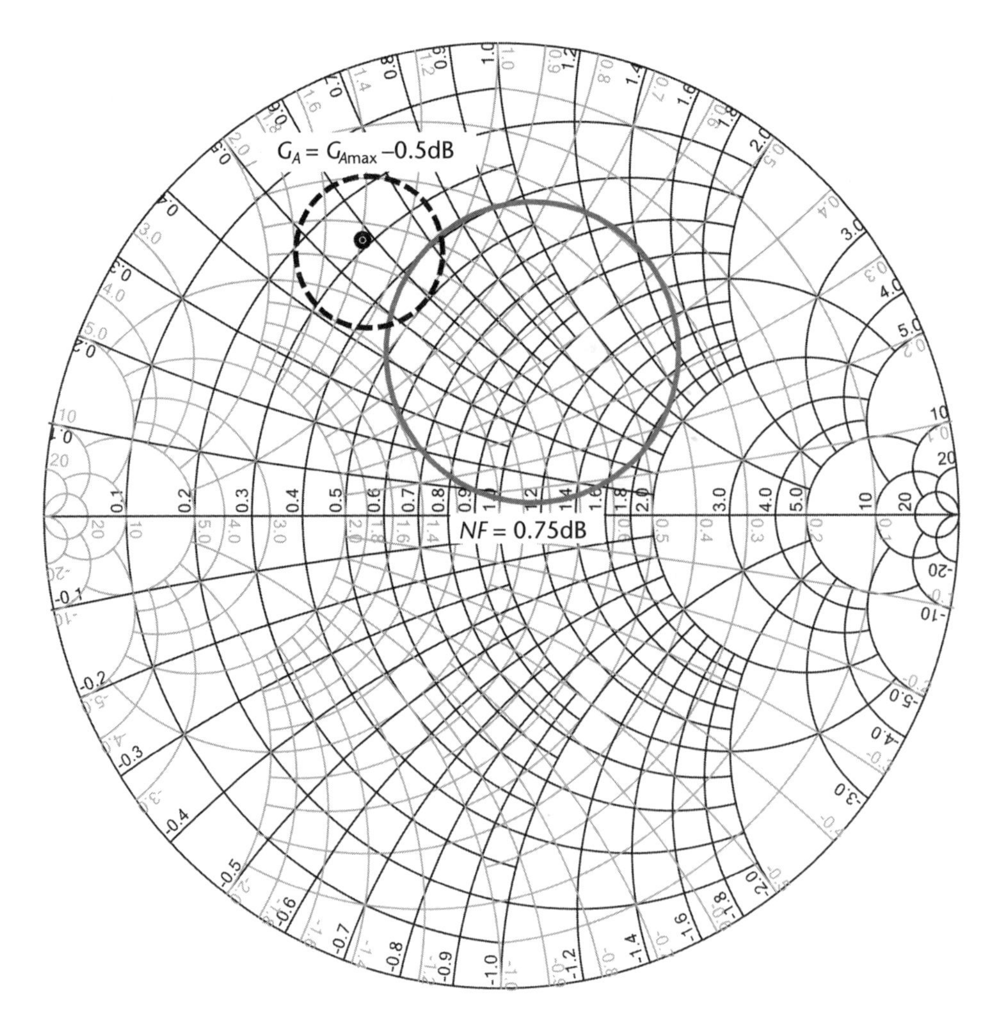

Figure 3.16 Constant G_A and NF circles for illustrating the values of source reflection coefficient that can be selected to meet the design specifications.

The overlap region between the two circles corresponds to Γ_S values that can be selected to simultaneously meet the design requirements for noise and gain performances. Since the overlap region is relatively large, one can also plot additional constant available gain circles to further optimize the gain of the device. For example, in Figure 3.17, two additional constant available gain circles were added ($G_A = G_{A\max} - 0.3$ dB and $G_A = G_{A\max} - 0.2$ dB). These two circles intersect with the constant noise circle. Hence, these circles can define better choices for Γ_S that will result in slightly higher gain.

The same approach can be used to optimize the noise performance of the amplifier while barely meeting the gain requirements. In this case, various constant noise circles are plotted in the Γ_S plane. Figure 3.18 reports the noise circles for $NF = 0.75$ dB, $NF = 0.70$ dB, and $NF = 0.65$ dB as well as the constant available gain circle for $G_A = G_{A\max} - 0.5$ dB. This figure shows that the noise figure can be further reduced below 0.75 dB while still meeting the gain requirement.

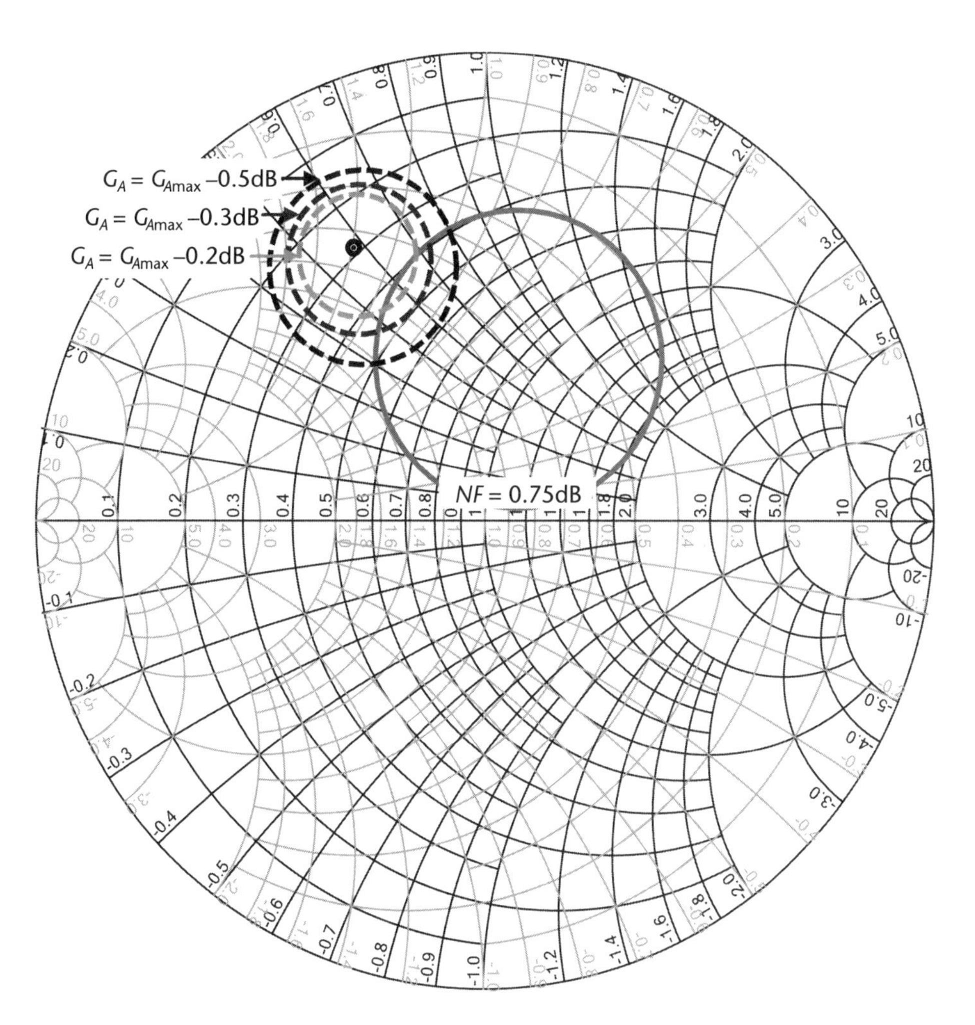

Figure 3.17 Constant G_A and NF circles for illustrating the values of Γ_S that can be selected to meet the design specifications with optimized gain performance.

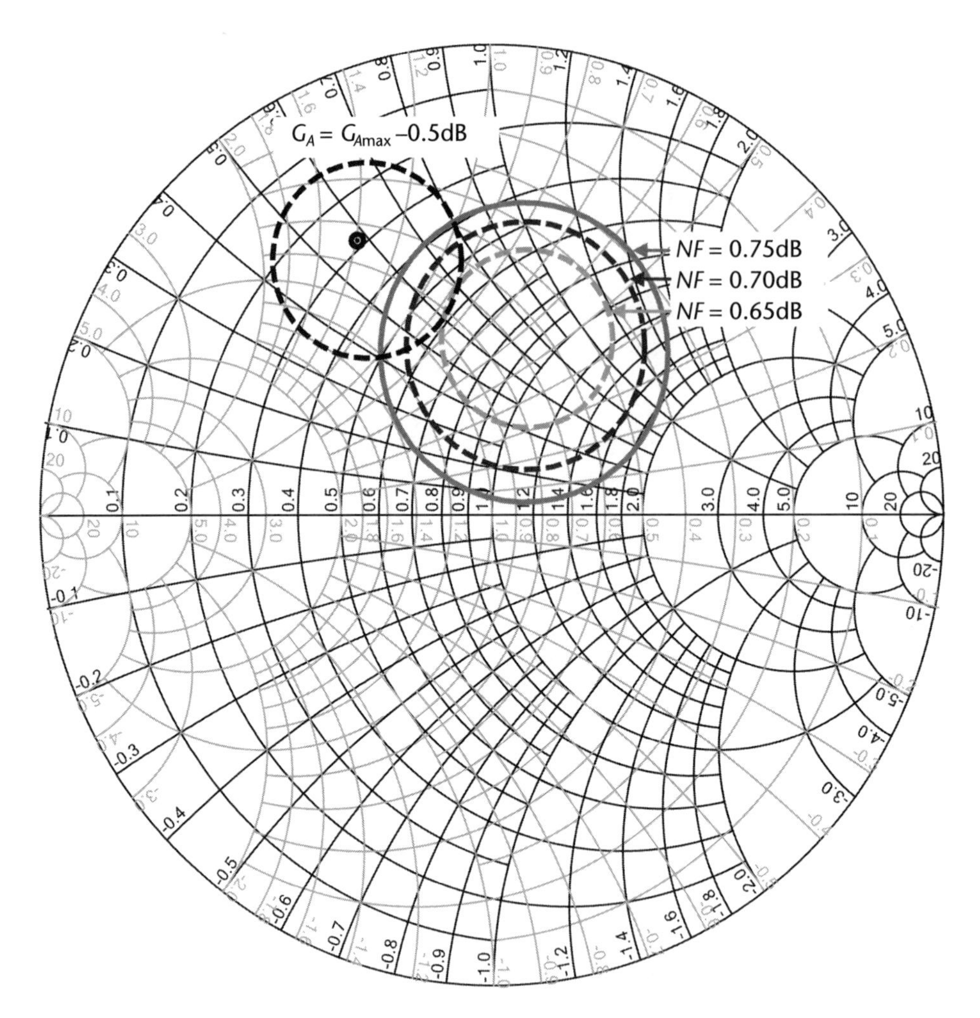

Figure 3.18 Constant G_A and *NF* circles for illustrating the values of Γ_S that can be selected to meet the design specifications with optimized noise performance.

In the general case, the design of a low noise and high gain amplifier has to take into consideration the *NF* of the components that will be placed after the amplifier being designed and hence select the best trade-off that would result in minimizing the noise figure of the cascade defined by (3.29).

3.5 PA Design

In PAs, the design concerns are different from those previously discussed for small-signal amplifiers. Since PAs are intended to operate in large-signal mode, output power, linearity, and efficiency are critical concerns. Hence, typical gain-oriented matching (similar to the approaches discussed in the previous sections) is not suitable for PA design. Indeed, in such applications, power match, which intends to maximize the output power, is more relevant. This matching approach will lead to a design with higher output power and larger 1-dB compression point which are crucial features for power amplifiers. Commonly, the design of PAs involves two steps: the choice of the class of operation, and the investigation of constant performance contours for the appropriate selection of the impedances to be presented to the device.

Conventional PA classes of operation include class A, AB, B, and C. As one moves from class A to class C, the quiescent current of the device is decreased to reduce the conduction angle and therefore limit the power consumption of the device to increase its power efficiency. Other classes include harmonically tuned PAs (such as class F and inverse class F) as well as switching mode amplifiers (classes D and E).

Since, in most modern applications, PAs are expected to be linear and power-efficient, single-ended amplifiers are often part of more advanced amplification systems such as Doherty amplifiers and envelope tracking amplifiers [1, 2].

3.5.1 Load-Pull Analysis

In a large-signal operation, the constant performance contours are not circles anymore. They are closed contours with a flattened oval-shaped due to the nonlinearity of the device and the parasitic effects of the package. It is a very tedious process to analytically derive these contours [1], which is why they are commonly obtained experimentally through load-pull measurements or by load-pull simulations using a comprehensive large-signal device model. In both cases, Γ_L is varied within the Smith chart boundary and the performance metrics of the amplifier are evaluated.

In an experimental setup, the load seen by the device is varied using impedance tuners. The impedance tuners can be either passive or active and are most of the time computer-controlled [11]. Important considerations when selecting tuners for load-pull systems include the frequency range, the resolution, the ability to sweep a wide area of the Smith chart, and harmonics capabilities [12]. The latter are needed when designing high-efficiency PAs that require presentation of a specific load at impedances at the harmonics (as is the case in class F amplifiers). However, with the advancement in computer aided-design (CAD) tools, and the availability of accurate large signal models of transistors, load-pull can be performed through simulations. One must keep in mind that, in this case, the performance of the designed amplifier will greatly depend on the accuracy of the models used to perform the load-pull simulations.

Whether performed through simulations or experimentally, the load-pull characterization enables the gathering of important data about the PA behavior including output power, gain, compression, and efficiency. This data is then analyzed to select the proper Γ_L to be presented to the transistor to achieve the desired performance. Source-pull, which consists in varying the source impedance seen by the device, is also performed to locate the appropriate Γ_S to be presented. In some cases, a few iterations might be needed to alternately optimize Γ_S and Γ_L to obtain the best possible trade-off, although the performance of the PA is more sensitive to Γ_L than Γ_S. Γ_S will mainly impact the gain of the amplifier and to some extent its amplitude-modulation to phase-modulation (AM-PM) characteristic [1].

3.5.2 Design Procedure

The design of PAs starts with the selection of the bias point, which defines the class of operation. Class A operation is obtained by selecting a quiescent point that maximizes the output signal swing and therefore results in conduction over the entire signal cycle. As the quiescent current is reduced, clipping will occur for negative AC signals, therefore reducing the conduction angle to less than the 360° of class A operation. In class B mode, the transistor's quiescent current is set to zero to conduct over only half the signal cycle. In this case, a push-pull configuration is commonly adopted to ensure linear operation [1, 2]. When the quiescent current is between that of class A and class B, class AB mode of operation is obtained as the device will conduct over more than half the signal cycle, although it will be unavoidably off during a fraction of the cycle. Class AB represents one of the most popular classes of operation as it achieves an attractive trade-off in terms of linearity and efficiency.

Once the bias point is set, load-pull-based experiments are carried out to identify the appropriate Γ_S and Γ_L to be presented to the device to achieve the desired specifications. This is conceptually similar to small-signal amplifiers optimized for gain and noise, while a few key differences must be noted:

- As mentioned above, there are no equations that can predict the constant performance contours for PAs. A large-signal device model or load-pull data is needed to derive these contours.
- Constant performance contours relevant to the case of power amplifiers are output power, efficiency and linearity.
- The results are power-dependent in the sense that the power level of the input signal affects the results. Therefore, it is essential to select the appropriate power level(s) at which the load-pull is to be performed.

In some specific classes of operations such as class E and class D, as well as class F, design equations are commonly used to determine the input and output reflections coefficients that need to be seen by the device [1].

3.6 Conclusions

The design of microwave amplifiers was thoroughly discussed in this chapter. Small-signal and large-signal amplifiers were covered. The design of small-signal amplifiers focuses on gain, noise, or a trade-off between both, whereas the design of PAs emphasizes more on output power, power efficiency, and linearity performances.

First, a generic amplifier design procedure was discussed. This allowed for introducing the concepts of unilaterality and stability common to all amplifiers designs. Then the basic noise analysis theory for cascaded networks was introduced to value the importance of LNAs and provide some guidelines on the optimization of their noise and gain performances for effective system-level designs. Maximum gain amplifiers were discussed for the cases of unilateral and bilateral transistors. Design equations and examples were presented for both cases. For a trade-off between gain and noise performances, gain and noise circles, along with their respective equations, were introduced, and the design procedure was thoroughly presented. Finally, the case of PA design was briefly addressed. In this case, the design is mainly based on load-pull data, since few, if any, theoretical equations can be used to derive the matching requirements for basic classes of operations in the large-signal mode.

Identifying the reflection coefficient to be presented to the transistor is an essential step for the design of AIAs in which the amplifier and antenna are integrated together. In receiver applications, AIAs are made of the receiving antenna and an LNA. Therefore, the antenna is to be connected to the input power of the LNA and hence appropriate matching is critical in order to achieve low noise performance. Conversely, in transmitting paths, the antenna is connected to the output of the power amplifier. Here also the proper matching between both will allow the transistor to see the adequate load reflection coefficient, which was shown to greatly influence the PA's performances.

References

[1] Cripps, S. C., *RF Power Amplifiers for Wireless Communications*, Second Edition, Norwood, MA: Artech House, 2006.

[2] Steer, M., *Microwave and RF Design: A System Approach*, Second Edition, Scitech Publishing, 2013.

[3] Gonzalez, G., *Microwave Transistor Amplifiers Analysis and Design*, Second Edition, Upper Saddle River, NJ: Prentice Hall, 1997.

[4] Bodway, G. E., "Two Port Power Flow Analysis Using Generalized Scattering Parameters," *Microwave Journal*, Vol. 10, No. 6, May 1967.

[5] Froehner, W. H., "Quick Amplifier Design with Scattering Parameters," *Electronics*, October 1967.

[6] Rollett, J. M., "Stability and Power-Gain Invariants of Linear Two-Ports," *IRE Transactions on Circuit Theory*, Vol. CT-9, March 1962, pp. 29–32.

[7] Pozar, D. M., *Microwave Engineering*, 4th ed., New York: Wiley, 2011.

[8] Chang, K., *RF and Microwave Wireless Systems*, New York: Wiley Interscience, 2000.

[9] Proakis, J. G., and M. Salehi, *Fundamentals of Communication Systems*, 2nd ed., New York: Pearson, 2013.

[10] Anderson, R. W., "S-Parameters Techniques for Faster, More Accurate Network Design," *Hewlett-Packard Journal*, Vol. 18, No. 6, February 1967.

[11] Simpson, G., "A Beginner's Guide to All Things Load Pull or Impedance Tuning 101," *Microwave and RF*, December 2014.

[12] Sevic, J. F., "Introduction to Tuner-Based Measurement and Characterization," Mauray Microwave Corporation, Technical data 5C-054, August 2004.

CHAPTER 4

Antenna Fundamentals

Antennas are transducers that convert electrical signals to electromagnetic waves (and vice versa) in the medium for wireless signal transmission. They are used in every wireless device. The design of antennas is a continuously developing discipline as it is a function of the wireless standards, the device shape, and the specifications required among many other factors. Although single antenna systems were widely used and are still being developed for various new devices, multi-antenna systems have also been required and are currently deployed. An example of a multi-antenna system is the one in all of our current fourth generation (4G)-enabled cell phones (multi-antenna systems also exist in antenna arrays for applications with beam/null steering capabilities). The topic of antenna design has been treated heavily in literature and dedicated books have been published on the topic [1–4]. In this chapter, we will provide an overview of the basic principles of the widely used antenna types and arrays (from which several variants can be designed) to give the reader a starting point in the design and provide the required basic principles for understanding its behavior. Additionally, several examples will be provided.

4.1 Antenna Features and Metrics

There are several performance metrics and features for various antennas. Based on the antenna type and material, such metrics can be directly affected and changed. Metallic-based antennas are the most widely used nowadays (i.e., dipoles, monopoles, and patches), but dielectric-based antennas have been suggested for millimeter-wave bands due to their lower losses and ease of integration within the same substrates compared to their metallic counterparts, although they occupy much larger volumes (there is always a trade-off) [5]. In this section, we will present the performance metrics of generic antennas and highlight the factors that control such parameters.

4.1.1 Input Impedance, Resonance, and Bandwidth

Any antenna can be represented by an equivalent RLC circuit. Series RLC circuits represent antennas of monopole and dipole types, while parallel RLC circuits are used to represent patch and planar inverted-F antennas (PIFAs) [6, 7]. A note here is that the input impedance of the antenna is frequency-dependent, and thus its behavior can change. Therefore, an antenna can behave as two different circuit equivalents during its resonance and antiresonance [8]. However, usually the first resonance is the one which we are concerned about. At resonance, the reactive part becomes almost zero leaving the real part only to be seen at the antenna input port. Depending on the antenna efficiency, a portion of the power delivered to this real valued impedance will be radiated (an imaginary impedance part will store the energy rather than radiate it).

Antennas are designed for a specific frequency band or multiple bands. The antenna material and geometry determine its resonance frequency and the bands covered (i.e., its operating frequency bandwidth). The resonance can be observed also from the reflection coefficient curve, as the point of minimum reflection will be the point of best match, corresponding to an almost pure real input impedance of the antenna. This antenna impedance will match the real impedance of the feeding line (care should be taken here, and the input impedance curves, real and imaginary, should be always checked to verify the resonance against the *S*-parameter curves) [1, 2, 7].

To determine the bandwidth of operation of the antenna, we define the frequency bandwidth to be the band covered by the two −10-dB points on the magnitude of the reflection coefficient ($|S_{11}|$) curve. These points correspond to the 2:1 voltage standing wave ratio (VSWR) points. For multiband (multiresonance) antennas, a more relaxed matching condition is adopted that defines the bandwidth between the −6-dB points corresponding to the 3:1 VSWR points. The relationships between the bandwidth (BW), $|S_{11}|$ and the VSWR are given by [1, 2]:

$$VSWR = \frac{1+|S_{11}|}{1-|S_{11}|} \tag{4.1}$$

$$|S_{11}| = \frac{VSWR-1}{VSWR+1} \tag{4.2}$$

$$BW = \frac{VSWR-1}{Q\sqrt{VSWR}} \tag{4.3}$$

where *Q* is the quality factor of the antenna under consideration.

An example is shown in Figure 4.1. A quarter-wavelength wire monopole antenna (modeled using HFSS; a complete tutorial on HFSS use and modeling is given in Appendix C) is shown with an ideal input matching and coaxial feed port. The antenna is resonating at 3 GHz (i.e., λ = 10 cm) and its input impedance and reflection coefficient curves are plotted. Note that at 3 GHz, the value of the real part of the impedance is close to 50Ω, while the imaginary part is 0Ω. The reflection coefficient minimum is at 3 GHz as well. The frequency bandwidth is 400 MHz for this antenna.

4.1.2 Radiation Pattern, Efficiency, Polarization, Gain, and MEG

The excitation of currents on the antenna surface translates into electromagnetic radiation [9]. The shape of the radiated fields, specifically the electric field (E-field), in the three-dimensional (3-D) space is called the radiation pattern of the antenna [1,

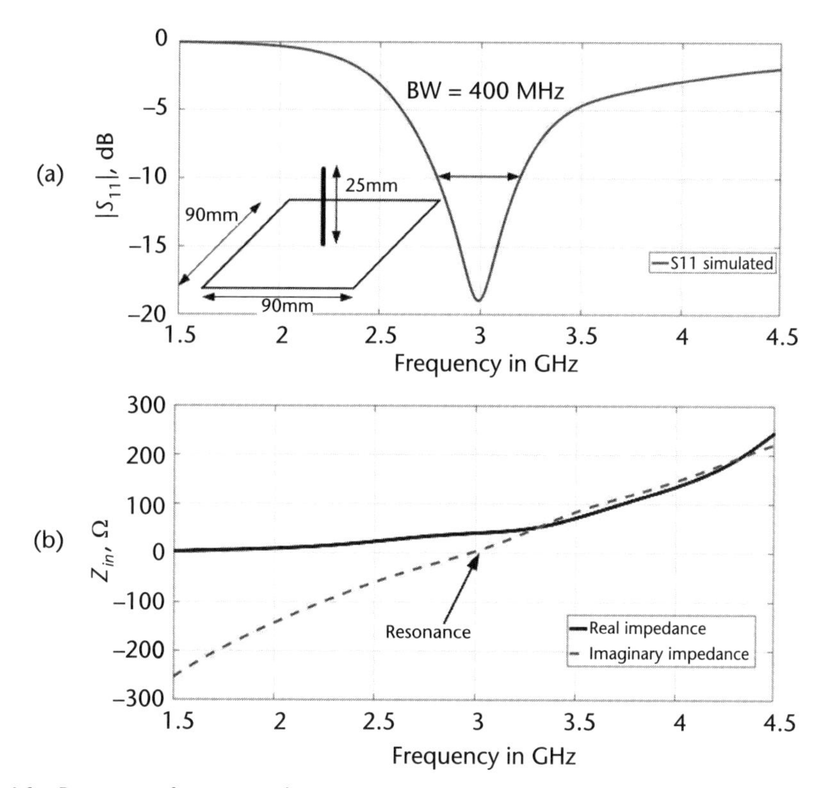

Figure 4.1 Response of a monopole wire antenna: (a) reflection coefficient with geometry at inset, and (b) the input impedance.

2, 9]. The radiation pattern of an antenna is a major performance metric as it shows the antenna radiated field behavior in a certain environment (i.e., if the antenna is placed on a surface, its pattern will change from that when it is isolated in space or placed in a measurement chamber). The radiation pattern has three subparameters that are of interest to the designer; the direction of maximum radiation (which translates to the maximum gain direction), the half-power beamwidth (HPBW) that identifies the points where the maximum radiated power falls to half of its maximum value (i.e., −3 dB from maximum), and finally the sidelobe level (SLL) that shows the ratio between the peak of the radiation and the first closest side peak. Figure 4.2 shows a 3-D as well as a 2-D rectangular and a 2-D polar plot gain patterns of a directional antenna (a printed Yagi antenna). The maximum gain is 5 dBi, the HPBW is 70°, and the SLL (back lobe for this single antenna) is −12 dB. Note that dBi is the gain unit with respect to the gain of a hypothetical isotropic radiator.

The polarization of an electromagnetic wave describes the time varying relative magnitude and direction of the electric field vector and its trace on a plane perpendicular to its direction of propagation at a fixed point in space. The polarization will show the locus of the sum of the components of the electric field in magnitude and phase on that plane at that point in space [1, 10]. Polarization is typically described as linear polarization, circular polarization and elliptical polarization. The conditions for having each are discussed in details in [10]. An important consideration

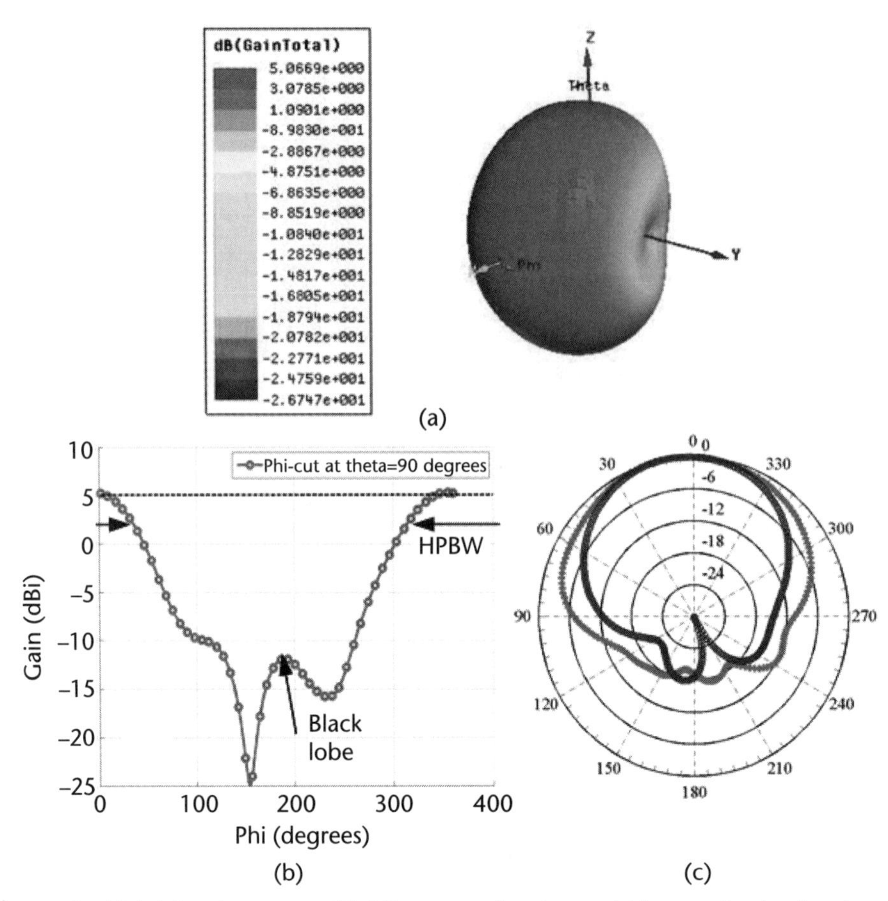

Figure 4.2 (a) A 3-D gain pattern, (b) 2-D rectangular plot, and (c) normalized polar plot.

in antenna design is to have the transmitting and receiving antennas with the same polarization to avoid power losses due to polarization mismatch.

Every antenna is connected to a feed line or network to transmit or receive energy. The efficiency with which the antenna radiates or collects electromagnetic energy depends on the internal losses and the matching in terms of impedance and polarization. In case of nonwire-like antennas, such as printed dipoles, monopoles, or patches, the substrate on which the antenna resides can store some energy and thus can reduce the radiation efficiency. Such losses are commonly referred to as substrate or material losses. In addition, the input impedance of the antenna should match that of the feeding transmission line or cable connected to it to achieve maximum power transfer and minimum reflection losses. Finally, the polarization of the antenna can deteriorate the efficiency if it does not match the polarization of the incoming waves. This is also called the polarization loss factor (PLF). Assuming the incoming wave and the antenna polarization are matched (i.e., $PLF = 1$, no loss due to polarization mismatch), the total efficiency (η_{Total}) of the antenna is usually expressed via the material losses and the matching efficiency as:

$$\eta_{\text{Total}} = \eta_{\text{material-loss}}\eta_{\text{matching}} = \eta_{\text{material-loss}}(1 - \Gamma) \tag{4.4}$$

where $\eta_{\text{maerial-loss}}$ is the efficiency due to material losses, η_{matching} is the efficiency due to matching losses, and Γ is the reflection coefficient.

Another important performance metric of an antenna is its gain. The power gain of an antenna is an indication of how much the transmitted or received signal is boosted due to the concentration of the radiated energy in a particular direction. Usually the gain is higher than 0 dBi, but it can be lower for low efficiency or highly miniaturized antennas [4]. The gain is obtained from the antenna directivity function (a spatial power distribution based on the radiated fields) and the antenna efficiency [1]. It is given by:

$$G(\theta,\phi) = \eta_{\text{Total}} D(\theta,\phi) \tag{4.5}$$

$$D(\theta,\varphi) = \frac{4\pi U(\theta,\varphi)}{P_{rad}} \approx \frac{2\pi r^2\left[\left|E_\theta(r,\theta,\phi)\right|^2 + \left|E_\phi(r,\theta,\phi)\right|^2\right]}{\zeta P_{rad}} \tag{4.6}$$

where $D(\theta,\phi)$ is the directivity of the antenna which is the ratio between the radiation intensity ($U(\theta,\phi)$) of an antenna in a given direction to its total radiated power (P_{rad}) averaged over a complete sphere surrounding it. r is the distance from the source, and ζ is the intrinsic impedance of the medium. The radiation intensity is a function of the E-field power components in the azimuth (ϕ) and elevation (θ) directions. The gain pattern will follow that of the directivity.

Another metric that is also of interest is the mean effective gain (MEG). MEG is a measure of the channel environment effect on the gain performance of the antenna [11]. This measure is important because the antenna gain measurements in an anechoic chamber are not usually representative of their values in a real wireless environment. Thus, MEG is used to predict the behavior of the antenna in a specific channel. In [12], a probabilistic model for the channel was obtained and combined with the gain patterns obtained from anechoic chamber measurements to yield MEG values. The expressions for finding MEG are

$$MEG = \int_0^{2\pi}\int_0^{\pi}\left[\frac{XPD}{1+XPD}G_\theta(\theta,\varphi)P_\theta(\theta,\varphi) + \frac{1}{1+XPD}G_\varphi(\theta,\varphi)P_\varphi(\theta,\varphi)\right]d\theta d\varphi \tag{4.7}$$

Satisfying the conditions,

$$\begin{cases} \int_0^{2\pi}\int_0^{\pi}\left[G_\theta(\theta,\phi) + G_\phi(\theta,\phi)\right]\sin\theta d\theta d\phi = 4\pi \\ \int_0^{2\pi}\int_0^{\pi}\left[P_\theta(\theta,\phi)\right]\sin\theta d\theta d\phi = \int_0^{2\pi}\int_0^{\pi}\left[P_\phi(\theta,\phi)\right]\sin\theta d\theta d\phi = 1 \\ XPD = \frac{P_V}{P_H} \end{cases} \tag{4.8}$$

where XPD is the cross-polarization discrimination (ratio between the vertical $[P_v]$ and horizontal $[P_H]$ power components), $G_\theta(\theta,\phi)$ and $G_\phi(\theta,\phi)$ are the antenna gain components, and $P_\theta(\theta,\phi)$ and $P_\phi(\theta,\phi)$ are the channel model power profiles. Depending on the channel under consideration, some simplifications for the values of the power profiles can be adopted as mentioned in [12].

4.2 Antenna Types

There are thousands of antenna designs in literature, each targeting a certain application and certain frequency bands. Fundamentally, they all can be considered modified versions of eight basic antenna types, specifically, the dipole, monopole, slot, loop, horn, reflector, dielectric resonator and patch antennas. Five of these antennas can be actively integrated with RF electronic components on the same substrate (board) due to their printed nature (placed on a dielectric substrate) and they will be the ones we focus on in the sections to come.

4.2.1 Dipole Antennas

A dipole antenna consists of two metallic arms with each having a length of $\lambda_0/4$ with a very small feeding gap to have a total length of almost $\lambda_0/2$, where λ_0 is the free space wavelength of the frequency of interest. The wire diameter should be much smaller than the operating wavelength. Figure 4.3(a) shows a wire dipole antenna and its feeding mechanism. To match the input to the specified load impedance and to have balanced currents (sinusoidal) on the wire structure, a balun is used [13]. The length of the resonant wire dipole can be a little less than $\lambda_0/2$ to provide proper resonance and matching (i.e., $0.48\lambda_0$). The same can be applied to a printed dipole antenna as shown in Figure 4.3(b). The feeding should be optimized to a 50Ω feeding line via a balun or proper matching technique. The length of the printed dipole will be close to $\lambda_g/2$ where λ_g is the guided wavelength (i.e., $\lambda_g = \lambda_0/\sqrt{\varepsilon_r}$, and ε_r is the relative dielectric constant of the substrate material).

The input impedance of a wire dipole antenna can be found based on the surface currents present on it [14], and for a half-wavelength ($\lambda_0/2$) dipole at its resonance frequency (and assuming no losses) is approximated by [1, 2]:

$$Z_{\text{in}} \approx 73 + j42.5 \tag{4.9}$$

The radiated E and H fields for the $\lambda_0/2$ wire dipole are expressed as

$$E_\theta \approx j\zeta \frac{I_0 e^{-jkr}}{2\pi r}\left[\frac{\cos(0.5\pi\cos\theta)}{\sin\theta}\right] \tag{4.10}$$

$$H_\varphi = \frac{E_\theta}{\zeta} = j\frac{I_0 e^{-jkr}}{2\pi r}\left[\frac{\cos(0.5\pi\cos\theta)}{\sin\theta}\right] \tag{4.11}$$

where ζ is the intrinsic impedance of the medium (120π for free space), k is the wave number, r is the radial distance from the antenna, and θ and ϕ are the elevation and azimuth angles, respectively. The radiation (and gain) patterns of a dipole are omnidirectional (doughnut shape) with its maximum pointing towards $\theta = 90°$. The directivity of a dipole is 1.64.

More accurate results can be computed using full-wave electromagnetic solvers that will take into account the effect of the wire radius relative to its length, as well as the feed gap effect (and when applicable, the effects that stem from the ground plane modeling). Equation (4.9) is an approximate value, and the imaginary part of the input impedance can be reduced to almost zero when the line dipole length is reduced a little as mentioned before. For example, to design a 900-MHz wire dipole antenna, each arm length should be around 8.33 cm.

Printed dipole antennas are widely used in small form factor devices and handheld terminals. Such printed antennas also get another size shrinkage factor from the value of the dielectric constant of the substrate as its size will be with respect to λ_g and not λ_0. The 3-D gain pattern of a printed dipole antenna is shown in Figure 4.3(c). Several examples of printed dipole antennas have appeared in literature such as those in [15–19]. In [15], a dual-band printed dipole antenna was presented on a $50 \times 5 \times 0.4$ mm^3 FR-4 substrate. The dual-band feature was obtained by etching a u-shaped slot within each dipole arm and it was fed via a 50Ω coaxial cable directly. The gain in the lower band (2.29–2.79 GHz) was 4 dBi, while it was around 2 dBi in the higher band (5.14–5.51 GHz). The geometry and reflection coefficient curve of this antenna are shown in Figure 4.4. A printed dipole antenna for wireless local area network (WLAN) applications, designed to resonate at 2.45 GHz was presented

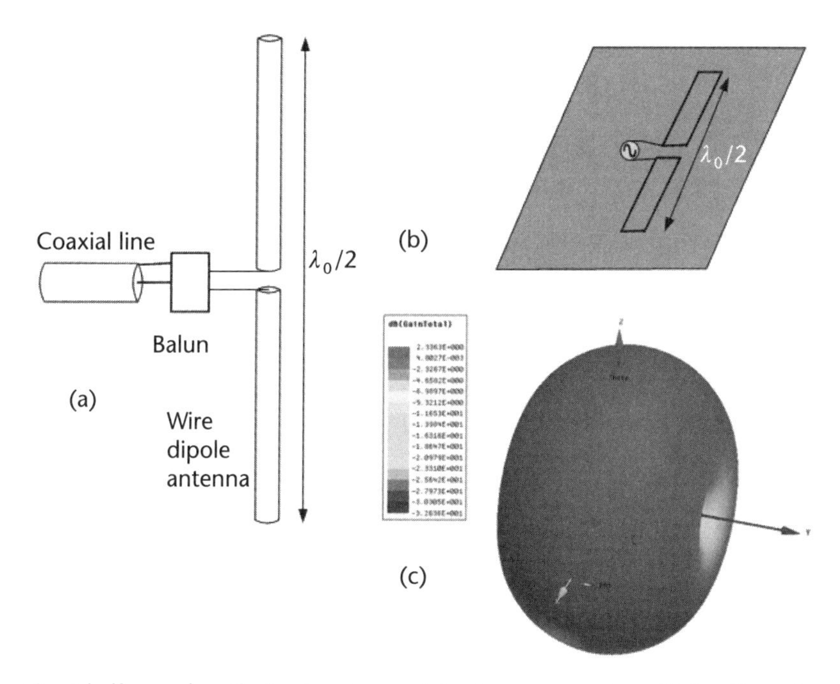

Figure 4.3 A half wave-length dipole antenna: (a) antenna geometry with feedlines of a wire dipole, (b) printed on a dielectric substrate, and (c) the radiation pattern of a printed dipole (omnidirectional).

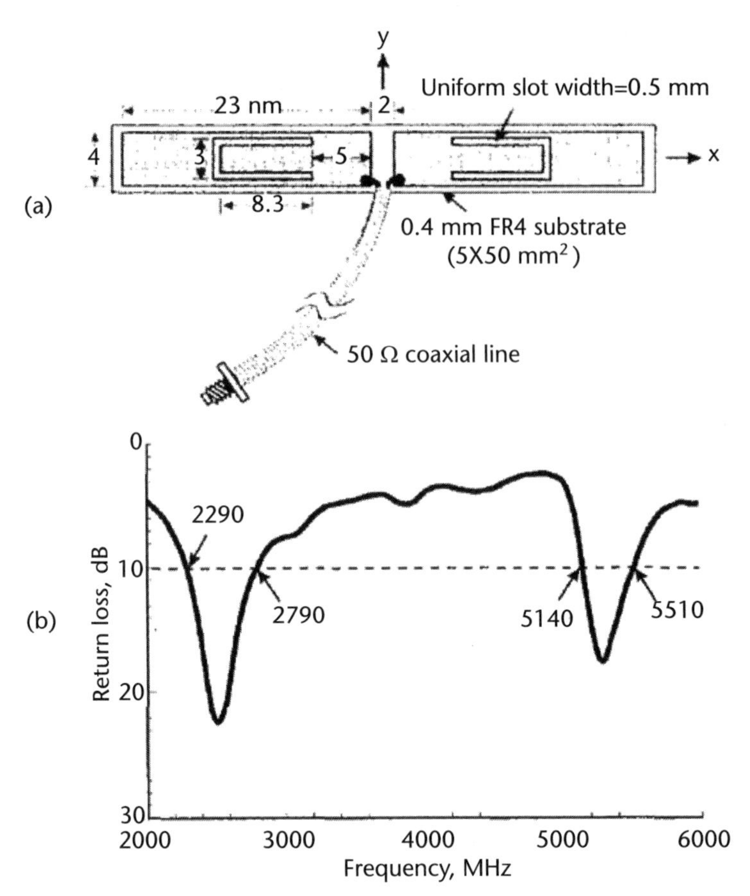

Figure 4.4 A printed dual-band dipole: (a) geometry, and (b) frequency response. (*From:* [15]. © 2002 IET. Reprinted with permission.)

in [16] with a tapered balun. The antenna had 260 MHz of bandwidth, and was etched on an FR-4 substrate with antenna size of 38 × 44 mm^2. An ultrawideband (UWB) printed dipole was presented in [18] covering the 3.1–10-GHz band. It was fabricated on an RT/Duroid 5880 substrate with 1.57-mm thickness. The geometry of the antenna is shown in Figure 4.5.

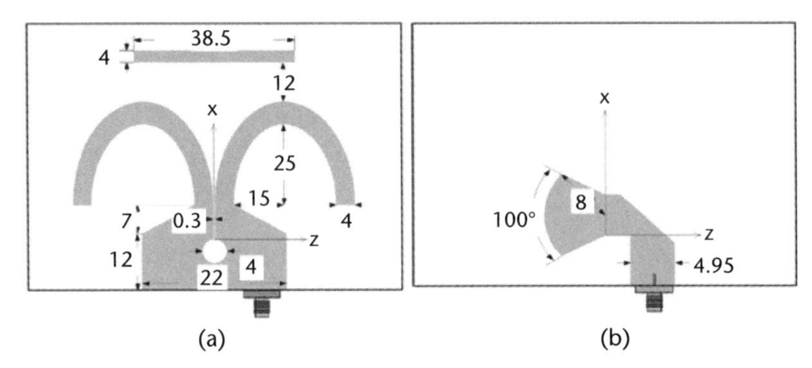

Figure 4.5 A UWB printed monopole antenna: (a) top view, and (b) bottom view. All dimensions are in millimeters. (*From:* [18]. © 2005 IEEE. Reprinted with permission.)

Several methods have been devised to miniaturize dipole antennas. Small size antennas are required for small form factor devices and sensor nodes. A direct wire dipole size reduction technique is via zigzagging or meandering the antenna arms [20]. This will shrink the electrical length without changing the physical length. The dipole input impedance is slightly affected in both its real and imaginary values. Another miniaturization technique is performed via placing it on a reactive impedance substrate (RIS) [21]. Such a substrate can provide dipole miniaturization ratios of 30% or more when designed carefully without reducing the operating bandwidth or efficiency (~96% radiation efficiency was obtained). In [22], a wideband miniaturized half bowtie printed antenna was presented with approximately 25% reduction in size compared to a regular one at 2.97 GHz. It achieved 47% bandwidth with an integrated balun within its sliced bowtie arm sizes and via optimizing the flare angle. Figure 4.6 shows the geometry of this miniaturized design.

4.2.2 Monopole Antennas

Taking one arm of a dipole antenna and placing it over a very large ground plane will result in a monopole antenna (i.e., one arm compared to two arms, dipole). A monopole has a radiation pattern that is similar to that of the dipole, as the ground plane assists in forming image currents of the excited arm above it, and thus similar radiation mechanism is obtained [1]. The input impedance of a monopole is half of that of its dipole equivalent, but its directivity is two times higher. A monopole consists of a $\lambda_0/4$ wire above a large ground plane, hence the name quarter-wave monopole [1, 2]. Monopoles are widely used in wireless terminals because of the presence of the system ground plane (a ground plane below a dipole will affect its impedance and radiation characteristics).

Printed monopole antennas or their modified versions are widely used in mobile devices (a complete design example is given in Section 4.5). A large number of examples have appeared in literature such as those in [23–33], covering most of the available communication standards and frequency bands, as well as various device sizes from cell phones, to tablets and radio frequency identification (RFID) tags. Dual-band printed monopole based antennas were presented in [23, 24], covering the WLAN bands of 2.4 GHz and 5.2/5.8 GHz. The former [23] was based on a

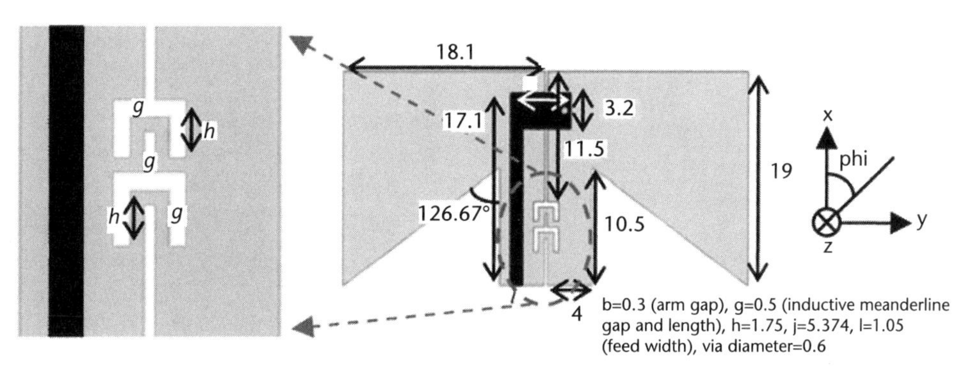

Figure 4.6 A wideband miniaturized half-bowtie dipole antenna. All dimensions are in millimeters. (*From:* [22]. © 2011 IEEE. Reprinted with permission.)

rectangular monopole with a trapezoidal conductor backing plane for wider bandwidth, while the latter [24] was based on a double-T simple monopole structure. Figure 4.7 shows the two geometries. Both provided omnidirectional patterns in all covered bands with gains of more than 2 dBi for the former and 1 dBi for the latter. To achieve wideband coverage, principles of wideband monopole operation were discussed in [25, 26].

Monopole antennas are widely used in UWB applications. In [27–29] several printed UWB monopole-based antennas were presented. The compact UWB monopole in [27] covered 2.7–6.2-GHz band, with a size of $30 \times 8 \times 0.8$ mm^3. The UWB designs in [28, 29] showed rounded edged rectangular patch and planar inverted cone monopole structures, respectively. Both provided large covered bandwidths and the one in [28] showed a band notch around 5.2-GHz WLAN to minimize interference between the two standards. This band-notch was achieved via a compact coplanar waveguide resonant cell within the feed line of the monopole. The fabricated models for the designs in [27, 29] are shown in Figure 4.8.

Miniaturization techniques can also be applied to printed monopole antennas similar to dipole ones. Electrically small monopole antennas (ESA) are widely used. Such antennas usually have narrow bandwidth and low efficiency levels unless they incorporate a technique for compensating for such drawbacks that are inevitable when the size becomes very small. For example, a $\lambda/40$ multi-element monopole antenna was presented in [30] for operation in the 460-MHz band with 3% bandwidth. The antenna was realized over a 100×100 mm^2 ground plane and was suspended 21 mm above it. The monopole arms were spiraled with round edges, and a gain of more than 2 dBi was obtained for the final optimized design. Several UWB compact-size antennas were also suggested in literature such as those in [31–33]. Some of the provided band-notch capabilities, such as the ones in [31, 33], relied on the use of stubs and slots within the radiating element or the introduction of defected ground structures (DGS) within the feeding lines to provide the band reject behavior at WLAN bands. The design in [32] showed UWB operation via a staircase structure at the side closer to the ground plane (from the feeding point of the monopole antenna).

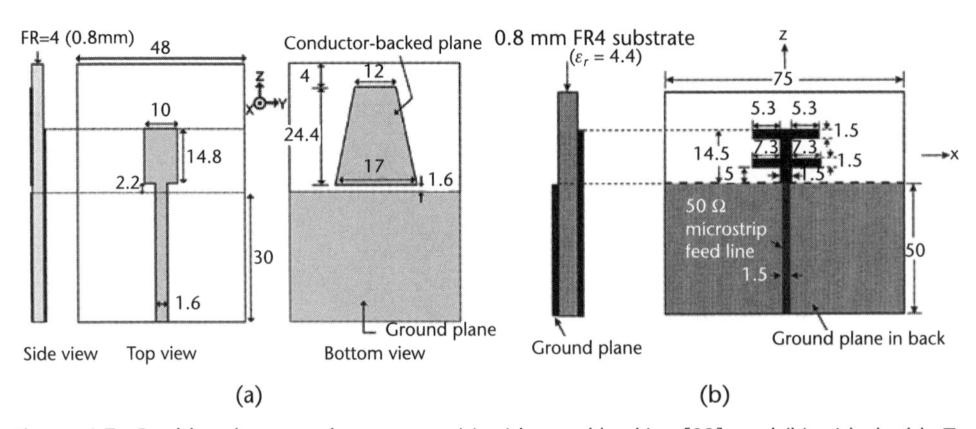

Figure 4.7 Dual-band monopole antennas: (a) with metal backing [23], and (b) with double-T structure. All dimensions are in millimeters. (*From:* [24]. © 2003 IEEE. Reprinted with permission.)

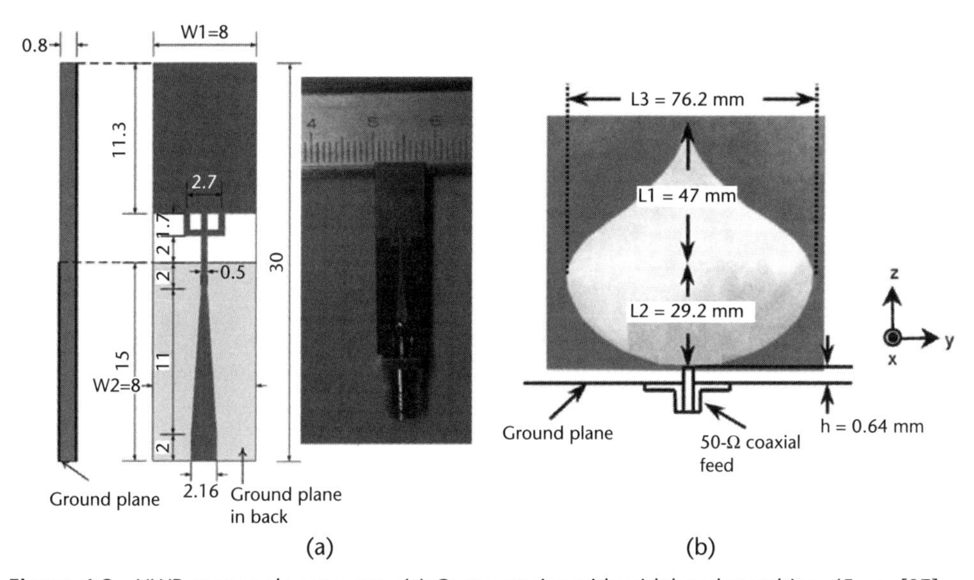

Figure 4.8 UWB monopole antennas. (a) Compact size with wideband matching. (*From:* [27]. © 2008 IEEE. Reprinted with permission.) (b) Inverted cone. (*From:* [29]. © 2004 IEEE. Reprinted with permission.)

A very widely used modified quarter-wavelength monopole antenna is the planar inverted-F antenna (PIFA). A PIFA antenna can be a bent monopole over a ground plane with a shorting stub to cancel out the capacitive reactance at resonance. PIFA antennas are widely used in wireless mobile devices because of their compact size and conformal structure. A PIFA can also be a rectangular patch like antenna with smaller size, where its length and width sum corresponding to about quarter of the wavelength of the resonance frequency of interest. The feeding and shorting pin locations have to be optimized for best matching [34]. In [35], an empirical equation was developed to come up with more accurate estimation of the resonance frequency taking into account factors such as the ground plane size and the antenna height. A PIFA design example is given in Section 4.5.

4.2.3 Patch Antennas

Patch antennas are low-profile, easy-to-fabricate elements that have seen a wide range of applications. The most common version of a patch antenna is the rectangular shape. Circular and triangular shapes are also available but not as common. The basic antenna geometry of a rectangular patch is shown in Figure 4.9. It consists of the top patch metallic part with dimensions of $L \times W$ placed on a dielectric material (substrate or even suspended with an air gap), with the other side size of the substrate containing a conducting sheet acting as the ground reference plane. The size of the patch can be estimated initially using a closed form set of equations based on the transmission line and cavity models [1, 2, 36–39], and then optimized using a full-wave electromagnetic solver to take into account the fringing, feeding, finite ground size, and other practical effects that are not considered in the initial closed-form calculations. The width and length of the patch based on the material

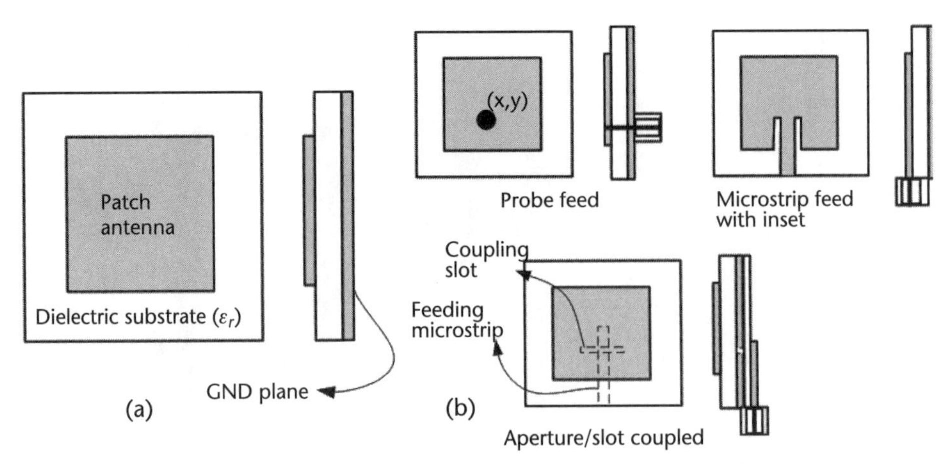

Figure 4.9 Patch antenna geometry: (a) rectangular patch antenna, and (b) possible feeding mechanisms.

of the substrate (ε_r), its height (h), and operating frequency can be estimated; for the dominant resonant mode and when $h << L$ and $h << W$ using [1, 36]:

$$W = \frac{v_0}{2f_r}\sqrt{\frac{2}{\varepsilon_{reff}+1}} \tag{4.12}$$

$$L = \frac{v_0}{2f_r\sqrt{\varepsilon_r}} = \frac{\lambda_g}{2} \tag{4.13}$$

$$\varepsilon_{reff} = \frac{\varepsilon_r+1}{2} + \frac{\varepsilon_r-1}{2}\left[1+\frac{12h}{W}\right]^{-0.5} \tag{4.14}$$

Equations (4.12) through (4.14) are the starting-point equations for a rough estimate of the patch dimensions for a given substrate. These values should be tuned later on in a simulation model (Appendix C shows a detailed design example of a patch antenna using HFSS with all its characteristics). The approximate input impedance (R_{in}) of a rectangular patch antenna at its resonance is found using (4.15) without taking into consideration the mutual coupling between its radiating slots (only the self-conductance G_1). To match the input impedance of a patch to that of the feeding line, an inset feed can be used (see Figure 4.9), or the feed is moved from the center towards the edges if a microstrip line is used for feeding. If probe feeding is considered, then several (x,y) feed points should be investigated to provide proper matching. Approximate analytical expressions can be found in [40] to determine the location of the feed on the patch surface. The resulting radiated electric fields for a typical linearly polarized rectangular patch excited with a voltage source with amplitude (V_0); and assuming it resembles two radiating slots; are given by (4.16) and (4.17) as a function of radial distance (r). The fields cover the top portion of the patch with a wide HPBW (top half sphere).

$$R_{\text{in}} = \frac{1}{2G_1} = \begin{cases} \frac{1}{180}\left(\frac{W}{\lambda_0}\right)^2, & W << \lambda_0 \\ \frac{1}{240}\left(\frac{W}{\lambda_0}\right), & W >> \lambda_0 \end{cases} \tag{4.15}$$

$$E_r = E_\theta \approx 0 \tag{4.16}$$

$$E_\varphi^t \approx \frac{j2V_0e^{-jkr}}{\pi r}\left[\sin\theta\frac{\sin\left(\frac{k_0W}{2}\cos\theta\right)}{\cos\theta}\right]\cos\left(\frac{k_0L}{2}\sin\theta\sin\varphi\right) \tag{4.17}$$

Regular patch antennas suffer from narrow bandwidth (typically of the order of 3% to 5%). Wideband patches can be achieved by inserting slot(s) within the patch. Such slot(s) will load the original resonant equivalent circuit of the ordinary patch with another for the slot(s) due to the longer current paths traveled around them, thus creating another resonance(s) that couple(s) to the original one. If multiple slots are carefully designed, the antenna can behave as a wideband one due to the creation of several additional resonances. This way, wider bandwidth or multiple bands can be created [41–44]. In [41], a wideband E-shaped patch antenna was presented. Its bandwidth was increased to about 30% compared to less than 5% of the original patch. A U-slot was introduced in [42, 43] to provide 27% and 15% improved impedance bandwidth, respectively. The former operated at 5.5 GHz while the latter operated at 1.7 GHz. Figure 4.10 shows their geometries and the fabricated prototypes. Loading patch antennas with metamaterial (MTM) unit cells is another approach for bandwidth enhancement as shown in [44] but not as wide as the slots, since MTM unit cells have narrow bandwidth, and thus the obtained bandwidth for the MTM loaded patch was only 6.8%.

There are numerous techniques used to miniaturize the size of patch antennas. A comprehensive review of such techniques is provided in [45]. A patch antenna can have its size reduced without affecting the band of interest by using one of these methods:

1. *Material loading:* Using high dielectric constant substrates can be applied to miniaturize the antenna size (reduce it) at the expense of reduced bandwidth, lower efficiency, and potentially higher cost. Various ceramic substrates and magneto-dielectric ones were proposed in literature. For example, in [46] a magneto-dielectric substrate was used to miniaturize a patch antenna by 65%, while maintaining a fractional bandwidth of 0.5% and radiation efficiency of 45% at 2.45 GHz.
2. *Shorting and folding:* Placing shorting pins or walls, as well as folding a patch, can reduce the antenna size. The field distribution below the patch is sinusoidal in nature, and thus placing a shorting wall at its middle where

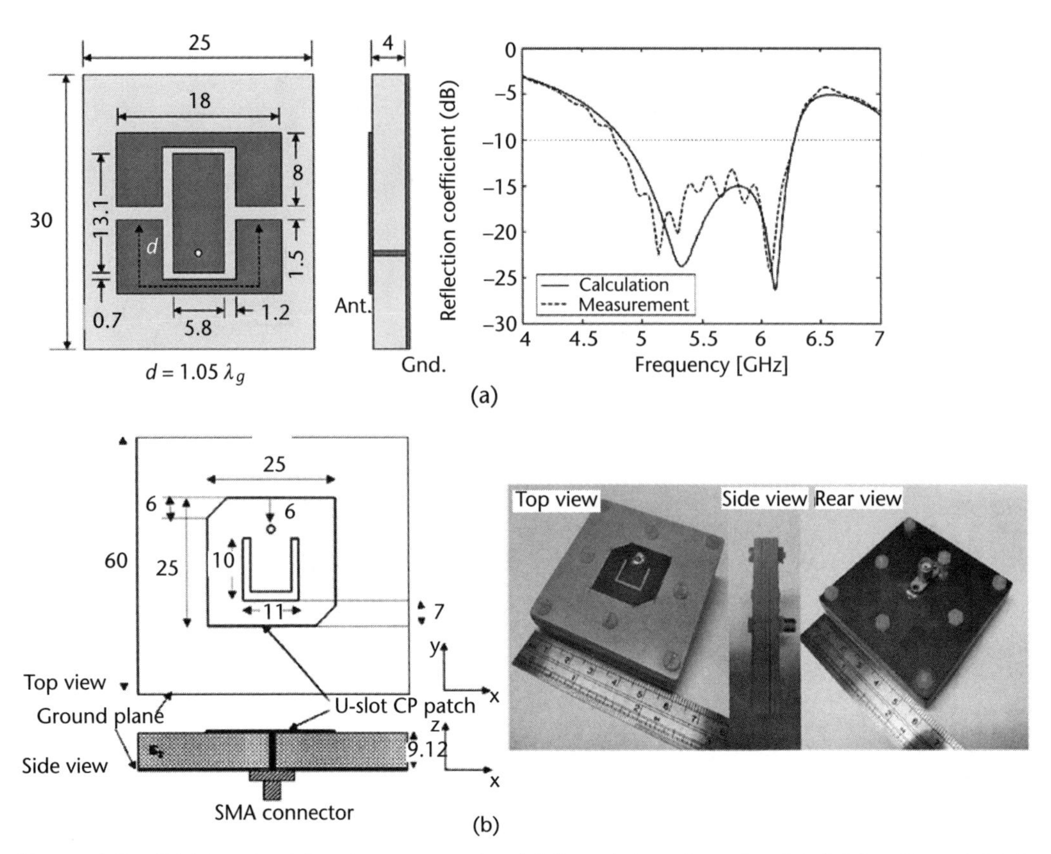

Figure 4.10 Wideband patch antenna examples. (a) Complete rectangular slot dividing the patch into two sections operating at 5.5 GHz. (*From:* [42]. © 2007 IEEE. Reprinted with permission.) (b) U-slot inside a circularly polarized patch for GPS (1.575 GHz) applications. (*From:* [43]. © 2011 IEEE. Reprinted with permission.)

the field diminishes to zero yields a $\lambda/4$ patch length, while maintaining its original resonance frequency (with a little degradation in directivity). Folding the patch on itself can yield $\lambda/8$ antennas as well, thus resulting in a size reduction of 75% [47].

3. *Reshaping and introducing slots:* We have seen earlier that introducing slots within the patch can excite other resonances and thus yield wide operating bandwidth. The same principle can be applied to make the original resonance of the patch get lower, yielding miniaturization when introducing loading slots that directly affect the original resonance of the antenna.
4. *Ground plane modifications:* The performance of patch antennas and their fields derived in most textbooks assume infinite ground planes. Finite ground planes allow the appearance of back lobes as well as affect the input impedance. Introducing defects in the ground plane such as complementary split-ring resonators (CSRR) can be used to load the patch and miniaturize it. For example, in [48] a CSRR was etched below a patch antenna to load it with its equivalent resonant and miniaturize it by 76% at 2.45 GHz. The efficiency achieved was 30% with 2% bandwidth.

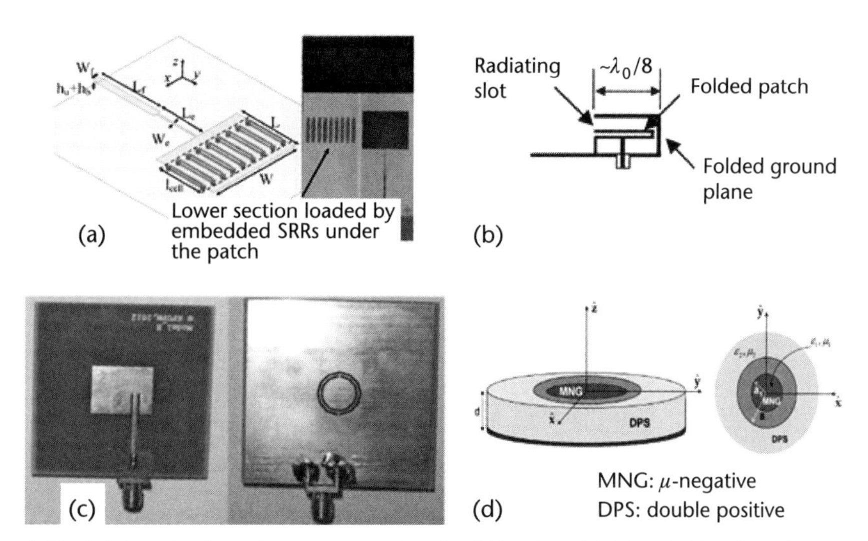

Figure 4.11 Miniaturized patch antennas using the following. (a) Material loading. (*From:* [46]. © 2011 IEEE. Reprinted with permission.) (b) Shorting and folding. (*From:* [47]. © 2004 IEEE. Reprinted with permission.) (c) CSRR loading. (*From:* [48]. © 2013 IEEE. Reprinted with permission.) (d) MTM substrate. (*From:* [49]. © 2007 IEEE. Reprinted with permission.).

5. *Engineered substrates and use of MTM:* The class of double-negative (DNG) MTM showed promising miniaturization ratios as the two negative values of permeability and permittivity will give a real index of refraction and thus when both values are high, a large miniaturization ratio can be obtained. Several examples in literature can be found such as [49].

Figure 4.11 shows several examples of miniaturized patch antennas using various techniques. Note that miniaturization usually affects efficiency, bandwidth, and input impedance.

4.2.4 Loop Antennas

Loop antennas are also used in wireless terminals. While circular loop shapes are used in wire based antennas, rectangular loops are used in printed versions. They cover wide range of frequencies and applications. The radiation pattern of a loop is omnidirectional and resembles that of a magnetic dipole. The fundamental theory of circular loops with uniform and nonuniform currents is discussed with some closed-form integrals in [50–53]. However, due to the complexity of their analytical analysis, loops are usually solved using numerical techniques [54] to take all geometrical and material effects into consideration and have good modeling accuracy. As a starting point, the basic loop size (circumference) of 1λ gives the fundamental resonance, although other loop lengths can be used for various mode excitations such as 0.25λ, 0.5λ, and 1.5λ. Figure 4.12 shows a multiband printed loop based antenna geometry and response [55]. The antenna resembles a 0.25λ loop at 900 MHz, and the matching circuit helps in making it work at the other bands.

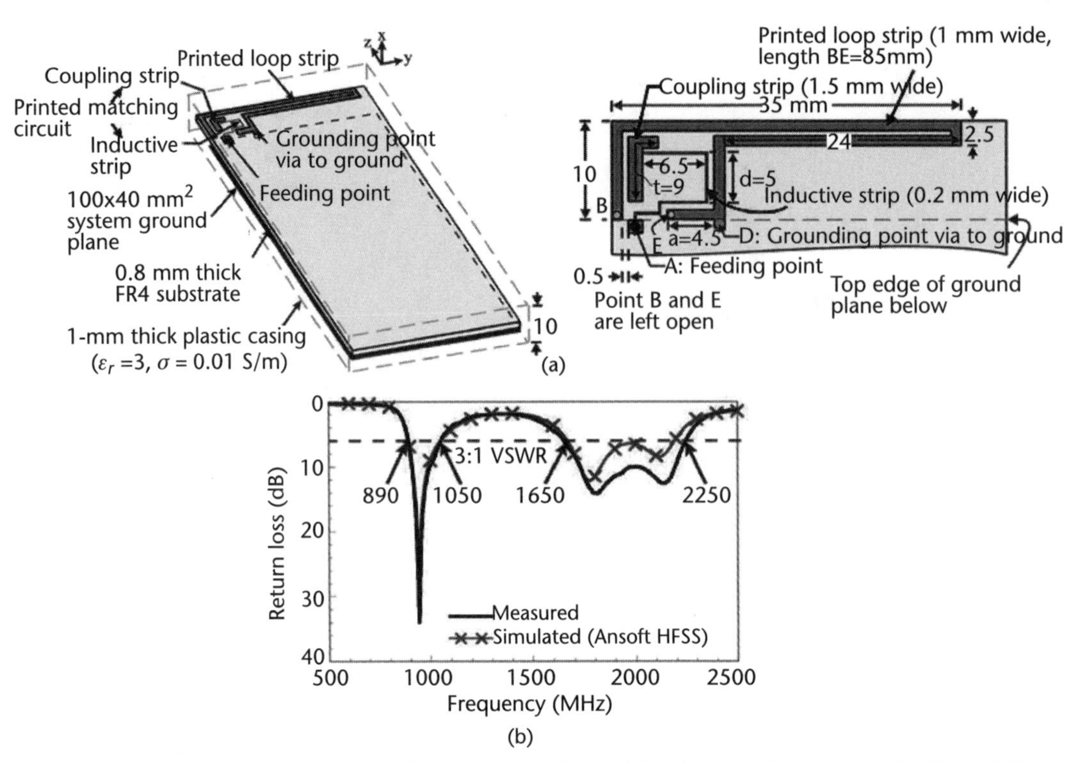

Figure 4.12 A printed multiband loop based antenna for mobile phones: (a) geometry details, and (b) frequency response and covered bands. (*From:* [55]. © 2009 IEEE. Reprinted with permission.)

4.2.5 Slot Antennas

Slot antennas are aperture antennas (small openings within a large ground plane) that are conformal to the structure and take the shape of its ground plane. They are used in designs where the aerodynamic behavior is important or very low-profile designs are desired (planes, missiles, compact wireless gadgets). The most widely used slot antenna is the rectangular half-wavelength long slot. The half-wavelength length will ensure that the voltage across it will add in phase causing radiation [1]. Its width should be much smaller than the operating wavelength, similar to the width of a printed dipole antenna, as the slot and dipole are complements of one another (the slot size can be covered by a dipole operating at the same frequency [i.e., to close the aperture], which is why they are complements). According to Babinet's principle, these two complementary radiating structures will have their impedances related according to [2]:

$$Z_{\text{Dipole}} \times Z_{\text{Slot}} = \frac{\zeta^2}{4} \tag{4.18}$$

thus giving an input impedance of approximately $363 - j211\Omega$ for the $\lambda/2$ slot (ideally). The length of the slot should be optimized numerically, and lengths slightly shorter than $\lambda/2$ will be obtained based on the material properties used for the substrate. The width of the slot should be optimized using a numerical solver for finite ground planes as well as to provide proper impedance matching with the feeding line [56]. The location of the feeding line, which is usually through aperture coupling from a microstrip line, should be closer to one of the two edges with distance not larger than $\lambda/4$ from that edge. This will make the feeding line act as an inductive load that will cancel out the capacitive input impedance of the slot. This location should be optimized according to the design under consideration, but small distances (around 0.05λ) usually give good matching. Figure 4.13 shows the geometry and feeding mechanism of a slot antenna operating at 28.5 GHz. The slot length and width are 3.75 mm and 0.5 mm, respectively. This slot is fed by a microstrip line that is 0.327 mm wide, and the substrate used was RO3003 with 0.13-mm thickness and $\varepsilon_r = 3$. The feed line was extended 0.3 mm beyond the slot edge for proper matching and was placed at 0.05λ from the edge. The ground plane dimensions were 16.8×18.8 mm^2. The radiated fields from a slot within an infinite ground plane are similar to those of the dipole, except that the **E** and **H** fields are interchanged (since the slot and dipole antennas are complements of one another). Thus, a rotated (90°) omnidirectional pattern is obtained.

Several examples of slot antennas have been reported in literature in single element and array formats. Some of the latest works appeared in [57–60]. A wideband annular slot antenna covering from 17–50 GHz was proposed in [57], while a reconfigurable wideband slot covering 4.5–6.5 GHz was proposed in [58] with two states of operation. In [59, 60], planar slot antenna arrays were proposed for the potential 5G frequency band of 28 GHz. In [59], a Butler matrix was used to feed the planar array to provide 4-beam switching patterns. In [60], a fixed radiation pattern was achieved from the planar slot antenna array via the use of a corporate feed network. The array was integrated in-between two 4G diversity systems with

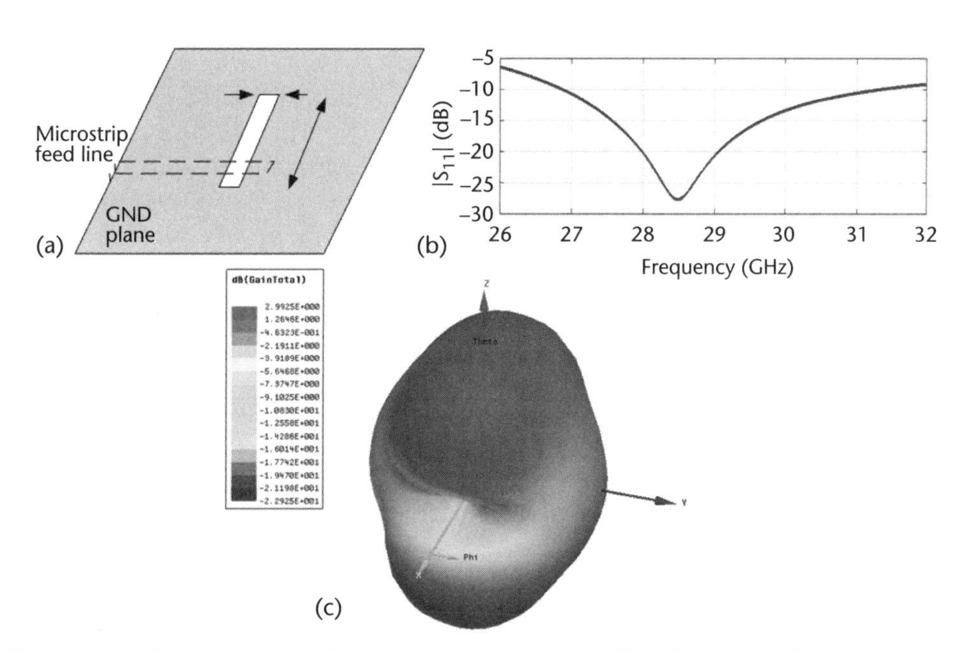

Figure 4.13 Slot antenna model at 28.5 GHz: (a) geometry, (b) reflection coefficient, and (c) radiation gain pattern.

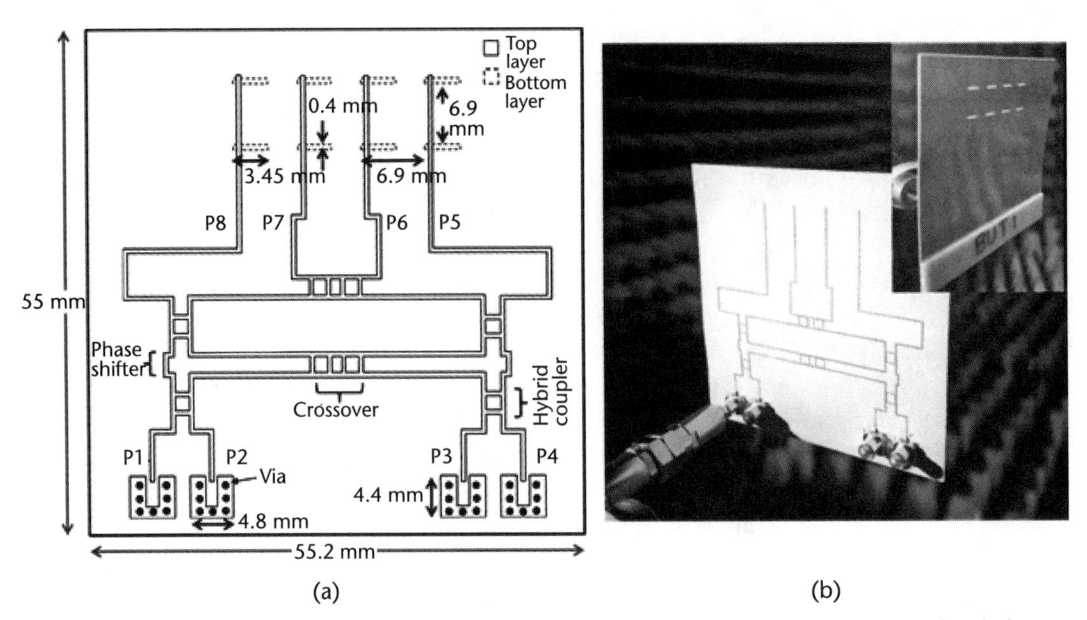

Figure 4.14 Planar slot antenna array with Butler matrix for beam switching at 28 GHz: (a) detailed geometry (RO3003 substrate with 0.13 mm was used), and (b) fabricated prototype. (*From:* [59]. © 2016 Wiley. Reprinted with permission.)

the first of its kind 4G/5G integration. Figure 4.14 shows the details of the design that appeared in [59].

4.3 Antenna Arrays

To improve the transmission or reception capability, to reduce the effect of interference, or to be able to steer the radiation patterns towards specific directions, an antenna array is usually required. An antenna array consists of several antennas that are placed at specific positions relative to one another, with specific amplitude and phase excitations to have a far-field radiation with certain characteristics. The fields from each radiating element are added to give the desired pattern. Usually, antenna arrays consist of multiple elements of the same type, and the degrees of freedom are the interelement spacing, the amplitude excitation, and the phase excitation. Any of these methods can be used individually or together to yield a certain pattern characteristics. An example from Section 4.2.5 was given to enhance the gain of the array at 28 GHz [59]. In this section, we will review the major array configurations and focus on their geometry and array factor they provide.

Once the array factor is obtained for a certain array geometry, the total radiated field is the array factor multiplied by the individual array pattern of a single antenna element within the array. This approximation is valid as long as the coupling between the elements is low, and is denoted as the array pattern multiplication principle [2]. Some more advanced methods have been devised to compensate for the errors encountered by this approximation such as those in [61, 62], but the technique is valid as long as the interelement spacing between the elements is around half a wavelength. For more accurate results, the actual antenna array should be modeled in a field solver to have the closest behavior to the real one to be obtained after fabrication. The array geometries discussed are linear, planar, circular, and conformal antenna arrays. A comprehensive review of antenna array theory is given in [63].

4.3.1 Linear Antenna Arrays

The detailed mathematical theory of linear antenna arrays has been discussed in details in [64–67], where the basic array along with other factors for obtaining phased arrays and SLL control have been extensively discussed. In this section, we will focus on the general operation, and we will provide expressions for the array factor of uniform as well as phased linear arrays.

Figure 4.15(a) shows a linear array of N-elements along the z-axis. All elements are identical. The array factor (AF) of such an equally spaced arrangement as a function of the interelement spacing (d), amplitude (I_i) and progressive phase excitations (α) is given by:

$$AF_{\text{linear}} = \sum_{i=1}^{N} I_i e^{j(i-1)\alpha} e^{[j(i-1)\psi]} = \sum_{i=1}^{N} I_i e^{j(i-1)\alpha} e^{[j(i-1)kd\cos\theta]} \tag{4.19}$$

where k is the wave number (i.e., $2\pi/\lambda$) and θ is the elevation angle. It is clear from (4.19) that when $\theta = 90°$, the exponent becomes zero and that will make all components add up showing a maximum in the AF. This is known as the broadside maximum, since it is perpendicular to the array axis. When the array is distributed along other axis (i.e., x or y), (ψ) will have different values, such that:

$$\psi = kd\sin\theta\cos\phi \quad , \quad \text{for } x\text{-axis array} \tag{4.20}$$

$$\psi = kd\sin\theta\sin\phi \quad , \quad \text{for } y\text{-axis array} \tag{4.21}$$

The amplitude and phase excitations can be used to lower the SLL or steer the beam maximum towards a specified direction. Different amplitude distributions have been proposed to lower the SLL and their effect on the HPBW and directivity were analyzed in literature [68]. Phased arrays with beam and null steering capabilities were also devised and studied in detail in the literature [67]. The simplest beam steering phased array can be constructed by using equal amplitude excitations (i.e., $I_i = 1$) and the progressive phases between adjacent elements are chosen such that the desired angle that the beam should be steered to is:

$$\psi + \alpha = kd\cos\theta_{\text{desired}} + \alpha = 0 \Rightarrow \alpha = -kd\cos\theta_{\text{desired}} \tag{4.22}$$

For example, if all the elements in the linear array (assume we have an 8-element array, that is, $N = 8$, with $d = \lambda/2$) of Figure 4.15(a) have unity amplitude and zero progressive phase ($\alpha = 0$), then the obtained broadside AF pattern will look like the one depicted in Figure 4.15(b). If the beam is to be directed toward $\theta = 45°$, then $\alpha = -127.28°$, and the obtained AF pattern will look like the one in Figure 4.15(c). It

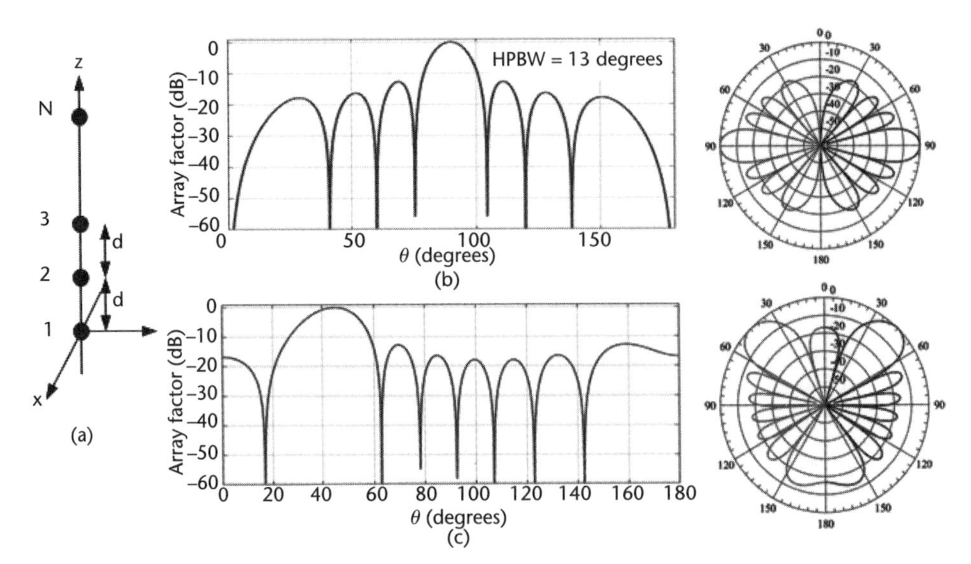

Figure 4.15 Linear array: (a) geometry, (b) AF for broadside case with $N = 8$ and $d = 0.5\lambda$ in rectangular and polar forms, and (c) AF with beam steering towards $\theta = 45°$ with $N = 8$ and $d = 0.5\lambda$ in rectangular and polar forms.

should be noted that grating lobes (sidelobes with amplitudes comparable to the main lobe, usually not desired as they spread the energy) are obtained if the interelement spacing exceed 1λ in equally distributed arrays (uneven element array distributions are more difficult to analyze).

4.3.2 Planar Antenna Arrays

We saw that in linear arrays, we have one degree of freedom when steering the beam. To have two degrees of freedom, we need planar arrays in which the beam can be steered towards specific θ and φ angles. Planar arrays have several applications in radars (particularly stationary and airborne tracking radars) and controlled radiation patterns antennas. The geometry of a planar array is shown in Figure 4.16(a). The AF for a generic planar antenna array at the x-y plane is given by:

$$
\begin{aligned}
AF_{\text{planar}} &= AF_x \times AF_y \\
AF_x &= \sum_{i=1}^{M} I_i e^{j(i-1)\alpha_x} e^{\left[j(i-1)kd_x \sin\theta\cos\phi\right]} \\
AF_y &= \sum_{k=1}^{N} I_k e^{j(i-1)\alpha_y} e^{\left[j(k-1)kd_y \sin\theta\sin\phi\right]}
\end{aligned}
\tag{4.23}
$$

To steer the beam towards a specific (θ_d, φ_d), we need to have progressive phases in the $x(\alpha_x)$ and $y(\alpha_y)$ directions as

$$
\begin{aligned}
\alpha_x &= -kd_x \sin\theta_d \cos\varphi_d \\
\alpha_y &= -kd_y \sin\theta_d \sin\varphi_d
\end{aligned}
\tag{4.24}
$$

To control the SLL, we need to adopt some amplitude distributions (amplitude tapers) according to the desired HPBW and directivity. Several of these tapers are discussed in details in [1, 2, 68]. As an example of beam steering, let us consider a planar antenna array with $M = 5$, $N = 6$ (i.e., 30 antenna elements), with uniform amplitude distribution (i.e., all amplitudes for individual elements are the same), and such that $d_x = d_y = \lambda/2$. Figure 4.16(b) shows the obtained AF pattern when we have broadside radiation (i.e., perpendicular to the array plane, toward $\theta = 0$), while Figure 4.16(c) shows the tilted AF beam towards ($\theta_d = 30°$, $\varphi_d = 45°$).

4.3.3 Circular Antenna Arrays

Circular antenna arrays are used in radar and direction finding applications. The antennas are placed on the periphery of a circular ring or fill a circular disk. Figure 4.17(a) shows a simple circular antenna array. The AF of a circular array is given by [1]:

$$AF_{\text{circular}} = \sum_{i=1}^{N} I_i e^{j\alpha_i} e^{\left[jka\sin\theta\cos(\phi-\phi_i)\right]} \tag{4.25}$$

where a is the radius of the circular array and $\phi_i = 2\pi_i/N$ is the angular position of the ith element. Equally spaced arrays are usually used. To be able to steer the beam in a specific direction in 3-D space (θ_d,φ_d), we need to have a progressive phase excitation according to

$$\alpha_i = -ka\sin\theta_d\cos\left(\varphi_d - \varphi_i\right) \tag{4.26}$$

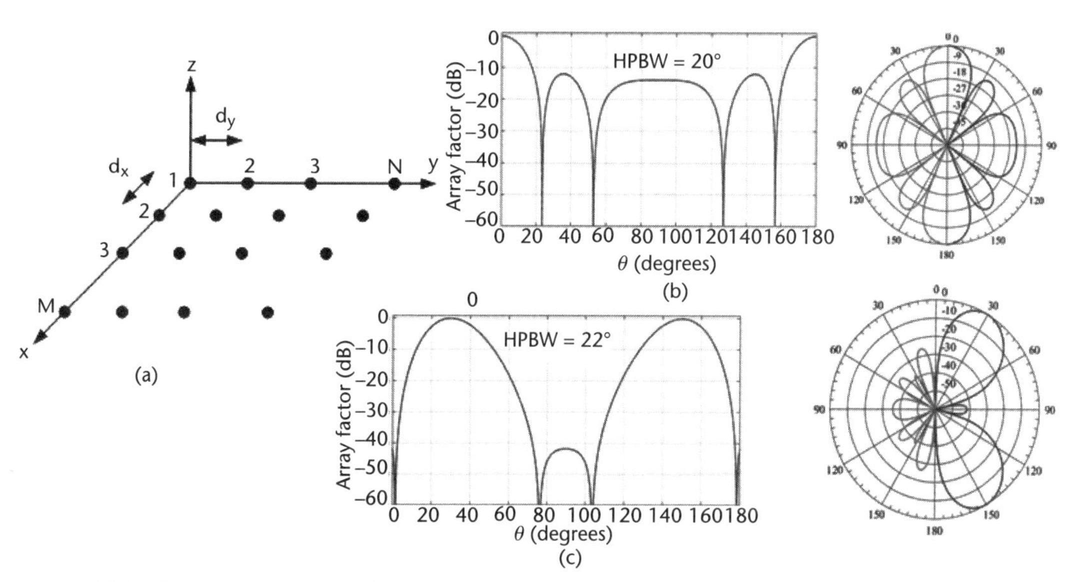

Figure 4.16 Planar antenna array: (a) geometry, (b) AF of a 30-element planar antenna array in rectangular and polar plots, broadside radiation, and (c) AF of a 30-element planar antenna array directed towards $\theta_d = 30°$, $\varphi_d = 45°$. The polar plot shows the elevation cut at $\varphi_d = 45°$.

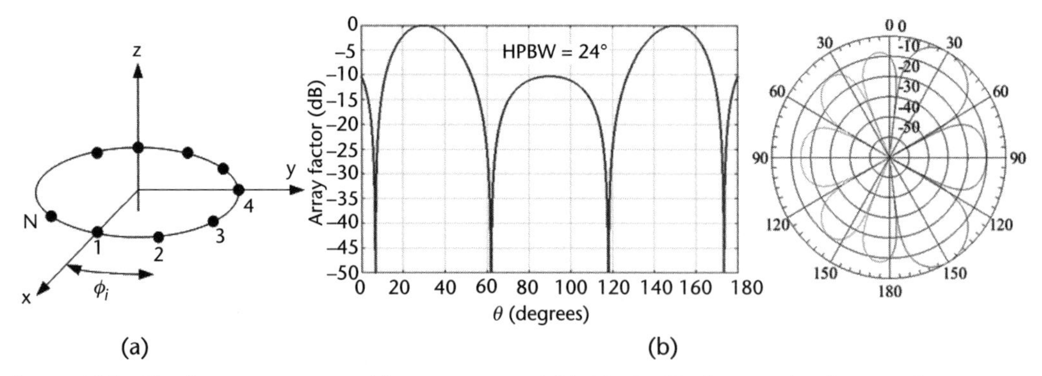

Figure 4.17 Circular antenna array: (a) geometry, and (b) AF of a 10-element circular array in rectangular and polar plots directed towards $\theta_d = 30°$, $\varphi_d = 45°$. The polar plot is showing the elevation cut at $\varphi_d = 45°$.

For example, for a circular antenna array with 10 elements and radius of 0.5m, if the beam is to be steered towards ($\theta_d = 30°$, $\varphi_d = 45°$), the obtained AF is shown in Figure 4.17(b).

4.4 MIMO Antenna Systems

The 4G standard of wireless communications marked the introduction of several new technologies, specifically adaptive modulation and coding, the use of new multiplexing techniques (such as orthogonal frequency division multiplexing [OFDM]), and multiple-input-multiple-output (MIMO) systems. MIMO systems provided a solution to the multipath fading problem that was degrading the performance of legacy wireless communication links by rather making multipath useful [69, 70]. By sending multiple streams of data, the reliability and data rates of the system can increase dramatically with some increase in complexity and processing. Such simultaneous multiple transmissions require multiple antenna systems at the transmitter and receiver sides. The design of such MIMO antenna systems to achieve the required increase in reliability and data rates needs careful attention and characterization, which will be the focus of the sections to come [69].

4.4.1 Features of MIMO Antennas and Systems

MIMO systems were developed to overcome the drawbacks of multipath from which single antenna systems suffered. MIMO systems can be operating in various modes based on the channel condition and the signal-to-noise ratio (SNR) levels. The three modes are the diversity mode, the spatial multiplexing mode, and the beam-forming mode [70]. In the diversity mode, the same information is sent over all the transmit antennas and is received by multiple receiving antennas, and then these multiple received signals are combined or the strongest is selected to retrieve the sent message. The same can be applied if only one message was sent, and multiple receivers were accepting the multipath signal. This will provide better reception in the presence of multipath. The spatial multiplexing mode will use different parallel streams at the transmitter and send them using the multiple antennas, and then, when the channels are isolated enough (low correlation), the multiple streams will be received by the multiple antennas simultaneously, thus providing a noticeable increase in the data rates. Spatial multiplexing is used in high SNR levels. The final mode of MIMO systems is the beam-forming mode that is used in low SNR scenarios. In this mode, the multiple elements at the transmitter and receiver sides are used as arrays to focus the energy and overcome low SNR levels. A diagram showing the multiple antennas at the transmitter and receiver ends is shown in Figure 4.18.

The use of MIMO antenna systems at the transmitter and the receiver parts of the wireless channel needs careful attention while designing them. Several extra performance metrics have been devised to assess the performance of MIMO antenna systems in addition to the regular metrics that are used for single antenna ones (see Section 4.1). The design of MIMO antenna systems for handheld portable devices is a challenging task, as the antennas are usually placed in close proximity. This will increase the field coupling as well as port coupling. These two features need to be minimized in MIMO antenna systems to achieve low correlation levels (important for achieving good channel capacity when the patterns are apart from one another helping to create isolated channels) and low-port isolation that will affect the efficiency of the antenna system. Also, it is essential to make sure that MIMO antennas share the same common system ground as they are not isolated. Indeed, not sharing the same system ground will cause errors in the levels detected as discussed in [71].

4.4.2 Performance Metrics of MIMO Antenna Systems

In addition to the performance metrics of single antenna systems such as resonance, bandwidth, gain, and efficiency as presented in Section 4.1, there are some specific metrics that need to be evaluated for MIMO antenna systems. These metrics are listed next.

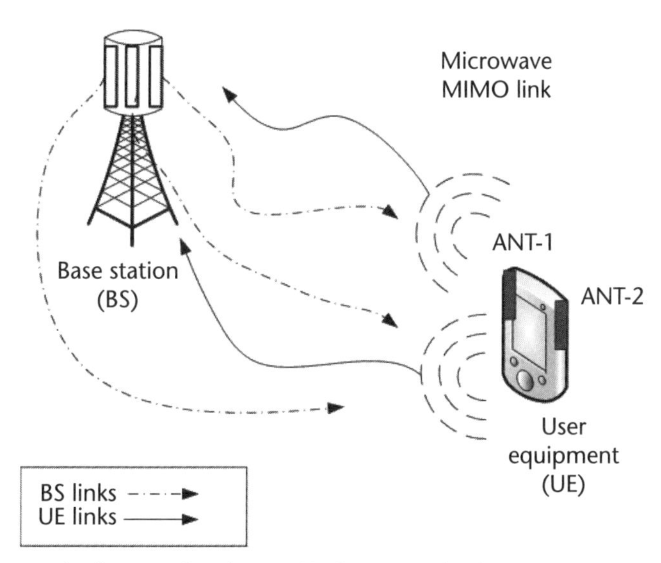

Figure 4.18 A generic diagram showing a MIMO communication system.

4.4.2.1 Port Isolation

Port isolation is important in order to achieve good radiation efficiency. Any energy lost in coupling to adjacent ports will affect the overall radiation efficiency of the antenna system (coupled energy to adjacent ports will reduce their efficiency). Port isolation is defined as the magnitude of the transmission coefficient ($|S_{ij}|$) between elements i and j in a multi-antenna system. A port isolation of 10 dB is typically the least acceptable value for MIMO antenna systems, but the higher the better.

Depending on the type of the antennas used in MIMO antenna systems as well as the proximity with respect to one another, low isolation (high port coupling) might be inevitable. Several techniques have appeared in the literature to improve port coupling such as DGS, neutralization lines, use of parasitic elements and structures, use of resonators, and many more. Details about each method can be found in [69]. Although some designs utilize separate grounds for their MIMO antenna systems (i.e., a separate ground for each antenna, and then placing them next to each other), the use of separate grounds is of no practical benefit and the increased port isolation achieved should not be acceptable, as in real systems, all signals should be referred to the system ground that is shared by all.

An important note should be mentioned here which is the misconception in several papers in the literature between field correlation and port coupling/isolation. There is no direct relationship between the two; thus, port isolation enhancement does not guarantee field correlation enhancement. However, it can enhance the radiation efficiency as the power will not couple to other ports and degrade their performance. This is a major issue that several researchers have encountered because they relied on the S-parameter equation for estimating the correlation coefficient which is a field based quantity and not a port based one (see Section 4.4.2.4).

4.4.2.2 Total Active Reflection Coefficient

In a multiport antenna system, the presence and activity of adjacent ports will affect the port under consideration in terms of its effective bandwidth and efficiency if not properly designed. To assess such interactions, the total active reflection coefficient (TARC) was introduced [72]. TARC is defined as the ratio between the square root of the sum of the reflected powers and their incident power counterpart. For an N-element antenna, TARC is found using:

$$\Gamma = \frac{\sqrt{\sum_{i=1}^{N} |b_i|^2}}{\sqrt{\sum_{i=1}^{N} |a_i|^2}} \tag{4.27}$$

where a_i and b_i are the incident and reflected waves, respectively. They are found from the measured S-parameters of the system. TARC values range between 0 and 1, where a 1 means that the incident power is totally reflected and not radiated, and 0 means that the incident power is totally radiated and nothing was reflected. TARC is usually represented in decibels.

For a two-port MIMO antenna system, TARC can be evaluated by fixing the angle of the incoming wave of one port and then sweeping the relative angle (θ) with the adjacent port to see the effect of such an angle change along with the mutual coupling on the effective bandwidth and efficiency of the multiport antenna system. This can be found using [73]:

$$\Gamma = \frac{\sqrt{\left(\left[\left|S_{11} + S_{12}e^{j\theta}\right|^2\right] + \left[\left|S_{21} + S_{22}e^{j\theta}\right|^2\right]\right)}}{\sqrt{2}} \tag{4.28}$$

4.4.2.3 Branch Power Ratio

The branch power ratio (BPR) shows the ratio between the lowest and highest power levels within the antenna system. For ideal operation of a multi-antenna system, we want the BPR to be 1 (0 dB), that is, all ports are receiving the same power level. BPR (K) can be found using [74]

$$K = \frac{P_{\min}}{P_{\max}} \tag{4.29}$$

Also, BPR can be found from the calculated MEG values for each antenna elements in a specific environment as

$$K_{i,j} = \min\left(\frac{MEG_i}{MEG_j}, \frac{MEG_j}{MEG_i}\right) \tag{4.30}$$

where i and j are the elements of the multi-antenna system. For acceptable diversity performance, the MEG values of various ports should not exceed 3 dB. The effect of the BPR on the channel capacity was investigated in [75]. It was found that the power levels of the incoming signals have more effect on the channel capacity compared to the BPR values that are typically below 1 dB.

4.4.2.4 Field Correlation

The correlation coefficient (CC) or the envelope correlation coefficient (ECC) is a very important MIMO antenna system metric that needs to be carefully evaluated. It shows the amount of channel correlation in a wireless communication link. When multiple MIMO channels are not correlated, the anticipated increase in the channel capacity can be achieved. Usually, we use ECC because we deal with power signals. Ideally, we want an ECC value of 0. While in reality, there is a nonzero correlation between the wireless channels due to the multipath environment power distribution as well as the radiation patterns being overlapped in a compact MIMO antenna system. The radiation patterns of the MIMO antenna system are the ones used in the evaluation of the CC (ρ) or ECC (ρ_e) since they directly affect the channel between the transmitter and the receiver. In an isotropic channel,[1] they can be evaluated according to [76]:

[1] An isotropic or uniform channel is the one that has $XPD = 1$ and uniform incoming wave distributions, that is, $P_\varphi = P_\theta = 1/4\pi$.

$$|\rho_{12}|^2 = \rho_e(ECC) = \frac{\left|\iint_{4\pi}\left[E_1(\theta,\phi) * E_2(\theta,\phi)\right]d\Omega\right|^2}{\iint_{4\pi}|E_1(\theta,\phi)|^2\, d\Omega \iint_{4\pi}|E_2(\theta,\phi)|^2\, d\Omega} \tag{4.31}$$

where $E_i(\theta,\phi)$ is the complex 3-D radiated field pattern for antenna i and the integral is evaluated over the whole sphere. The expression in (4.31) is based on the assumption of having an isotropic wireless environment, and this should always be highlighted. The relation between CC and ECC is not exactly the square function, but they follow approximately a square one in a Rayleigh distributed narrowband environment [77]. A more general expression that takes into account the environment effect is:

$$|\rho_{12}|^2 \approx \rho_{e12} = \left|\frac{\oiint XPD \times E_{\theta i}(\Omega)E^*_{\theta j}(\Omega)P_\theta(\Omega) + E_{\varphi i}(\Omega)E^*_{\varphi j}(\Omega)P_\varphi(\Omega)\,d\Omega}{\sqrt{\oiint XPD \times G_{\theta i}(\Omega)P_\theta(\Omega) + G_{\varphi i}(\Omega)P_\varphi(\Omega)\,d\Omega} \times \sqrt{\oiint XPD \times G_{\theta j}(\Omega)P_\theta(\Omega) + G_{\varphi j}(\Omega)P_\varphi(\Omega)\,d\Omega}}\right| \tag{4.32}$$

where XPD is the cross-polarization discrimination factor, which gives the ratio between the vertically polarized and horizontally polarized field components in the environment, $P_{\theta,\varphi}(\Omega)$ is the wave distribution of the specific (φ,θ) angular directions, and $G_{\theta,\varphi}(\Omega)$ is the gain (i.e., $G_\theta[\Omega] = [E_\theta(\Omega)E^*_\theta(\Omega)]$). In [78], a simplified relationship to find (4.31), based on the port parameters (S-parameters) for lossless antenna systems was provided based on equating the powers in and out of the system. For a two-element MIMO antenna system, this relationship is given by:

$$\rho_{12} = \left|\frac{|S^*_{11}S_{12} + S^*_{21}S_{22}|}{\left\{\left[1-\left(|S_{11}|^2+|S_{21}|^2\right)\right]\left[1-\left(|S_{22}|^2+|S_{12}|^2\right)\right]\right\}^{\frac{1}{2}}}\right| \tag{4.33}$$

Several works in literature advocate the use of (4.33) due to the simplicity of obtaining the port parameters. However, several issues should be pointed out. The first is the assumption of using lossless antennas in an isotropic environment. Unless the antenna system under consideration is lossless, (4.33) should not be used. Since all printed antennas are lossy, and unless the efficiency of the antenna is very high, then (4.33) should not be used to evaluate CC or ECC in any work, because it is not valid [79, 80].

Another important issue when using (4.33) is the fact that port parameters have nothing to do with channel behavior that is important for evaluating the channel capacity. This means that (4.33) should not be used to evaluate CC or ECC even for 100% efficient MIMO antenna systems, but rather (4.31) or (4.32). Considering similar antennas that are spatially placed in close proximity, (4.31) or (4.32) will predict high ECC values compared to (4.33) that depend on the port coupling alone between the adjacent elements and not their radiating patterns.

To improve (lower) CC/ECC values in a MIMO antenna system, the designer should try to spatially isolate/tilt/separate the radiation patterns. This can be done by using different polarizations, or tilted beams (via beam steering, use of reflectors within the ground, use of parasitic directors). Some examples can be seen in [81–83]. Figure 4.19 shows one way of achieving tilted patterns for low ECC values.

4.4.2.5 Diversity Gain

Diversity schemes were devised to combat multipath effects. When the signals are not correlated, their combination at the receiver will provide higher SNR levels and thus better signal reception. A measure of diversity performance in a communication system is the diversity gain (DG). It is defined as the difference between the averaged SNR of the combined signals in the diversity system and the single antenna system in a specified channel, provided that the SNR is above a threshold value. Mathematically, it is given by [84]:

$$DG = \left[\frac{SNR^{i}_{\text{div}}}{SNR^{m}_{\text{div}}} - \frac{SNR^{i}_{\text{sin}}}{SNR^{m}_{\text{sin}}} \right]_{P\left(SNR^{i}_{\text{div}} < SNR_{\text{ref}}\right)} \tag{4.34}$$

where SNR^{i}_{div} and SNR^{m}_{div} are the instantaneous and average (mean) SNR for the diversity (MIMO) system, respectively. SNR^{i}_{sin} and SNR^{m}_{sin} are the instantaneous

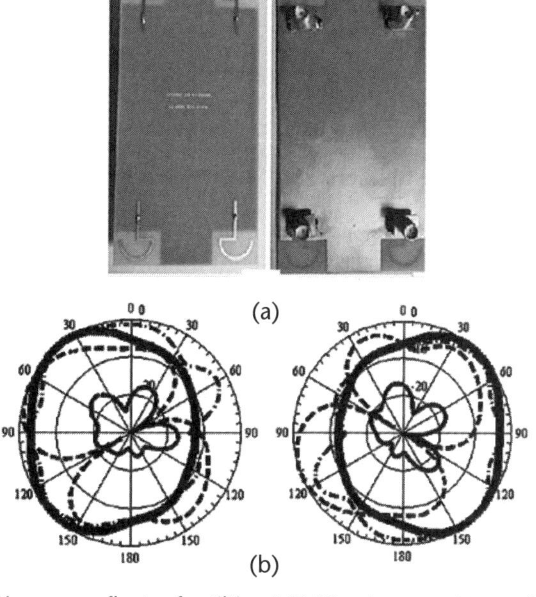

Figure 4.19 GND acting as a reflector for tilting MIMO antenna patterns: (a) four-element MIMO antenna fabricated prototype, and (b) tilted patterns of the top antennas. Note the effect of the GND sheet behind them. (*From:* [82]. © 2016 Wiley. Reprinted with permission.)

and average (mean) SNR for the single antenna system, respectively. SNR_{ref} is the reference level. The increase in the number of antennas should provide higher diversity gain, but the gain saturates after 8 antennas or so according to (4.34). Cumulative distribution functions (CDF) are used to show the DG versus the SNR. Two main DG metrics are usually used: apparent DG (ADG) and effective DG (EDG). The former defines the improvement between the single antenna and the combined signals at a reference level of 1% on the CDF (marking 99% reliability), while the latter compares the improvement of the combined signal with respect to the ideal theoretical Rayleigh CDF. Figure 4.20 shows the CDF curves for a two-port dipole antenna system with 0.5λ separation. The ADG and EDG are clearly marked. The combined curve is based on maximum ratio combining scheme (MRC) [85]. The total efficiency of the two ports can be found from the difference between the port (ANT) CDF and that of the theoretical Rayleigh one.

An approximate formula for antennas of the same type (i.e., similar radiation efficiencies) that relates the ADG to CC/ECC is given by [86]:

$$ADG = 10.5\sqrt{1 - |0.99\rho|^2} \tag{4.35}$$

4.4.2.6 Channel Capacity

The general channel capacity equation for M transmitting and N receiving antennas, with no channel state information, Gaussian distributed signals, and identity covariance matrix is given by [70]:

$$C = BW\log_2\left[\det\left(\mathbf{I}_N + \frac{P_T}{\sigma^2 M}\mathbf{H}\mathbf{H}^H\right)\right] \tag{4.36}$$

where C is the channel capacity (in bits/second), BW is the channel bandwidth (in hertz), P_T is the equally distributed input power among the elements, σ^2 is the noise

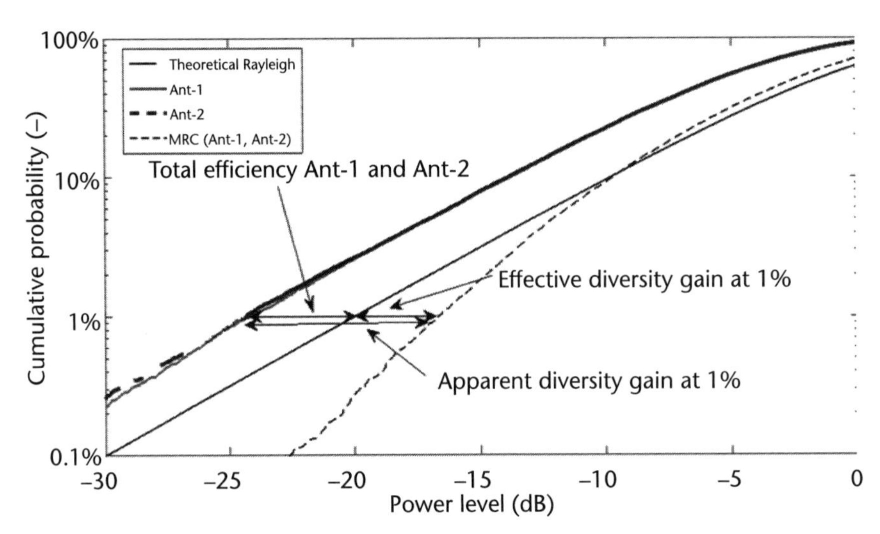

Figure 4.20 CDF curves for finding ADG and EDG.

power, $\mathbf{I_N}$ is the $N \times N$ identity matrix, and $\mathbf{H}$ is the complex channel matrix. From (4.36), it is clear that an increase in BW or power (i.e., SNR) will yield a direct increase in C. However, due to the limited spectrum bands, as well as predefined power level transmissions set by governmental agencies and operators, these two factors are rarely available for any modifications. This leaves us with increasing M or N or both to achieve an increase in C. This is the main driver behind MIMO systems.

Increasing the number of antenna elements needs to be performed at the base station as well as the user terminal (UT) sides. This increase is easy to accomplish on the base station side, while it is very challenging when targeting small form factor UT, as placing antennas closer together will increase port coupling (decrease isolation) and increases field correlation. Evaluating the channel capacity is an essential step to verify that the designed antenna is going to provide the expected MIMO advantage. Coming up with the capacity performance is usually done via actual measurements in a defined environment to evaluate the complex channel matrix $\mathbf{H}$. Several procedures appeared in the literature such as the ones mentioned in [69, 74, 85]. Figure 4.21 shows the channel capacity curves for a dual-band MIMO antenna system (2.45 and 5.5 GHz) that appeared in [74] and compares the measured actual curves with the theoretical ones estimated based on ideal channel conditions.

4.4.3 MIMO Antenna System Examples

There is a large collection of MIMO antenna systems using various antenna types and for a vast number of applications covering a wide spectrum of frequencies. In this section, we will present few demonstrative examples of some recent applications. Several references discussing other applications will be pointed out as well.

MIMO antenna systems for mobile terminals and cell phones are among the most designed due to the stringent requirements of placing several antennas within a small volume in such terminals as well as due to the large coupling and interactions that take place. Figure 4.22 shows the most widely used locations for MIMO

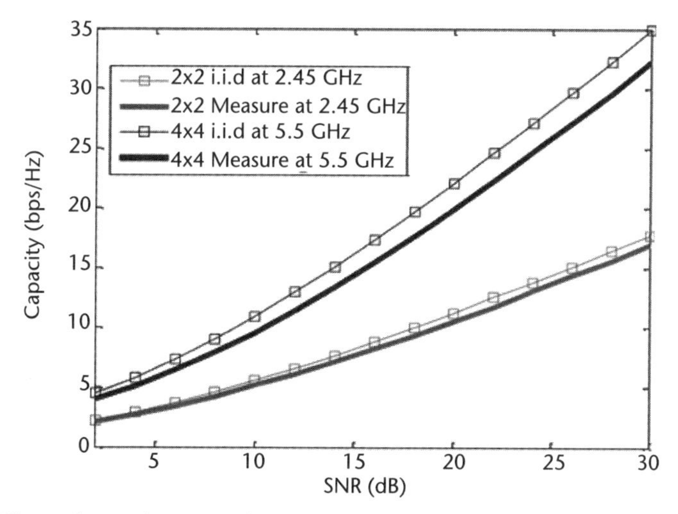

Figure 4.21 Channel capacity curves for a 2 × 2 and 4 × 4 MIMO antenna system. (*From:* [74]. © 2016 IEEE. Reprinted with permission.)

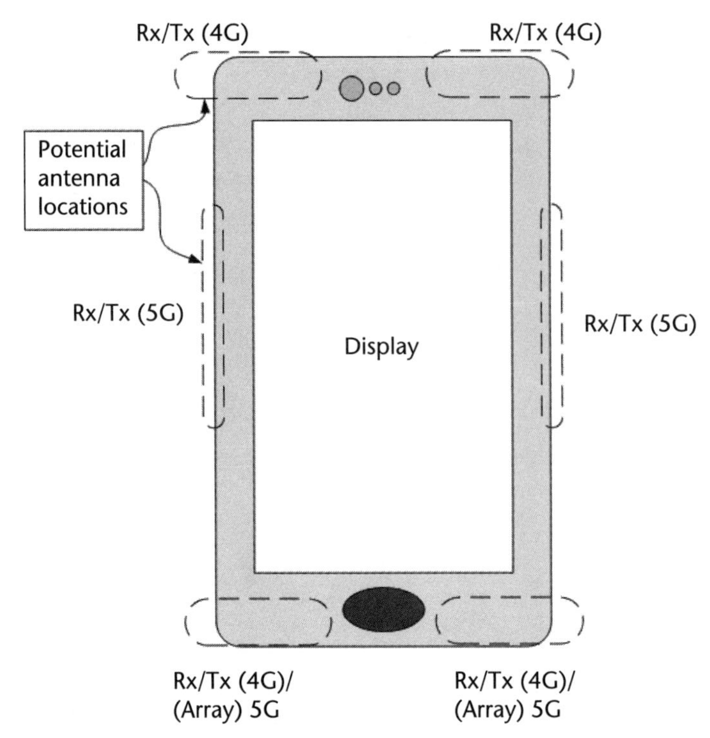

Figure 4.22 Most widely used locations for MIMO antenna systems in current 4G and future 5G bands.

antennas in smart phones. The most widely used locations for 4G based antennas are the top and bottom sides of the mobile device. Most 4G-enabled smart phones have two MIMO antennas. For the upcoming 5G standard, higher frequency bands are targeted such as 3.6–6 GHz and the millimeter-wave bands in 27–29.5 GHz and 36–47 GHz among many others such as 64–71 GHz (several other bands that have not yet been approved are listed in Table 6-1 in [87]). For these bands, the waves will encounter high attenuation when propagating from the mobile terminal to the base station, and thus antenna arrays are to be used to compensate for such transmission loss. Locations of such antennas are towards the bottom of the phone. For the lower bands, the edges of the mobile terminal can be used to stack 8 to 10 elements for an improved channel capacity.

Several recent works have provided multiband MIMO antenna system solutions for mobile terminals such as those presented in [88–92]. For covering the low-frequency bands of the 4G standard (i.e., 700–900 MHz), the periphery of the mobile phone is used to place such antennas because they usually need long current paths to cover lower bands such as the solution in [86]. Two metal rings separated by 3.5 mm that encircle the GND plane with a total volume of $150 \times 70 \times 7.5$ mm^3 are used. The rings were connected to the GND plane at optimized locations to provide the resonance frequencies of the two-element cellular band (820–900 MHz) MIMO antenna and the two-element Wi-Fi (2,400–2,500 MHz and 5,200–5,800 MHz) MIMO antenna system. One of the cellular antennas covers an extra band as well (due to the length differences of the radiators). This is shown in Figure 4.23(a). Satisfactory MIMO performance was measured in three used positions in an anechoic chamber. The four corners of a standard mobile phone backplane are used in [91] for a four-element frequency reconfigurable MIMO antenna system with wide tunability range and wide bandwidth in each step. Each tuned band provides at least 450 MHz of bandwidth, and the reconfigurability covers the 1,610–2,710-MHz band. Varactor diodes were used for frequency sweeps. The patterns were tilted because of the presence of the GND plane (the antennas were surrounded by the GND from three sides). The fabricated prototype is shown in Figure 4.23(b). The beams from each antenna were tilted with respect to one another because of the GND plane, thus ensuring low ECC values.

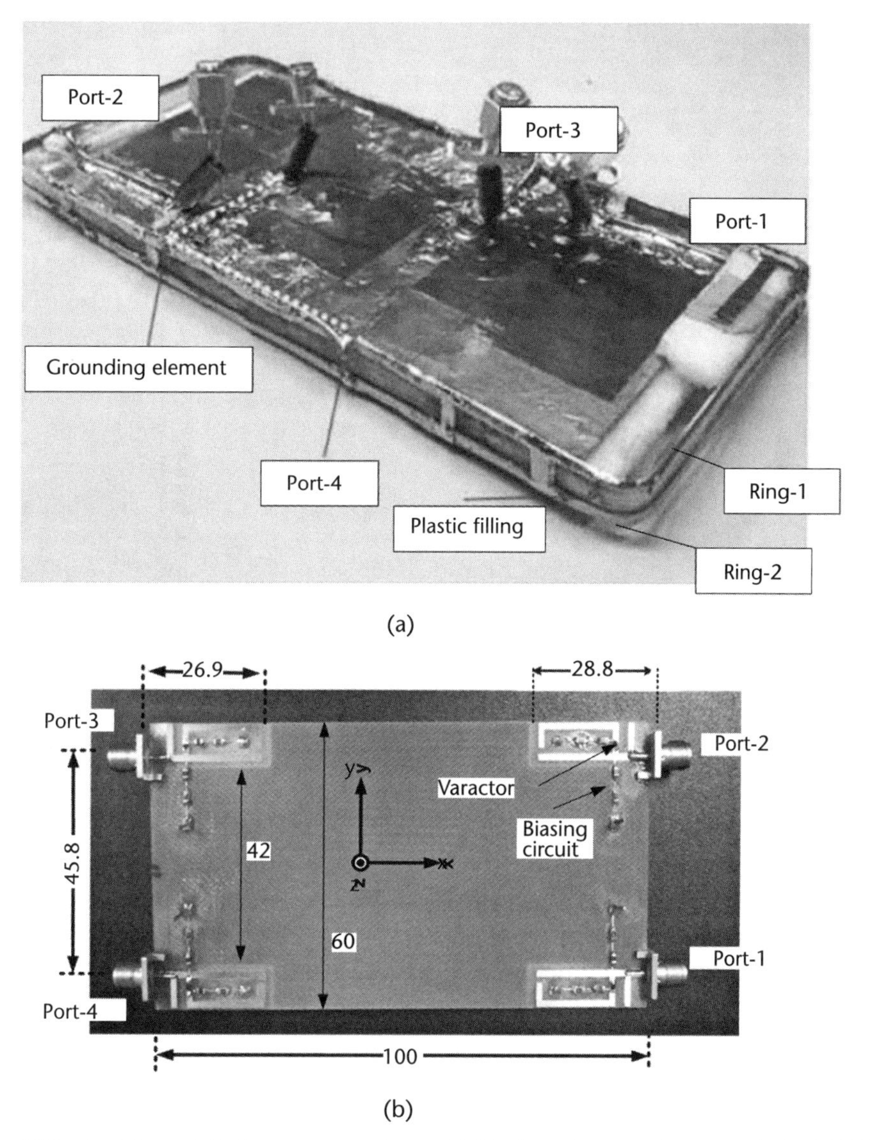

Figure 4.23 Multiband MIMO antenna systems. (a) Mockup of a two-element cellular MIMO antenna (ports 1 and 2) and 2-element WLAN (ports 3 and 4) one using shorted rings. (*From:* [88]. © 2015 IEEE. Reprinted with permission.) (b) Four-element frequency reconfigurable MIMO antenna using varactor diodes. (*From:* [91]. © 2016 Wiley. Reprinted with permission.)

Monopole-based antennas are used in UWB MIMO antenna systems as those in [93–95]. Such antennas can be used in mobile terminals supporting UWB operation and standards. USB dongle size MIMO antenna systems were proposed in [96, 97] using slotted patch 3-D antennas as well as helix based ones. The former targeted WLAN bands in the 2.4-GHz and 5-GHz bands, while the latter targeted low 4G bands in the 700-MHz range. Figure 4.24(a) shows the USB dongle design in [96]. Wireless access point MIMO antenna systems for WLAN have relied on adjacent antenna covering various sectors in 3-D space such as those in [98–100]. A 4-beam directional access point design with 4 elements in each of its 4 sectors was proposed in [100]. Each of the 4 elements in each of the 4 sectors had two concentric loops for 2.45-GHz WLAN and 3.6-GHz WiMAX bands and a dipole in its center for 5-GHz WLAN. The MIMO and single-element geometries are shown in Figure 4.24(b).

The final examples on MIMO antenna systems are the ones that were recently proposed targeting the 5G potential bands and standard. Two examples are given here. The first example is a 3.4–3.6-GHz MIMO antenna system with 8 elements that can provide high channel capacities for future 5G terminals targeting low frequency bands. The design integrates the 8-element monopole-based 5G MIMO antenna system on the sides of the mobile terminal main board while its top is used for a 4G two-element MIMO antenna system. The details of this proposed design are shown in Figure 4.25(a) [101]. Note that a neutralization line is used to enhance the isolation between the 4G antennas. The 28-GHz band is licensed as one of the 5G potential bands. Another integrated MIMO antenna system consisting of two 4G-based MIMO antenna systems and a 5G planar slot antenna array operating at 28 GHz is presented in Figure 4.25(b) [102]. The 4G dual-element MIMO antenna system covered 1.87–2.53 GHz, and the 5G antenna array covered 26.8–28.4 GHz. A modified two-arm-based monopole was used for the 4G antennas while a

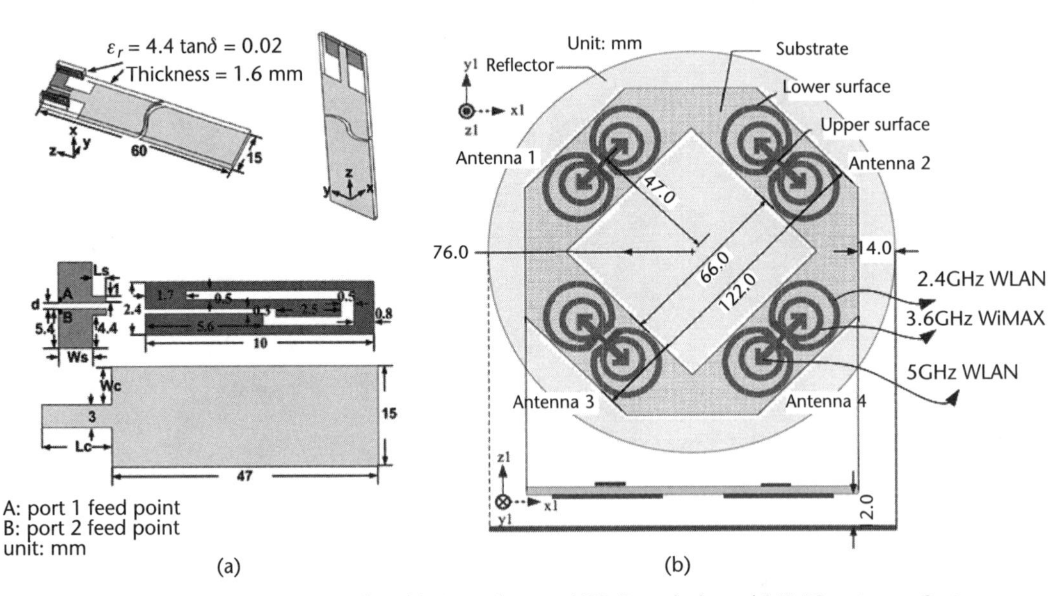

Figure 4.24 MIMO antenna examples. (a) Two-element USB Dongle-based MIMO antenna System covering WLAN bands. (*From:* [96]. © 2014 IEEE. Reprinted with permission.) (b) Multi-element multiband 4-element per one sector (4-sector) design. (*From:* [100]. © 2016 IEEE. Reprinted with permission.)

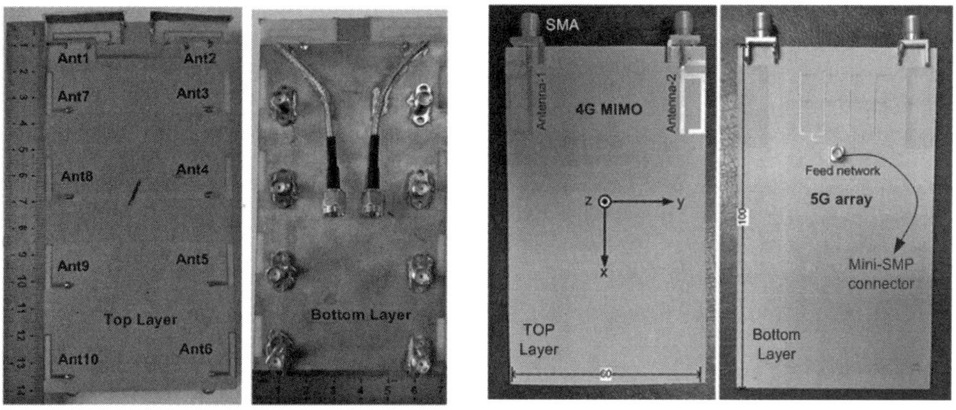

(a) (b)

Figure 4.25 5G antenna systems. (a) Integrated 4G/5G with 8-element MIMO at 3.6 GHz. (*From:* [101]. © 2016 IEEE. Reprinted with permission.) (b) Integrated 4G/5G with two-element wideband MIMO and a 5G slot-based array at 28 GHz. (*From:* [102]. © 2005 IET. Reprinted with permission.)

half-wavelength slot was used as the basic 5G array element. The array consisted of 4 × 2 slots with a corporate feeding network. The design was constructed on a three-layer printed circuit board (PCB), with the GND plane in the middle, and the top layer had the 4G elements while the bottom layer had the 5G feed network.

While several examples have been presented in this section on various MIMO antenna systems and implementations for a wide range of applications and wireless standards, there are much more in open literature targeting specific designs, geometries, device sizes and standards. The reader is encouraged to check the latest designs in various well-known repositories online.

4.5 Computer-Aided Antenna Design

Analyzing antenna performance and radiation mechanisms analytically can be impossible for complex antenna geometries. That is why closed-form approximate expressions for the radiated fields and impedances were developed for only the basic antenna elements discussed in Section 4.2. All other antennas are analyzed numerically (with the help of computer-aided design [CAD] tools); using one of the three well-known computational electromagnetic methods; method of moments (MoM), finite-difference time domain (FDTD), or finite element method (FEM). These numerical methods discretize the antenna geometry and find the currents on it from which the radiated fields and behavior are obtained.[2] This shows how

[2] Other tools are available. Some are free of charge such as NEC, OpenEMS, while some basic packages exist within some commercial versions of MATLAB, but without the detailed capabilities that the three methods (i.e., MoM, FDTD, FEM) provide, which make them capable of analyzing arbitrary geometries. Computational time becomes more dramatic when 3-D structures are meshed, and thus simplified models can be adopted using the tool boxes mentioned earlier.

important it is to be able to use modern tools to model and design antennas for the application and specifications under consideration. In this section, we present two examples using two well-known antenna design software packages; High Frequency Structure Simulator (HFSS) from Ansys Inc. and Computer Simulation Technology (CST Inc.). Detailed, step-by-step tutorials are given in Appendices C and D for the use of these two packages and the way to apply them in modeling various antenna types.

4.5.1 Printed Monopole Antenna Modeling Example Using HFSS

Printed monopole antennas are among the most widely used in mobile terminals nowadays due to their ease of modeling and attractive features, starting from their simple structure and behavior. Several techniques are used to enhance the bandwidth and radiation properties of such antennas, but we will focus here on a simple example of designing a two-element MIMO antenna system operating at the 2.45-GHz WLAN band. The backplane size is that of a standard smart phone terminal with $110 \times 60 \times 0.8$ mm^3. Figure 4.26 shows the geometry of the antenna under consideration. This geometry is modeled using HFSS with all dimensions to be used as shown in Figure 4.26.

The length of a single monopole is around $\lambda_g/4$ at 2.45 GHz which is about 19 mm ($\lambda_0 = 122.5$ mm at 2.45 GHz, and thus $\lambda_g = 61.2$ mm), since the substrate used is FR-4 with $\varepsilon_r = 4$, and loss tangent of $\delta = 0.02$. The length should be tuned (increased/decreased around the optimal length) to provide proper matching at the frequency of operation. After the simulation is run over the frequency band 1.5–3

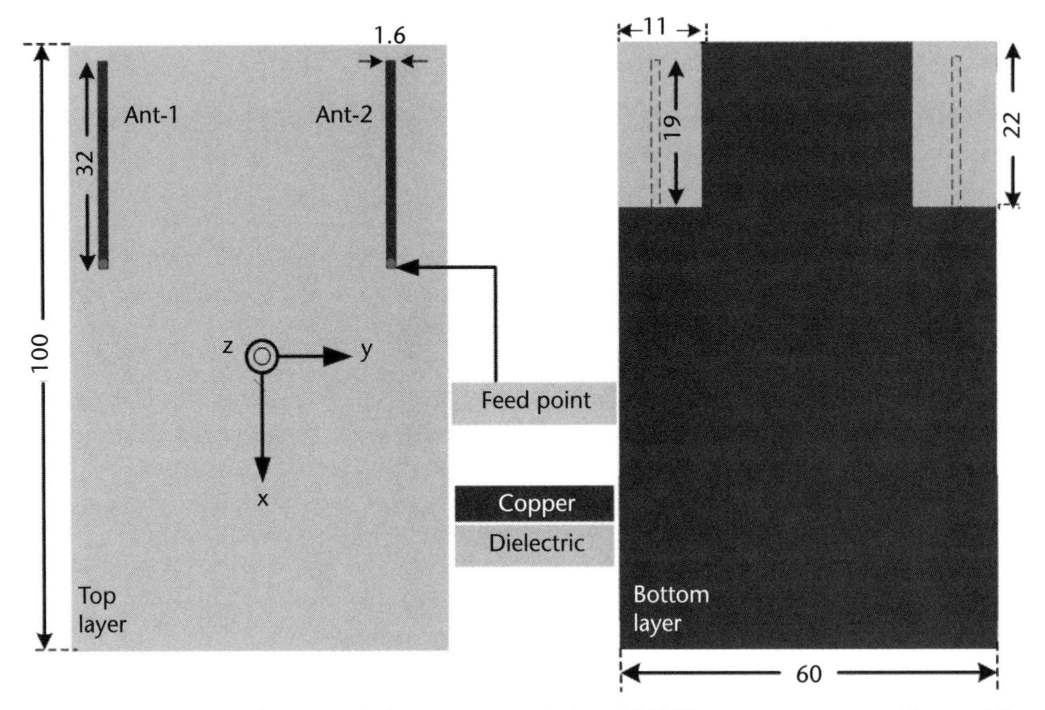

Figure 4.26 Geometry of a WLAN 2-element monopole based MIMO antenna system within a mobile phone-sized backplane.

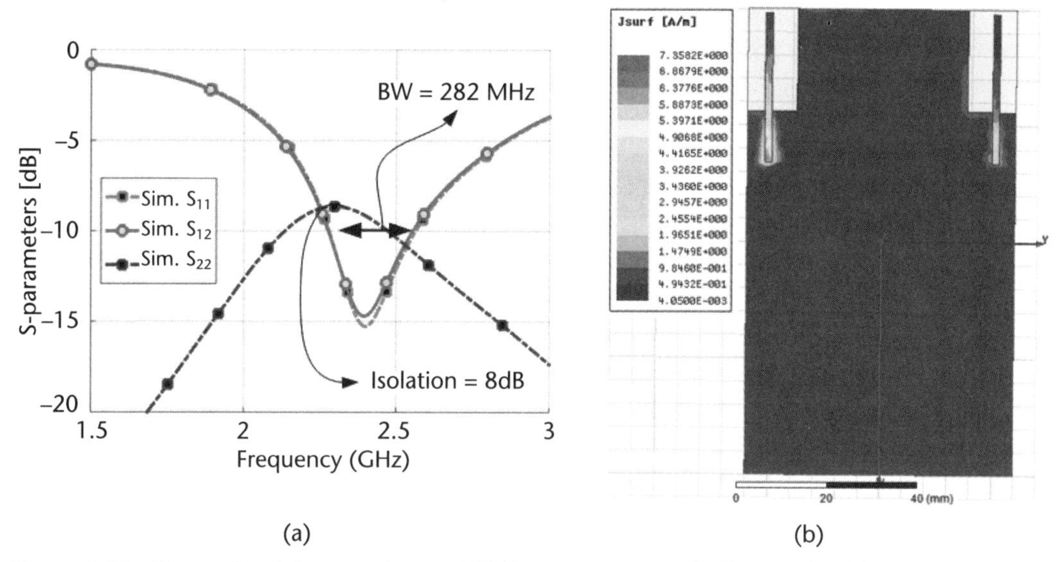

Figure 4.27 The results of the two-element MIMO antenna system in Figure 4.26: (a) *S*-parameters, and (b) current distribution when antenna 1 is active and antenna 2 is terminated with 50Ω.

GHz, the *S*-parameters are plotted as shown in Figure 4.27(a). As can be seen, the antenna is resonating at 2.45 GHz with a −10-dB bandwidth of 282 MHz. The coupling between the two antennas is better than 8.5 dB, but needs improvement for good port efficiency. We will not improve it further in this example, but the reader is referred to one of the methods in [69] to further enhance the performance of this antenna. The current distribution at the center frequency when antenna 1 is activated is shown in Figure 4.27(b). As can be seen, the current is distributed along antenna 1 representing a current path of approximately $\lambda_g/4$. Thus confirming the monopole operation, and some of the current is coupled to antenna 2 as shown by the coupling curves.

The 3-D gain patterns when each of the two elements is excited are shown in Figure 4.28. It is clear from this figure that each pattern is tilted (its maximum) opposite to the other due to the presence of the GND plane on the monopole side.

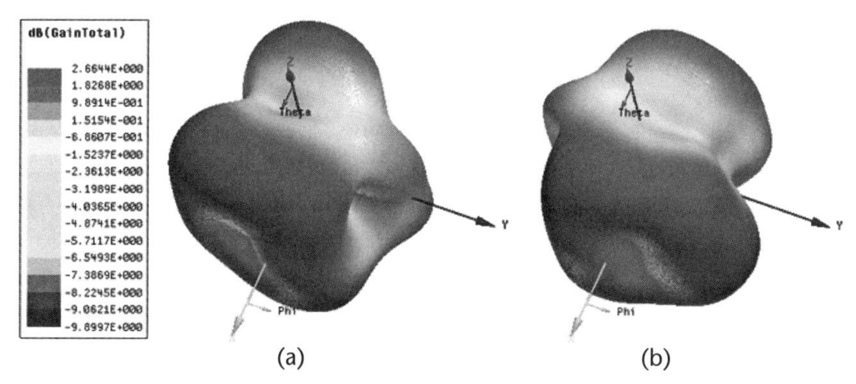

Figure 4.28 Gain patterns of the two-element MIMO antenna system at 2.45 GHz: (a) antenna 1 is active, and (b) antenna 2 is active.

This will ensure low values of ECC as the patterns do not overlap much. The evaluated ECC value based on (4.31) is 0.07, which is way below the maximum limit, while the efficiency of the antennas was found to be higher than 90%.

4.5.2 Printed PIFA Antenna Modeling Example Using CST

The second design example is conducted using CST. We model a two-element PIFA-based MIMO antenna system as shown in Figure 4.29. Again, the main PIFA arm length (from feed to tip of antenna) is approximately $\lambda_g/4$ at 2.45 GHz like its monopole counterpart. The length should be tuned along with the shorting strip location to provide appropriate matching and impedance bandwidth. In this model, we have incorporated the effect of the SubMiniature version A (SMA) connector that is usually used to feed the antenna from the backside of the board. The SMA model was created by using concentric cylinders resembling the actual geometry of a real SMA connector with the same dimensions and using the same metal and dielectric properties of the inner feeding pin and isolation material (i.e., Teflon). This is an important step to have close match between modeling and prototyped results.

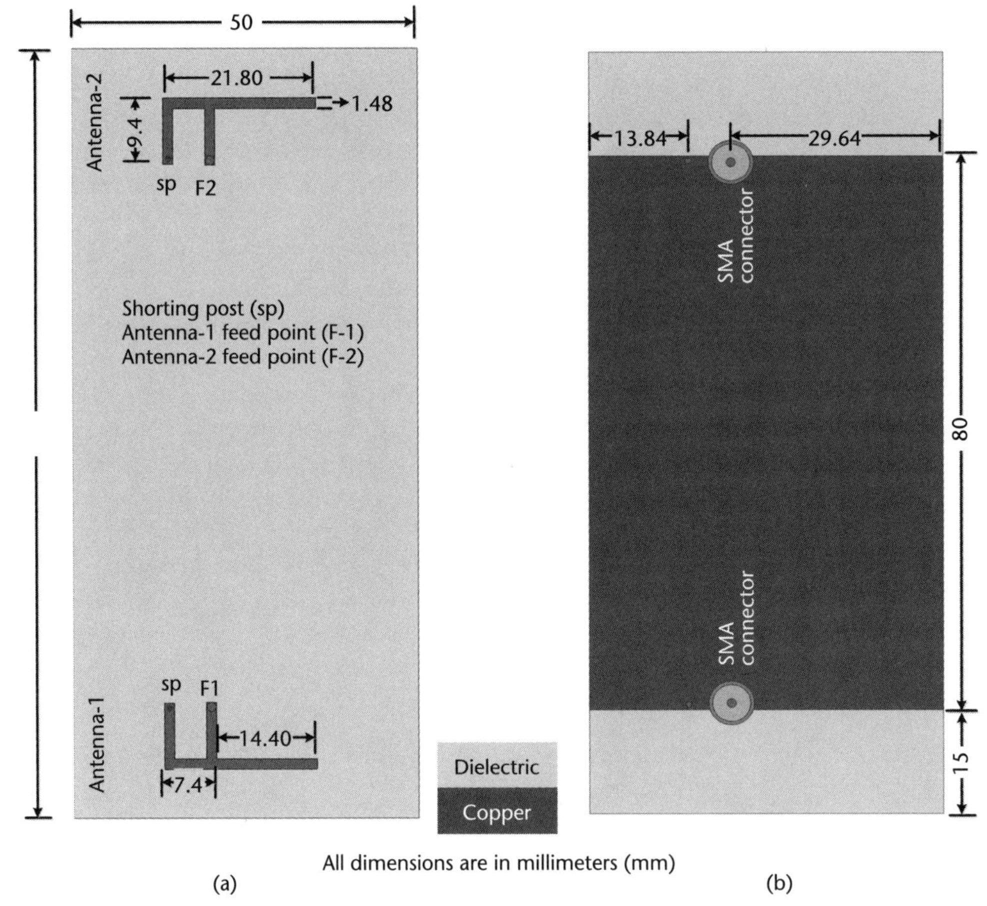

Figure 4.29 Geometry of the two-element PIFA-based MIMO antenna: (a) top layer, and (b) bottom layer.

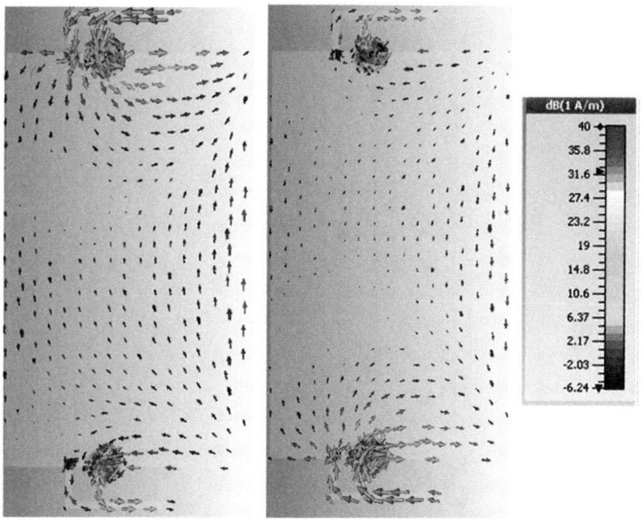

(a) (b)

Figure 4.30 Current vector distribution on the surface of the antenna and GND plane: (a) antenna 1 is active, and (b) antenna 2 is active.

To understand the behavior of the current on the antenna surface as well as the GND plane, Figure 4.30 shows the vector currents when both antennas are activated, one at a time. Notice that the current is concentrated on the surface of the active antenna, giving a current path of almost $\lambda_g/4$ at 2.45 GHz. Some small amounts of currents are coupled to the other antenna (although not excited) through the GND plane that is why the isolation is not that high (i.e., 12 dB). Figure 4.31 shows the *S*-parameters, where the resonance is around 2.45 GHz with a shared bandwidth of at least 300 MHz between the two MIMO antennas. Notice that there is a slight deviation between the reflection coefficient curves of the two antennas due to the coupling and manual placement (they are created using a mirror copy operation).

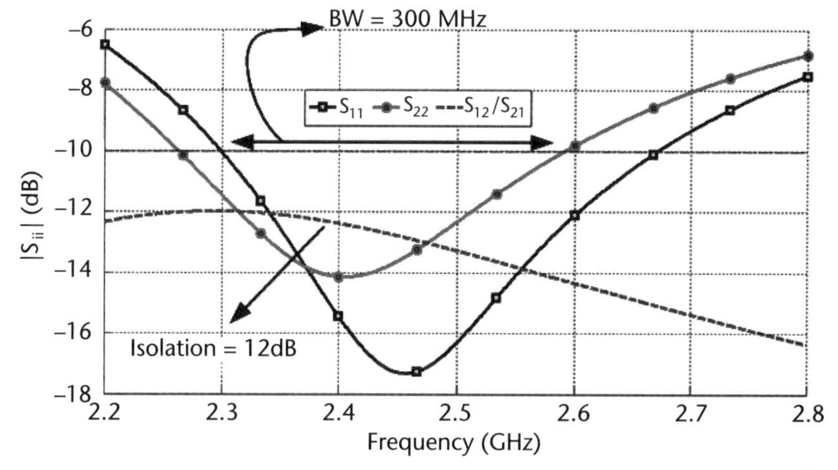

Figure 4.31 *S*-parameter curves including reflection coefficients of elements 1 and 2 ($|s_{11}|$, $|s_{22}|$) and mutual coupling ($|s_{12}|$).

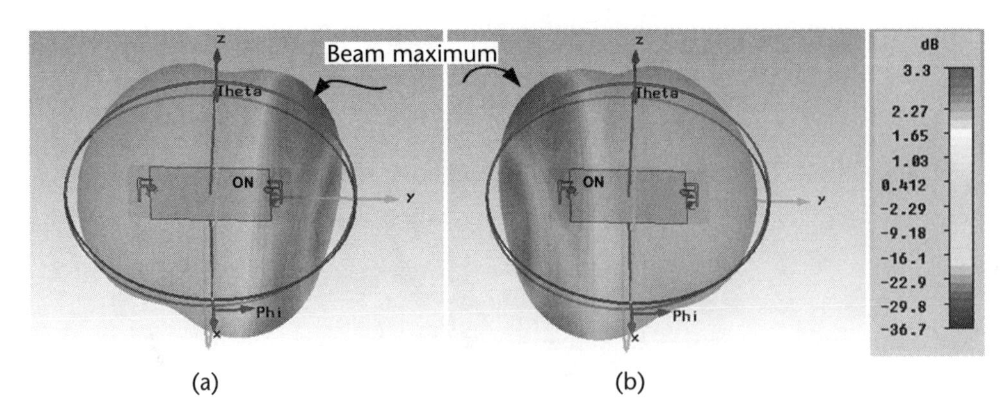

Figure 4.32 Gain patterns of the PIFA MIMO antenna system: (a) Antenna 1 is active, and (b) Antenna 2 is active.

The 3-D gain patterns are shown in Figure 4.32. It is clear that the beams when either antenna is excited are pointing in opposite directions, which will ensure good MIMO performance due to low field coupling and low ECC values. The maximum gain versus frequency for both antennas as well as the ECC curve is shown in Figure 4.33. The antennas have an efficiency of more than 80% across the band of operation with a gain of at least 3 dBi. ECC values evaluated using (4.31) do not exceed 0.07 in the whole band of operation.

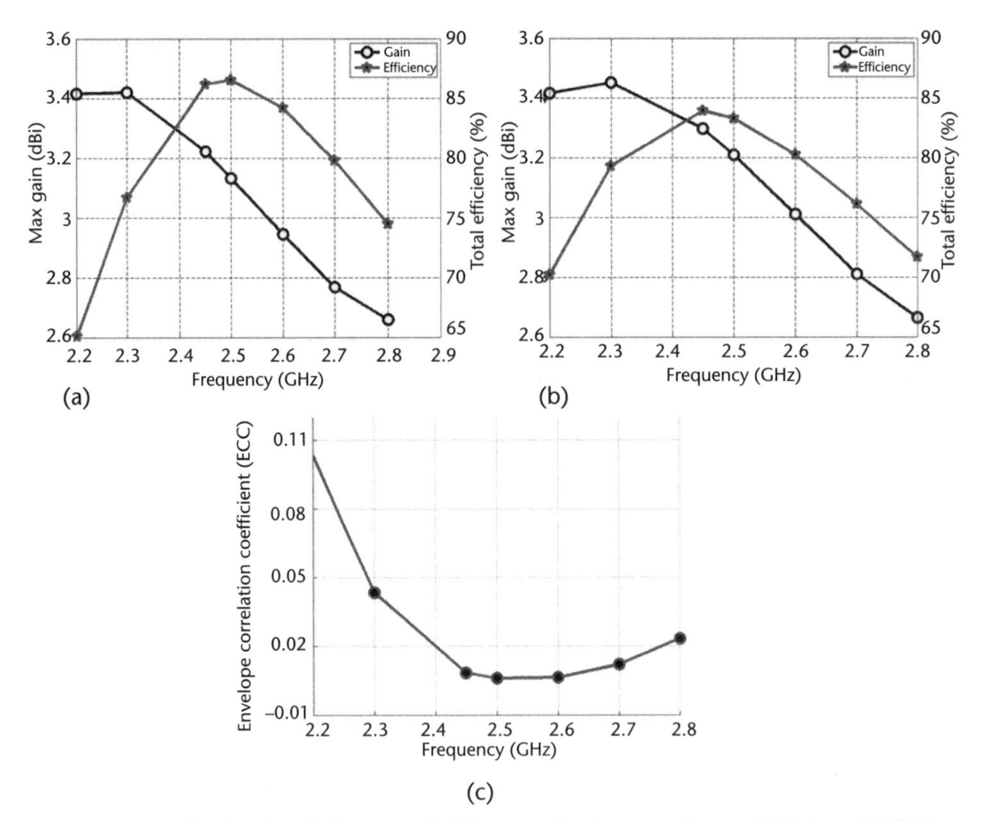

Figure 4.33 Realized gain, efficiency and ECC curves for the two-element PIFA-based MIMO antenna system: (a) maximum and efficiency of antenna 1, (b) maximum and efficiency of antenna 2, and (c) ECC curve.

4.6 Conclusions

Proper antenna design is essential to any successful wireless system. In this chapter, we reviewed antenna fundamental operating principles and metrics. The features and geometries of basic antenna elements that are widely used in wireless systems were discussed and touched upon. Several examples of printed antennas covering various wireless standards and bands were presented. Antenna design for the recent MIMO technology was also discussed along with its new performance metrics that are required for achieving the anticipated advantages from this technology. Examples of 4G and potential 5G MIMO antenna solutions were provided. The chapter ended with two detailed design examples using two CAD tools that can help the antenna designers start their project and understand the various metrics that they need to evaluate to ensure close to first-run success with the minimum amount of fine-tuning.

References

[1] Balanis, C. A., *Antenna Theory: Analysis and Design*, Fourth Edition New York: Wiley, 2015.

[2] Kraus, J. D., and R. J. Marhefka, *Antennas for All Applications*, Third Edition, New York: McGraw-Hill, 2002.

[3] Stutzman, W. L., and G. A. Thiele, *Antenna Theory and Design*, Second Edition, New York: Wiley, 1998.

[4] Volakis, J. L., *Antenna Engineering Handbook*, 4th Edition, New York: McGraw Hill, 2007.

[5] Mongia, R. K., and P Bhartia, "Dielectric Resonator Antennas: A Review and General Design Relations for Resonant Frequency and Bandwidth," *International Journal on RF and Microwave Computer Aided Engineering*, Vol. 4, No. 3, July 1994, pp. 230–247.

[6] Garg, R., et al., *Microstrip Antenna Design Handbook*, Norwood, MA: Artech House, 2001.

[7] Hamid, M., and R. Hamid, "Equivalent Circuit of a Dipole Antenna of Arbitrary Length," *IEEE Transactions on Antennas and Propagation*, Vol. 45, No. 11, November 1997, pp. 1695–1696.

[8] Best, S. R., and B. C. Kaanta, "A Tutorial on the Receiving and Scattering Properties of Antennas," *IEEE Antennas and Propagation Magazine*, Vol. 51, No. 5, October 2009, pp. 26–37.

[9] Balanis, C. A., "Antenna Theory: A Review," *IEEE Proceedings*, Vol. 80, No. 1, January 1992, pp. 7–23.

[10] Balanis, C. A., *Advanced Engineering Electromagnetics*, Second Edition, New York: Wiley, 2012.

[11] Nielsen, J. O., et al., "Computation of Mean Effective Gain from 3D Measurements," *Proc. IEEE 49th Vehicular Technology Conference*, Vol. 1, July 1999, pp. 787–791.

[12] Toga, T., "Analysis of Mean Effective Gain of Mobile Antennas in Land Mobile Radio Environments," *IEEE Transactions on Vehicular Technology*, Vol. 39, No. 2, February 1990. pp. 117–131.

[13] Lewallen, R. W., "Baluns: What Do They Do and How They Do It," *American Radio Relay League*, Vol. 1, pp 1991–2004.

[14] Shen, L. C., T. T. Wu and R. W. King, "A Simple Formula of Current in Dipole Antennas," *IEEE Transactions on Antennas and Propagation*, Vol. AP-16, No. 5, September 1968, pp. 542–547.

[15] Su, C. M., H. T. Chen, and K. L. Wong, "Printed Dual-Band Dipole Antenna with U-Slotted Arms for 2.4/5.2 GHz WLAN Operation," *Electronics Letters*, Vol. 38, No. 22, October 2002, pp. 1308–1309.

[16] Chen, G. Y., and J. S. Sun, "A Printed Dipole Antenna with Micro Strip Tapered Balun," *Microwave and Optical Technology Letters*, Vol. 40, No. 4, February 2004, pp. 344–346.

[17] Chuang, H. R., and L. C. Kuo, "3-D FDTD Design Analysis of a 2.4-GHz Polarization-Diversity Printed Dipole Antenna with Integrated Balun and Polarization-Switching Circuit for WLAN and Wireless Communication Applications," *IEEE Transactions on Microwave Theory and Techniques*, Vol. 51, No. 2, February 2003, pp. 374–381.

[18] Ma, T. G., and S. K. Jeng, "A Printed Dipole Antenna with Tapered Slot Feed for Ultrawide-Band Applications," *IEEE Transactions on Antennas and Propagation*, Vol. 53, No. 11, November 2005, pp. 3833–3836.

[19] Floch, J. M., and H. Rmili, "Design of Multibrand Printed Dipole Antennas Using Parasitic Elements," *Microwave and Optical Technology Letters*, Vol. 48, No. 8, August 2006, pp. 1639–1645.

[20] Nakano, H., et al., "Shortening Ratios of Modified Dipole Antennas," *IEEE Transactions on Antennas and Propagation*, Vol. AP-32, No. 4, April 1984, pp. 385–386.

[21] Mosallaei, H., and K. Sarabandi, "Antenna Miniaturization and Bandwidth Enhancement Using a Reactive Impedance Substrate," *IEEE Transactions on Antennas and Propagation*, Vol. 52, No. 9, September 2004, pp. 2403–2414.

[22] Yeoh, W. S., K. L. Wong, and W. S. T. Rowe, "Wideband Miniaturized Half Bowtie Printed Dipole Antenna with Integrated Balun for Wireless Applications," *IEEE Transactions on Antennas and Propagation*, Vol. 59, No. 1, January 2011, pp. 339–342.

[23] Pan, C. Y., et al., "Dual Wideband Printed Monopole Antenna for WLAN/WiMAX Applications," *IEEE Antennas and Wireless Propagation Letters*, Vol. 6, 2007, pp. 149–151.

[24] Kuo, Y. L., and K. L Wong, "Printed Double-T Monopole Antenna for 2.4/5.2 GHz Dual-Band WLAN Operations," *IEEE Transactions on Antennas and Propagation*, Vol. 51, No. 9, September 2003, pp. 2187–2192.

[25] Agrawall, N. P., G. Kumar, and K. P. Ray, "Wide-Band Planar Monopole Antennas," *IEEE Transactions on Antennas and Propagation*, Vol. 46, No. 2, February 1998, pp. 294–295.

[26] Ammann, M. J., and Z. N. Chen, "Wideband Monopole Antennas for Multi-Band Wireless Systems," *IEEE Antennas and Propagation Magazine*, Vol. 45, No. 2, April 2003, pp. 146–150.

[27] Wu, Q., et al., "Printed Omni-Directional UWB Monopole Antenna with Very Compact Size," *IEEE Transactions on Antennas and Propagation*, Vol. 52, No. 3, March 2008, pp. 896–899.

[28] Qu, S. W., J. L. Li, and Q. Xue, "A Band-Notched Ultra Wideband Printed Monopole Antenna," *IEEE Antennas and Wireless Propagation Letters*, Vol. 5, 2006, pp. 495–498.

[29] Suh, S. Y., W. L. Stutzman and W. A. Davis, "A New Ultra Wideband Printed Monopole Antenna: The Planar Inverted Cone Antenna (PICA)," *IEEE Transactions on Antennas and Propagation*, Vol. 52, No. 5, May 2004, pp. 1361–1365.

[30] Hong, W., and K. Sarabandi, "Low-Profile, Multi-Element, Miniaturized Monopole Antenna," *IEEE Transactions on Antennas and Propagation*, Vol. 57, No. 1, January 2009, pp. 72–80.

[31] Abdollahvand, M., G. Dadashzadeh, and D. Mostafa, "Compact Dual Band-Notched Printed Monopole Antenna for UWB Application," *IEEE Antennas and Wireless Propagation Letters*, Vol. 9, 2010, pp. 1148–1151.

[32] Zaker, R., and A. Abdipour, "A Very Compact Ultra Wideband Printed Omnidirectional Monopole Antenna," *IEEE Antennas and Wireless Propagation Letters*, Vol. 9, 2010, pp. 471–473.

[33] Ahmadi, B., and R. F. Dana, "A Miniaturized Monopole Antenna for Ultra-Wide Band Applications with Band-Notch Filter," *IET Microwaves, Antennas and Propagation*, Vol. 3, No. 8, 2009, pp. 1224–1231.

[34] Virga, K. L., and Y. R. Samii, "Low-Profile Enhanced-Bandwidth PIFA Antennas for Wireless Communications Packaging," *IEEE Transactions on Microwave Theory and Techniques*, Vol. 45, No. 10, October 1997, pp. 1879–1997.

[35] Chattha, H. T., et al., "An Empirical Equation for Predicting the Resonant Frequency of Planar Inverted-F Antennas," *IEEE Antennas and Wireless Propagation Letters*, Vol. 8, 2009, pp. 856–860.

[36] Carver, K. R., and J. W. Mink, "Micro Strip Antenna Technology," *IEEE Transactions on Antennas and Propagation*, Vol. AP-29, No. 1, January 1981, pp. 2–24.

[37] Richards, W. F., U. T. Lo, and D. D. Harrison, "An Improved Theory for Micro Strip Antennas and Applications," *IEEE Transactions on Antennas and Propagation*, Vol. AP-29, No. 1, January 1981, pp. 38–46.

[38] Lo, Y. T., D. Solomon and W. F. Richards, "Theory and Experiment on Micro Strip Antennas," *IEEE Transactions on Antennas and Propagation*, Vol. AP-27, No. 2, March 1979, pp. 137–145.

[39] Pozar, D. M., "Microstrip Antennas," *Proceedings of the IEEE*, Vol. 80, No. 1, January 1992, pp. 79–91.

[40] Lee, K. F., and K. M. Luk, *Microstrip Patch Antennas*, London: Imperial College Press, 2011.

[41] Yang, F., et al., "Wide-Band E-Shaped Patch Antennas for Wireless Communications," *IEEE Transactions on Antennas and Propagation*, Vol. 49, No. 7, July 2001, pp. 1094–1100.

[42] Wi, S. H., Y. S. Lee, and J. G. Yook, "Wideband Micro Strip Patch Antenna with U-Shaped Parasitic Elements," *IEEE Transactions on Antennas and Propagation*, Vol. 55, No. 4, April 2007, pp. 1196–1199.

[43] Lam, K. Y., et al., "Small Circularly Polarized U-Slot Wideband Patch Antenna," *IEEE Antennas and Wireless Propagation Letters*, Vol. 10, 2011, pp. 87–90.

[44] Ha, J., et al., "Hybrid Mode Wideband Patch Antenna Loaded with a Planar Metamaterial Unit Cell," *IEEE Transactions on Antennas and Propagation*, Vol. 60, No. 2, February 2012, pp. 1143–1147.

[45] Khan, M. U., M. S. Sharawi, and R. Mittra, "Microstrip Patch Antenna Miniaturization Techniques: A Review," *IET Microwaves, Antennas and Propagation*, Vol. 9, No. 9, 2015, pp. 913–922.

[46] Farzami, F., K. Forooraghi, and M. Norooziarab, "Miniaturization of a Micro Strip Antenna Using a Compact and Thin Magneto-Dielectric Substrate," *IEEE Antennas and Wireless Propagation Letters*, Vol. 10, 2011, pp. 1540–1542.

[47] Li, R., et al., "Development and Analysis of a Folded Shorted-Patch Antenna with Reduced Size," *IEEE Transactions on Antennas and Propagation*, Vol. 52, No. 2, February 2004, pp. 555–562.

[48] Sharawi, M. S., et al., "A CSRR Loaded MIMO Antenna System for ISM Band Operation," *IEEE Transactions on Antennas and Propagation*, Vol. 61, No. 8, August 2013, pp. 4265–4274.

[49] Alu, A., et al., "Sub Wavelength, Compact, Resonant Patch Antennas Loaded with Metamaterials," *IEEE Transactions on Antennas and Propagation*, Vol. 55, No. 1, January 2007, pp. 13–25.

[50] Rao, B. R., "Far Field Patterns of Large Circular Loop Antennas: Theoretical and Experimental Results," *IEEE Transactions on Antennas and Propagation*, Vol. 16, No. 2, March 1968, pp. 269–270.

[51] Storer, J. E., "Impedance of Thin-Wire Loop Antennas," *Transactions of the American Institute of Engineers*, Vol. 75, No. 5, November 1956, pp. 606–619.

[52] Wu, T. T., "Theory of the Thin Circular Loop Antenna," *AIP Journal of Mathematical Physics*, Vol. 3, No. 6, 1962, pp. 1301–1304.

[53] Werner, D. H., "An Exact Integration Procedure for Vector Potentials of Thin Circular Loop Antennas," *IEEE Transactions on Antennas and Propagation*, Vol. 44, No. 2, February 1996, pp. 157–165.

[54] Li, L. W., and M. S. Leong, "Method-of-Moments Analysis of Electrically Large Circular-Loop Antennas: Nonuniform Currents," *IEEE Proceedings Microwaves Antennas and Propagation*, Vol. 146, No. 6, December 1999.

[55] Chi, Y. W., and K. L. Wong, "Quarter-Wavelength Printed Loop Antenna with an Internal Printed Matching Circuit for GSM/DCS/PCS/UMTS Operation in the Mobile Phone," *IEEE Transactions on Antennas and Propagation*, Vol. 57, No. 9, September 2009, pp. 2541–2547.

[56] Das, B. N., and K. K. Joshi, "Impedance of a Radiating Slot in the Ground Plane of a Microstripline," *IEEE Transactions on Antennas and Propagation*, Vol. AP-30, No. 5, September 1982, pp. 922–926.

[57] Ha, J., M. A. Tarifi, and D. S. Filipovic, "Design of Wideband Combined Annular Slot-Monopole Antenna," *IEEE Transactions on Antennas and Propagation*, Vol. 64, No. 9, September 2016, pp. 4138–4143.

[58] Fakharian, M. M., et al., "A Wideband and Reconfigurable Filtering Slot Antenna," *IEEE Antennas and Wireless Propagation Letters*, Vol. 15, 2016, pp. 1610–1613.

[59] Alreshaid, A. T., et al., "Compact Millimeter-Wave Switched-Beam Antenna Arrays for Short Range Communications," *Microwave and Optical Technology Letters*, Vol. 58, No. 8, August 2016, pp. 1917–1921.

[60] Hussain, R., et al., "A Compact 4G MIMO Antenna Integrated with a 5G Array for Current and Future Mobile Handsets," *IET Microwaves, Antennas and Propagation*, No. 2, 2017, pp. 1–17.

[61] Baumgartner, P., et al., "Limitations of the Pattern Multiplication Technique for Uniformly Spaced Linear Antenna Arrays," *International Conference on Broadband Communications for Next Generation Networks and Multimedia Applications (CoBCom)*, September 2016, pp. 1–7.

[62] Kelley, D. F., and W. L Stutzman, "Array Antenna Pattern Modeling Methods That Include Mutual Coupling Effects," *IEEE Transactions on Antennas and Propagation*, Vol. 41, No. 12, December 1993, pp. 1625–1632.

[63] Kummer, W. H., "Basic Array Theory," *Proceedings of the IEEE*, Vol. 80, No. 1, January 1992, pp. 127–140.

[64] Schelkunoff, S. A., "A Mathematical Theory of Linear Arrays," *Bell System Technical Journal*, Vol. 22, No. 1, 1943, pp. 80–107.

[65] Lo, Y. T., "A Mathematical Theory of Antenna Arrays with Randomly Spaced Elements," *IEEE Transactions on Antennas and Propagation*, Vol. 12, No. 3, May 1964, pp. 257–268.

[66] Murthy, P. K., and A. Kumar, "Synthesis of Linear Antenna Arrays," *IEEE Transactions on Antennas and Propagation*, Vol. 24, No. 6, November 1976, pp. 865–870.

[67] Mailloux, R. J., *Phased Array Antenna Handbook*, Second Edition, Norwood, MA: Artech House, 2005.

[68] Haupt, R., *Antenna Arrays: A Computational Approach*, New York: Wiley-IEEE Press, 2010.

[69] Sharawi, M. S., *Printed MIMO Antenna Engineering*, Norwood, MA: Artech House, 2014.

[70] Goldsmith, A., *Wireless Communications*, Cambridge, UK: Cambridge University Press, 2005.

[71] Sharawi, M. S., "Printed MIMO Antenna Systems: Current Misuses and Future Prospects," *IEEE Antennas and Propagation Magazine*, February 2017.

[72] Manteghi, M., and Y. R. Samii, "Multiport Characteristics of a Wide-Band Cavity Backed Annular Patch Antenna for Multipolarization Operations," *IEEE Transactions on Antennas and Propagation*, Vol. 53, No. 1, January 2005, pp. 466–474.

[73] Su, S. W., C. T. Lee, and F. S. Chang, "Printed MIMO-Antenna System Using Neutralization-Line Technique for Wireless USB-Dongle Applications," *IEEE Transactions on Antennas and Propagation*, Vol. 60, No. 2, February 2012, pp. 456–463.

[74] Soltani, S., P. Lotfi, and R. D. Murch, "A Port and Frequency Reconfigurable MIMO Slot Antenna for WLAN Applications," *IEEE Transactions on Antennas and Propagation*, Vol. 64, No. 4, April 2016, pp. 1209–1217.

[75] Nielsen, J. O., et al., "On Antenna Design Objectives and the Channel Capacity of MIMO Handsets," *IEEE Transactions on Antennas and Propagation*, Vol. 62, No. 6, June 2014, pp. 3232–3241.

[76] Vaughan, R. G., and J. B Andersen, "Antenna Diversity in Mobile Communications," *IEEE Transactions on Antennas and Propagation*, Vol. 36, No. 4, November 1987, pp. 149–172.

[77] Pierce, J. N., and S. Stein, "Multiple Diversity with Nonindependent Fading," *Proceeding of the IRE*, Vol. 48, No. 1, 1960, pp. 89–104.

[78] Blanch, S., J. Romeu, and I. Corbella, "Exact Representation of Antenna System Diversity Performance from Input Parameter Description," *Electronics Letters*, Vol. 39, No. 9, May 2003, pp. 705–707.

[79] Sharawi, M. S., "Current Misuses and Future Prospects for Printed Multiple-Input-Multiple-Output Antenna Systems," *IEEE Antennas and Propagation Magazine*, Vol. 59, No. 2, April 2017, pp. 162–170.

[80] Sharawi, M. S., A. Hassan, and M. U. Khan, "Correlation Coefficient Calculations for MIMO Antenna Systems: A Comparative Study," *International Journal on Microwaves and Wireless Technology*, Cambridge Press, Vol. 9, No. 10, December 2017, pp. 1991–2004.

[81] Park, S., and C. Jung, "Compact MIMO Antenna with High Isolation Performance," *Electronics Letters*, Vol. 46, No. 6, March 2010.

[82] Hassan, A. T., and M. S. Sharawi, "Four Element Half Circle Shape Printed MIMO Antenna," *Microwave and Optical Technology Letters*, Vol. 58, No. 12, December 2016, pp. 2990–2992.

[83] Valdes, J. F. V., et al., "Evaluation of True Polarization Diversity for MIMO Systems," *IEEE Transactions on Antennas and Propagation*, Vol. 57, No. 9, September 2009, pp. 2746–2755.

[84] Vaughan, R., and J. B. Andersen, *Channels, Propagation and Antennas for Mobile Communications*, London: IET Press, 2003.

[85] Kildal, P. S., *Foundations of Antenna Engineering*, Norwood, MA: Artech House, 2015.

[86] Rosengren, K., and P. S. Kildal, "Radiation Efficiency, Correlation, Diversity Gain and Capacity of a Six-Monopole Antenna Array for a MIMO System: Theory, Simulation and Measurement in Reverberation Chamber," *IEEE Proceedings on Microwave and Antenna Propagation*, Vol. 152, No. 1, February 2005, pp. 7–16.

[87] International Wireless Industry Consortium, "Evolutionary & Disruptive Visions Towards Ultra High Capacity Networks," *IIWPC Ultra High Capacity Networks, White Paper Version 1.1,* April 2014, pp. 1–89.

[88] Zhao, K., et al., "Body-Insensitive Multimode MIMO Terminal Antenna of Double-Ring Structure," *IEEE Transactions on Antennas and Propagation*, Vol. 63, No. 5, May 2015, pp. 1925–1936.

[89] Cihangir, A., et al., "Neutralized Coupling Elements for MIMO Operation in 4G Mobile

Terminals," *IEEE Antennas and Wireless Propagation Letters*, Vol. 13, 2014, pp. 141–144.

[90] Shoaib, S., et al., "Design and Performance Study of a Dual-Element Multiband Printed Monopole Antenna Array for MIMO Terminals," *IEEE Antennas and Wireless Propagation Letters*, Vol. 13, 2014, pp. 329–332.

[91] Hussain, R., and M. S. Sharawi, "Wide-Band Frequency Agile MIMO Antenna System with Wide Tunability Range," *Microwave and Optical Technology Letters*, Vol. 58, No. 9, September 2016, pp. 2276–2280.

[92] Hussain, R., and M. S. Sharawi, "Integrated Reconfigurable Multiple-Input-Multiple-Output Antenna System with an Ultra-Wideband Sensing Antenna for Cognitive Radio Platforms," *IET Microwaves, Antennas and Propagation*, Vol. 9, No. 9, 2015, pp. 940–947.

[93] Li, J. F., et al., "Compact Dual Band-Notched UWB MIMO Antenna with High Isolation," *IEEE Transactions on Antennas and Propagation,* Vol. 61, No. 9, September 2013, pp. 4759–4766.

[94] Srivastava, G., and A. Mohan, "Compact MIMO Slot Antenna for UWB Applications," *IEEE Antennas and Wireless Propagation Letters*, Vol. 15, 2016, pp. 1057–1060.

[95] Dhar, S. K., and M. S. Sharawi, "A UWB Semi-Ring MIMO Antenna with Isolation Enhancement," *Microwave and Optical Technology Letters*, Vol. 57, No. 8, August 2015, pp. 1941–1946.

[96] Liao, W. J., et al., "Compact Dual Band WLAN Diversity Antennas on USB Dongle Platform," *IEEE Transactions on Antennas and Propagation*, Vol. 62, No. 1, January 2014, pp. 109–118.

[97] Krewski, A., W. L. Schroeder and K. Solbach, "2-Port DL-MIMO Antenna Design for LTE-Enabled USB Dongles," *IEEE Antennas and Wireless Propagation Letters*, Vol. 12, 2013, pp. 1436–1439.

[98] Su, S. W., "High-Gain Dual-Loop Antennas for MIMO Access Points in the 2.4/5.2/5.8 GHz Bands," *IEEE Transactions on Antennas and Propagation*, Vol. 58, No. 7, July 2010, pp. 2412–2419.

[99] Han, W., et al., "A Six-Port MIMO Antenna System with High Isolation for 5-GHz WLAN Access Points," *IEEE Antennas and Wireless Propagation Letters*, Vol. 13, 2014, pp. 880–883.

[100] Pan, Y., Y. Cui, and R. L. Li, "Investigation of a Triple-Band Multibeam MIMO Antenna for Wireless Access Points," *IEEE Transactions on Antennas and Propagation*, Vol. 64, No. 4, April 2016, pp. 1234–1241.

[101] Ban, Y. L., et al., "4G/5G Multiple Antennas for Future Multi-Mode Smartphone Applications," *IEEE Access*, Vol. 4, June 2016, pp. 2981–2988.

[102] Hussain, R., et al., "Compact 4G MIMO Antenna Integrated with a 5G Array for Current and Future Mobile Handsets," *IET Microwaves, Antennas and Propagation*, Vol. 11, No. 2, January 2017, pp. 271–279.

CHAPTER 5

Active Integrated Antennas

In a wireless communication system, the signals to be transmitted are upconverted and then amplified before they are transmitted into the air by the antennas (see Figure 1.14). The received signals by the antennas are filtered and amplified before being downconverted and processed. Active radio frequency (RF) circuits are connected to antennas, and the design of these active parts along with their antenna systems is denoted as active antennas. Such active antennas appeared some time ago, but the term active integrated antennas (AIAs) started to surface after the success of microwave integrated circuits (MICs) to denote the design and placement of the active part (i.e., amplifiers and mixers) as well as the antenna on the same substrate [1–5]. Then it was used to denote the integration of the active circuits with the passive antenna systems to yield an overall integrated module with enhanced characteristics (i.e., in terms of gain, matching, and operating bandwidth when properly designed). This type of integration can provide several benefits such as smaller overall size, antenna miniaturization (reduce system size), enhanced bandwidth, improved gain, and improved noise performance [4–6].

This chapter will cover the basic types of AIAs, their features, design steps, and performance metrics. Detailed designs from literature will be used to clarify the concepts and provide working examples. These types include oscillator AIAs, low noise amplifier (LNA) and power amplifier (PA)-based AIAs, mixer AIAs, transceiver AIAs, and other types including reconfigurable and on-chip AIAs.

5.1 Performance Metrics of AIAs

Assessing the performance of AIAs depends on their structure and application. Since we are talking about antennas, the standard antenna parameters such as impedance matching, bandwidth, radiation patterns, and gain are of fundamental importance. Additional parameters related to the integration of active components with the antenna must be taken into account, and these are driven by the requirements of the application. Some of these parameters are discussed in the following sections.

5.1.1 Frequency Bandwidth

Once an active element is connected to a passive antenna, the overall frequency bandwidth will be changed and will typically be limited by the narrower bandwidth of the two. It is thus very important that the designer assesses the total bandwidth after the integration and not to rely on the individual bandwidth of each stage alone

[7]. The designer should also be careful and aware that cascading the active circuits (i.e., amplifier, mixer) and the antenna will also affect the matching between them and thus the output impedance of the active circuit and the input impedance of the antenna should not have steep variations around the center frequency of operation. This is of paramount importance especially when using multiband or wideband systems.

5.1.2 Power Gain

One of the main advantages of AIAs is their enhanced gain due to the presence of an active element. Whether we are considering a transmitting or receiving AIA, the gain of the AIA will be boosted by that of the active element in addition to the original antenna gain. The amount of gain obtained from the AIA should be according to the standard limits if we are considering a transmitter, while the receiver sensitivity and following stage gains in the receiver chain will determine the amount of gain from the LNA [8, 9]. For example, if the maximum gain of an antenna is 3 dBi, and it is integrated with an amplifier with 20 dB of gain, the overall gain of the AIA will be around 23 dBi if the design is well integrated (matched) across the operating frequencies.

5.1.3 Total Efficiency

Amplifier and antenna efficiencies are very important metrics as was discussed in Chapters 3 and 4. The total efficiency of an AIA determines how much power is radiated for a given input power. Thus, to achieve high efficiency, both the antenna and the amplifier should operate with high efficiency with minimum insertion loss and high power-added efficiency (PAE). For a mixer-type AIA, we want high conversion efficiencies from the mixer circuits. The total efficiency is an important metric that is usually not fully evaluated due to some difficulties in relating the radiated power levels to the input power levels considering all the other factors such as impedance mismatch, amplifier nonidealities, and the antenna losses.

5.1.4 Stability

When active circuits are considered, the issue of stability is of great importance and should be checked each time the design is changed. As discussed in Chapter 3, amplifier stability should be maintained for its proper operation. The Rollett stability factor (k) should be kept above 1 even beyond the edges of the operating bandwidth to give some room for some possible fluctuations, and avoid any instability conditions [7, 9, 10]. The designer should give some room and make sure that the circuit is stable with good frequency margins (depending on the application, we need to factor in any frequency variation from the fabrication process or operating temperatures; thus, the design should satisfy the requirements beyond the bandwidth of interest). In the case of an oscillator-type AIA, the circuit should be kept in the unstable region to maintain the oscillation within that narrow band of interest. Proper limiting circuitry should be used to avoid saturating the active element and ending up with nonlinear distortion, but the overall design should have a stable behavior.

5.1.5 Noise Performance

Since all practical components are noisy, the noise effect should not be overlooked when designing RF circuits as mentioned in Chapter 3. The noise factor (F) and the noise figure (NF) are the two important parameters to be assessed. Thus, the cascading of multiple RF circuits will give an NF that is mainly dependent on the first-stage RF circuit [7], which is why the LNA noise figure is critical in receiver designs.

5.1.6 Example

Let us consider an active antenna design example and try to check the previously mentioned performance metrics. Start by using a simple rectangular patch antenna resonating at 2.39 GHz, with an inset feed for 50Ω matching. An RO4350B substrate with $\varepsilon_r = 3.48$ and thickness of 1.52 mm is used and the antenna is modeled using CST. The antenna bandwidth obtained is 45 MHz. The fabricated antenna geometry and its S-parameter curves are shown in Figure 5.1. Note that Figure 5.1(b) shows the two-port model of the antenna using the method provided in [11], as this is needed to evaluate the integrated design in ADS when the amplifier is connected to the antenna (ADS requires a two-port circuit model for the antenna if we want to check the output gain using it). The two-port equivalent model of the antenna is created using the S-parameters of the single-port antenna in addition to its radiation efficiency.

An RF amplifier is designed (using an ATF 53198 transistor) and analyzed over its valid operating range and then optimized for operation between 1.6 and 2.6 GHz. The transistor model is imported into ADS, and is examined via the proper biasing ($V_{GS} = 0.7$V, $V_{DS} = 3.5$V, and $I_{DS} = 110$ mA) and is stabilized using a gate resistor $R_{\text{stab}} = 100$ and capacitor $C_{\text{stab}} = 4.7$ pF. Two RF chokes (RFC) are used for biasing and isolating high-frequency power supply noise ($L1 = L2 = 25$ nH) and bypass capacitors are used for voltage regulation ($C3$ = 47 pF, $C4$ = 470 pF, and $C5$ = 4,700 pF). $C1$ and $C2$ are decoupling capacitors used at the input and output of the transistor with 1-nF values. The schematic and response of the amplifier are

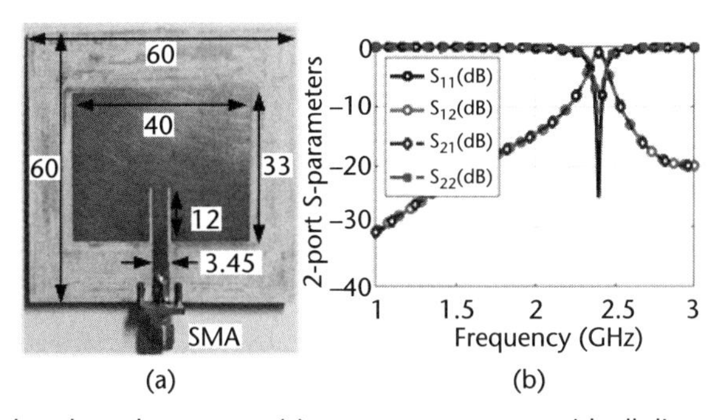

Figure 5.1 Narrowband patch antenna: (a) antenna geometry with all dimensions in millimeters, and (b) S-parameters of the two-port model of the antenna.

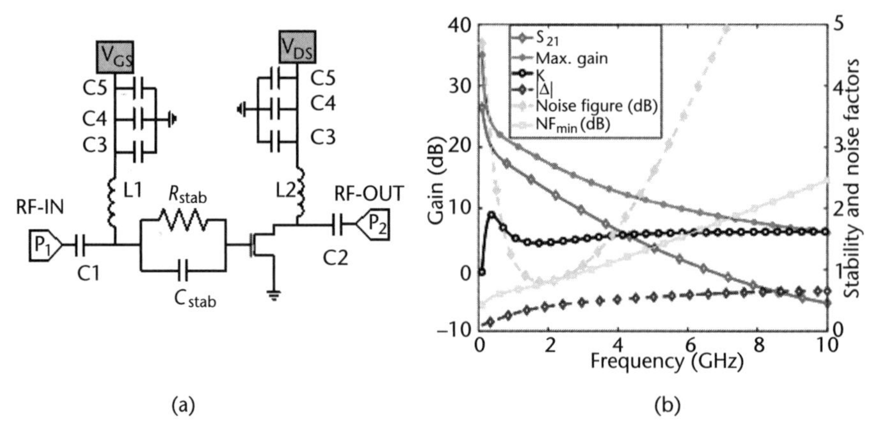

Figure 5.2 RF amplifier design: (a) schematic of the amplifier, and (b) response of the various amplifier parameters.

shown in Figure 5.2. The response shows the maximum possible gain (in decibels) versus frequency, the transmission coefficient (S_{21}) (in decibels), the stability metrics (K and $|\Delta|$—unitless) as well as the noise figure (NF) of the designed amplifier and $NF_{\min}$ of the transistor (both in decibels).

The amplifier can be optimized to operate over a wide frequency range. Since the antenna is intended to operate around 2.39 GHz, and because the matching networks are fundamentally band-limited, the amplifier is optimized via designing its input matching network (IMN) and output matching network (OMN) for flat operating power gain covering 1.6–2.6 GHz. The details of matching network design are discussed in Chapters 2 and 3.

The resulting optimized RF amplifier along with its IMN and OMN is shown in Figure 5.3. Lumped element-based IMN and OMN are used in this example (using three inductors and three capacitors for each network). The values of the matching network components are shown in boxes on the top of Figure 5.3(a). As can be seen, the amplifier response shows maximum flat gain between 1.6 and 2.6 GHz, with values ranging between 16.2 and 17.7 dB. The noise figure values in the band of interest were between 1.4 and 1.5 dB. Backward gain was less than –20 dB, and the matching response was acceptable and less than −10 dB.

The two-port model of the antenna was imported into ADS and simulated with the amplifier complete circuit. Figure 5.4 shows the results obtained when cascading the amplifier with the antenna to build an active antenna system. As is clearly shown, the frequency response of the complete system is now following that of the antenna, and thus the operating 3-dB gain bandwidth is now around 100 MHz as shown in the gain response (S_{21}), and the −10 dB impedance bandwidth is around 40 MHz (S_{22}, output of the antenna model), following that of the antenna (the lower bandwidth element will dictate the overall system response). The NF was around 1.6 to 1.7 dB in the region of operation, a marginal change compared to the amplifier NF alone (as indicated before).

To predict the radiation characteristics of the antenna with the amplifier, a combined simulation is performed. In order to do so, the amplifier two-port (S-parameters) model is exported to CST, and then an integrated model of the active antenna is created to simulate the patterns and total efficiencies. Figure 5.5

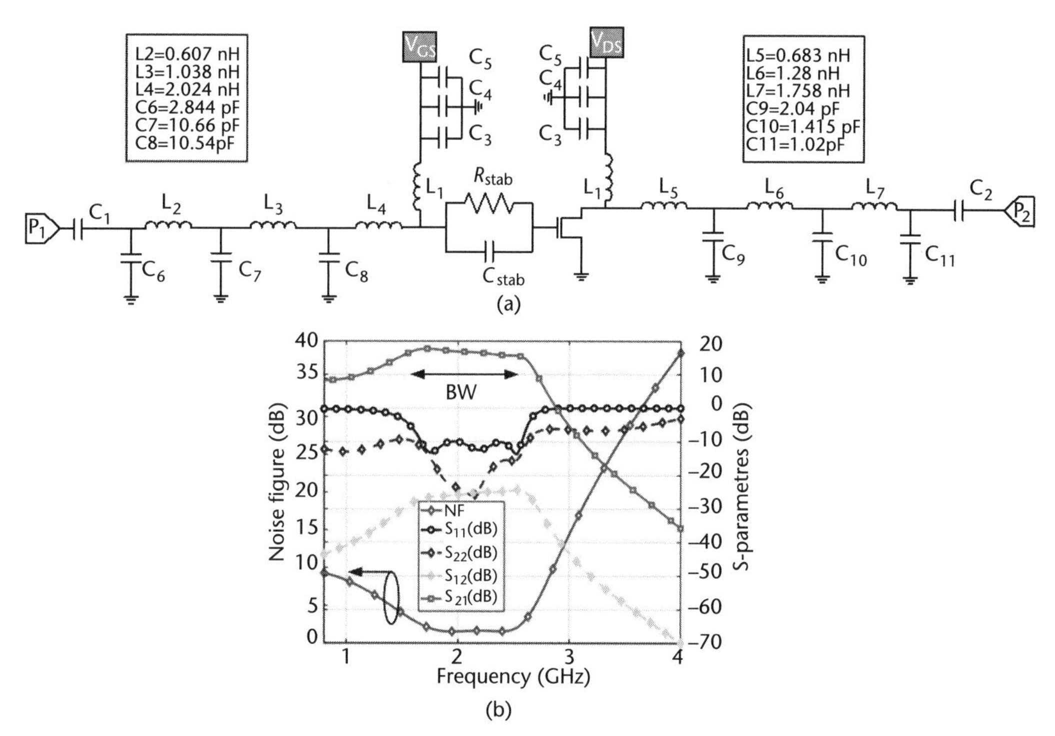

Figure 5.3 Optimized amplifier design for operation between 1.6 and 2.6 GHz: (a) circuit schematic, and (b) complete response.

shows the CST model of the active antenna and its response. Figure 5.5(a) depicts the active antenna model showing the amplifier connected to the antenna geometry. Figure 5.5(b) shows the radiation efficiency of the system, which does not consider the matching effects, as well as the total radiation efficiency that factors in any mismatch losses. Note that at the center frequency, the two are almost the same (as the matching values are very good), but beyond that, the total efficiency shows a much narrower efficiency bandwidth due to the presence of the active circuit and mismatch losses from the antenna response. The radiation gain patterns are also

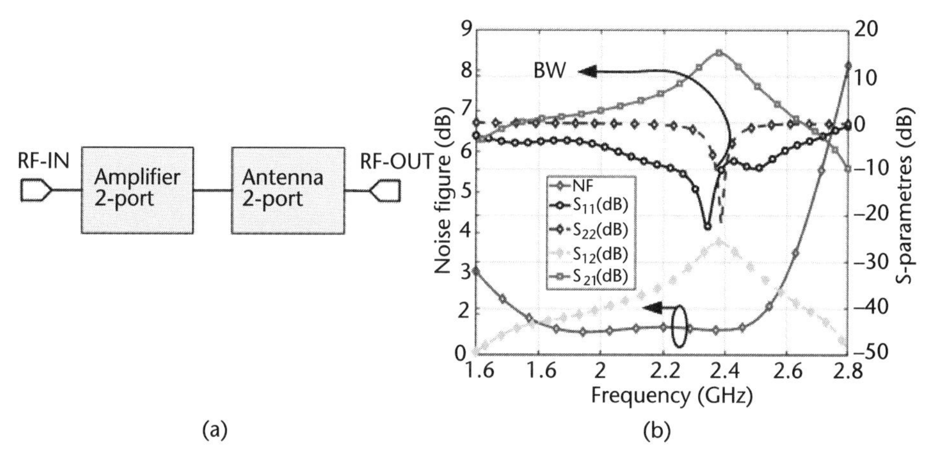

Figure 5.4 Results of the active antenna system within ADS: (a) block diagram of the complete system, and (b) response curves.

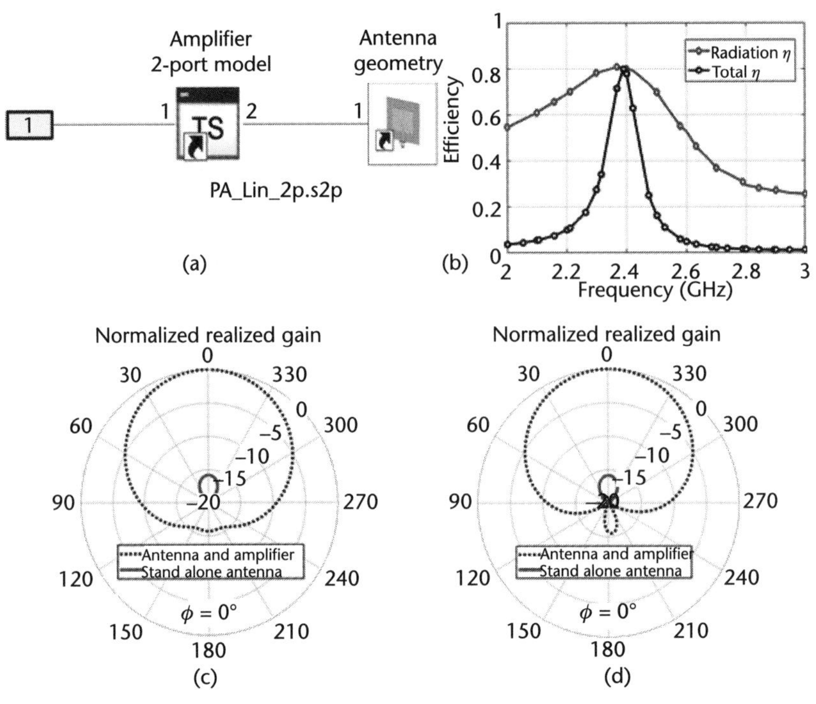

Figure 5.5 The active antenna integrated model in CST: (a) active antenna model, (b) obtained radiation efficiencies, (c) $\phi = 0°$ cut of the gain pattern of the active antenna and standalone one (polar plot in degrees), and (d) $\phi = 90°$ cut of the gain pattern of the active antenna and standalone one (polar plot in degrees).

found for this simple active antenna system for two elevation-plane (θ) cuts at $\phi = 0°$ and $\phi = 90°$. Figure 5.5(c, d) shows the realized gain patterns normalized with respect to the active antenna response. A boost of about 15 dB is obtained from the active antenna system compared to the passive one.

5.2 Oscillator-Based AIAs

Oscillator-based AIAs can be used in broadcasting, wireless power transmission, and wireless charging applications. They are very good candidates for continuous-wave (CW) sources with a possible application in radio frequency identification (RFID) systems. Since they depend on an oscillator circuit, they can modulate an incoming signal directly and transmit it through the integrated antenna. The antenna can be placed in the feedback loop of the oscillator circuit to act as a resonator contributing to the oscillator operation, thus reducing its overall size. Several oscillator-based AIAs have appeared in literature [12–25].

Table 5.1 shows several antenna types used in oscillator AIAs. The antenna is connected to follow two configurations as will be outlined in the following section. From the table, it can be easily seen that most of the works used patch or ring-based antennas in their feedback loops. Very few works utilized slot, meander-line monopoles, printed Yagi, or printed loop antennas. Some of the proposed designs have a self-mixing capability, thus having the ability to downconvert or upconvert using fewer circuit elements and components in an integrated design as proposed in [13].

Table 5.1 Antenna Types Used in Oscillator AIAs

Patch	*Ring*	*Slot*	*Meandered Monopole*	*Printed Yagi*	*Printed Loop*
[12, 14, 15, 20]	[13, 17, 19, 23, 25]	[16, 18]	[21]	[22]	[24]

5.2.1 Design Outline

The design of oscillator AIAs can be conducted in two configurations as illustrated in Figure 5.6. Figure 5.6(a) shows a fully integrated oscillator AIA where the antenna is acting as part of the resonating circuit in the transistor feedback circuit. The second configuration is larger in size as the antenna is connected to the oscillator output and thus acts more as a reactive load. Note that the transistor is shown in a common source configuration; this is not necessarily the case, as any other transistor configuration can be used as along as the oscillation conditions are maintained. Some standard oscillator circuits can be used such as Clapp oscillators, among others. The two bias voltages are needed to guarantee proper transistor biasing via the use of an RFC to guarantee isolating DC from RF paths and to start and maintain the oscillation criterion (Barkhausen criterion) [7]. This criterion states that an oscillation is maintained if the feedback loop gain is unity and the angle is 360°.

For the configuration in Figure 5.6(a), oscillation can be guaranteed if the two conditions in (5.1) and (5.2) are satisfied simultaneously:

$$\left|G_{\text{ant}}\left(f_{\text{osc}}\right)\right|\left|G_{\text{osc}}\left(f_{\text{osc}}\right)\right| = 1 \tag{5.1}$$

$$\angle G_{\text{ant}}\left(f_{\text{osc}}\right) + \angle G_{\text{osc}}\left(f_{\text{osc}}\right) = 360° = 2\pi \text{ (rad)} \tag{5.2}$$

The conditions in (5.1) and (5.2) are direct translation of the Barkhausen criterion at the oscillation frequency (f_{osc}), where G_{ant}, G_{osc}, are the antenna and oscillator gains, respectively. These also mean that the antenna impedance (Z_{ant}) cancels out the oscillator input impedance (Z_{osc}) at the feedback path, resulting

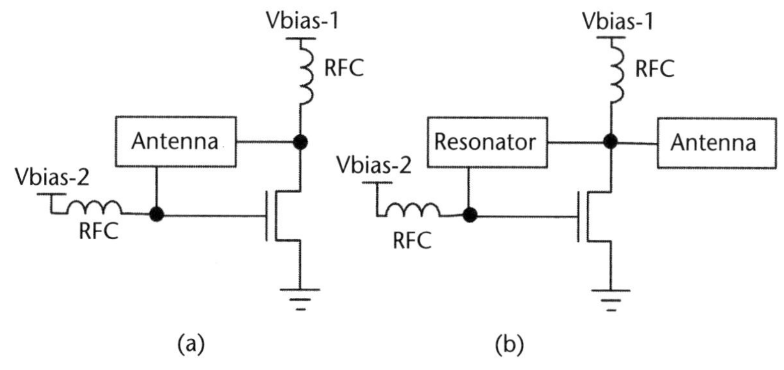

Figure 5.6 Possible configurations of oscillator AIAs: (a) antenna in feedback loop and part of resonance circuit, and (b) antenna as a load at oscillator output.

in zero impedance at the gate of the transistor in Figure 5.6(a) due to the negative resistance of the unstable transistor. This will maintain the oscillator operation. This can be written as:

$$R_{\text{ant}}\left(f_{\text{osc}}\right) + R_{\text{osc}}\left(f_{\text{osc}}\right) = 0 \tag{5.3}$$

$$X_{\text{ant}}\left(f_{\text{osc}}\right) + X_{\text{osc}}\left(f_{\text{osc}}\right) = 0 \tag{5.4}$$

where R_{ant} and X_{ant} are the real and imaginary parts of the antenna impedance, and R_{osc} and X_{osc} are the real and imaginary parts of the oscillator circuit (seen towards the gate/base of the transistor in a common source/emitter configurations, respectively). Since the antenna is connected at two ports for the configuration in Figure 5.6(a), the two-port parameters can be obtained using a full-wave electromagnetic simulator (i.e., HFSS, CST, or any available tool). For the configuration in Figure 5.6(b), the antenna is connected to the circuit via a single port, and thus its equivalent input impedance can be connected to the oscillator circuit to load it and maintain proper oscillation conditions. It should be noted that as shown in Section 5.1.6, a single-port antenna model can be converted to a two-port equivalent circuit following the procedure in [11]. Once the two-port model of the antenna is incorporated with the oscillator model, they can be analyzed and simulated using a harmonic balance based tool (i.e., ADS or MWO) to check its performance and behavior and to perform fine-tuning.

As for regular oscillators, the major figures of merit that describe its behavior are the phase noise, the DC to RF conversion efficiency, the DC power consumption in addition to the obtained gain, radiation patterns, and overall AIA size. It should be noted that integrating the antenna with the active circuit can yield a miniaturized antenna design, as the loading of the antenna with other parasitic elements (i.e., inductors and capacitors in the feedback loop of the oscillator as well as the capacitors between the terminals of the transistor) reduces its resonance frequency (remember that an antenna can be modeled as an RLC resonance circuit) and thus yields a miniaturized antenna size for that new lower frequency covered.

5.2.2 Examples

Two examples will be given in this section to cover each of the two configurations in Figure 5.6. The first example shows the design presented in [19]. This represents the configuration in Figure 5.6(a). An annular ring antenna is integrated in the feedback loop of a transistor to maintain oscillations. The antenna geometry is shown in Figure 5.7(a). The inner and outer radii of the antenna are $r1 = 2$ mm and $r2 = 3.9$ mm. The gap where the transistor is placed is 2.6 mm and the source plate of the transistor is placed in the gap. The length of the RFC microstrip line is 4.4 mm with a width of 0.15 mm. The three high impedance lines are serving as RFC for basing the transistor ($V_{DD} = 2.3$V, $V_{GG} = -0.6$V, and $V_{SS} = 0$). The size of the ground plane is $W \times L = 25 \times 25$ mm^2. The design is made on an RO4003C substrate with $\varepsilon_r = 3.38$, thickness of 0.508 mm, and loss tangent ($\tan\delta$) of 0.0027. An NEC NE3512S02 n-channel hetero-junction field effect transistor (FET) is used.

The open-circuit condition at the edges of the C-shaped makes the current path length half a guided wavelength ($\lambda_g/2$) at the unloaded resonance frequency. The DC-blocking capacitor is used to isolate the gate and drain bias voltages. Figure 5.7(b) shows the fabricated prototype.

Figure 5.8 shows the impedances of the unloaded (Figure 5.8[a, b]) and loaded (Figure 5.8[c, d]) configurations of the C-shaped AIA. It should be noted that the unloaded antenna resonates around 6.95 GHz, while the loaded one around 5.65 GHz. The modeling is conducted for the antenna alone using HFSS (other EM solvers can be used as well) while the integrated design with the amplifier using ADS. The two-port parameters of the c-shaped ring are extracted from HFSS and loaded

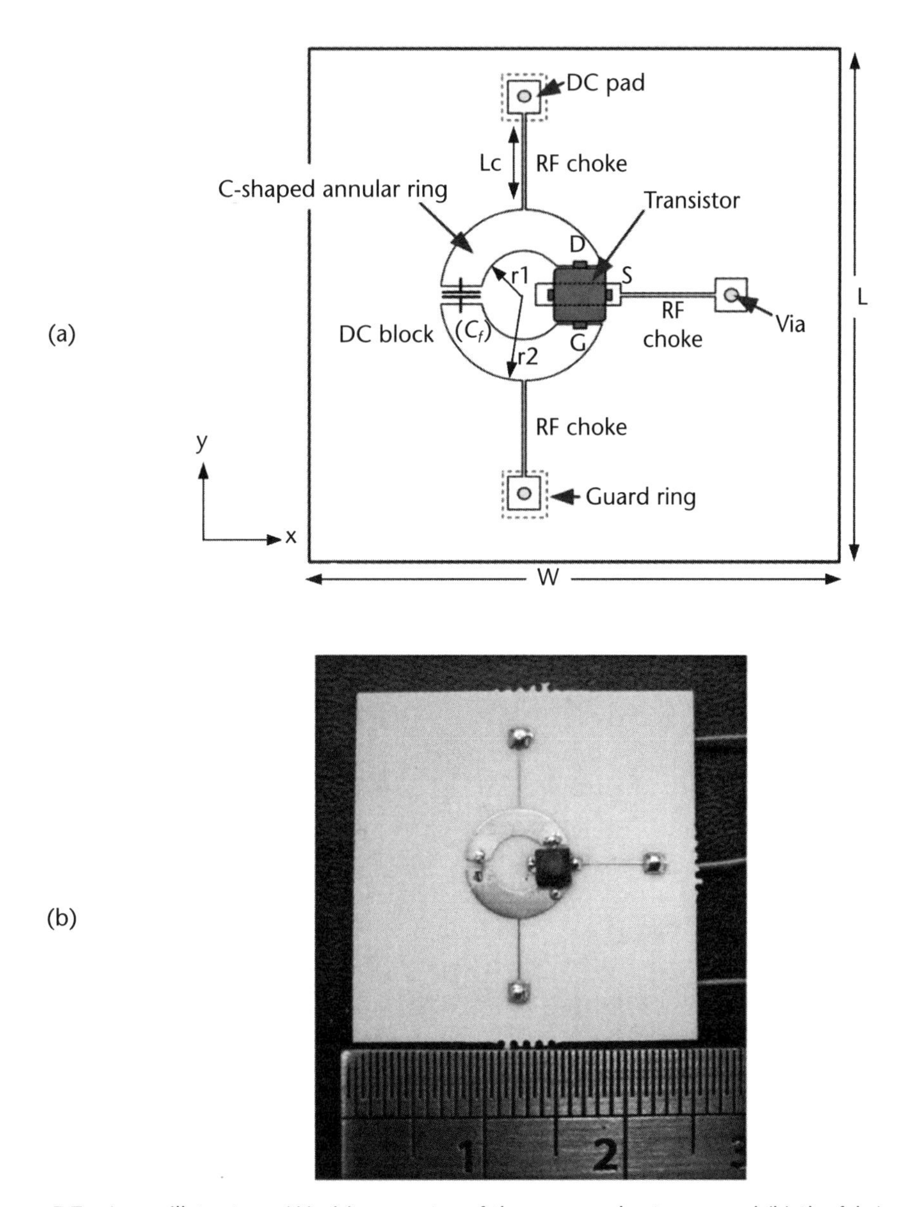

Figure 5.7 An oscillator type AIA: (a) geometry of the proposed antenna, and (b) the fabricated prototype. (*From:* [19]. © 2011 IEEE. Reprinted with permission.)

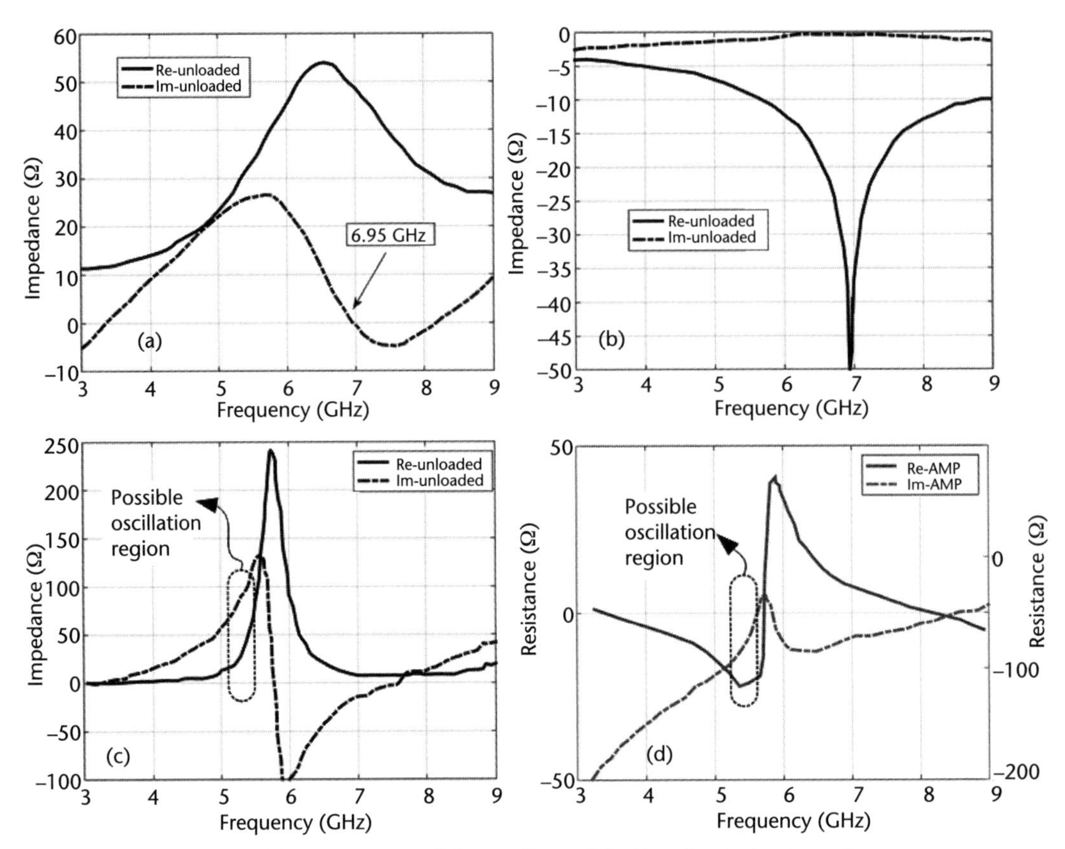

Figure 5.8 Responses of the two stages of the oscillator AIA: (a) unloaded c-shaped ring antenna impedance, (b) *S*-parameters of the c-shaped unloaded antenna, (c) transistor loaded c-shaped antenna impedance, and (d) the input impedance of the transistor with the c-shaped antenna in its feedback loop.

into ADS for analysis. The condition for oscillation is clear around 5.4 GHz as the impedance of the antenna cancels out the input impedance of the transistor looking into its gate (note the negative impedance values for the amplifier in the region of oscillation). It is clear how loading the antenna results in miniaturizing its size as well. The effective isotropic radiated power (EIRP) of the miniaturized oscillator AIA was measured to be 7.3 dBm at 5.383 GHz. The phase noise was −91.9 dBc/Hz at 100 kHz with a DC-RF efficiency of 25.9% and DC power consumption of 59.8 mW.

The second example follows the configuration in Figure 5.6(b). In this configuration, the antenna is not used in the feedback path of the oscillator circuit, and thus no DC blocking capacitors are needed, but the size of the circuit is larger compared to the first configuration. For this configuration, the active GaAs metal-semiconductor field-effect transistor (MESFET) (ATF-13786) is biased with V_{DS} = 4.1V and V_{GS} = 0.1V to provide the negative resistance, while a resonator consisting of a microstrip line loaded with a varactor diode for voltage tuning is used. A voltage controlled oscillator (VCO) is obtained with this configuration that covers 5.73–5.97 GHz, with 5.85 GHz as its center frequency. A slot antenna is used to radiate the energy with 20% impedance bandwidth. A tuned open stub at the source of the transistor is used to provide the negative resistance needed to maintain the oscillations. Figure 5.9(a, b) shows the geometry of the design as presented in [16]. The design was implemented on an RT/duriod 5880 substrate with 0.508-mm thickness and $\varepsilon_r = 2.2$. The design achieved 16.29 dBm of EIRP with and RF-DC efficiency of 24.6%. The phase noise was −104.1 dBc/Hz at 100 kHz. Figure 5.9(c) shows the measured output power level indicating −20 dBm at a distance of 2m and using a horn antenna with 17 dBi of gain.

5.3 Amplifier-Based AIAs

Most of the AIAs are of the amplifier type. Amplifier-based AIAs are mostly used in wireless terminals and devices and can be in the transmitting as well as the receiving paths. In transmitters, AIAs are mainly used to boost the power gain of the transmitting path. Integrating the antenna with the amplifier to form an AIA has the advantages of antenna miniaturization and improved transmitted gain. The amplifiers of such antennas are optimized for higher gain. Several AIA-based transmitting amplifier designs have appeared, such as those in [26–36]. Nonlinear PAs are also used in AIAs, and an example is shown in [32]. Details of optimizing PA using load-pull method are discussed in Section 3.4 of Chapter 3.

At the receiving end, an LNA is required as a first stage of amplification to reduce the overall *NF* of the system (since the first stage in the RF chain determines the *NF* value [7–9]). AIA-based LNA designs are also widely available in literature such as those in [37–46]. As with oscillator-type AIAs, the main antenna structure used in amplifier-based AIAs is the patch antenna. This is due to its ease of integration. Other antenna types such as slots, Yagi-based, and dipole/monopole are occasionally used. Table 5.2 shows the antenna types used in various references in amplifier-based AIA designs to give the reader an idea of the most common antenna types used in this type of AIA.

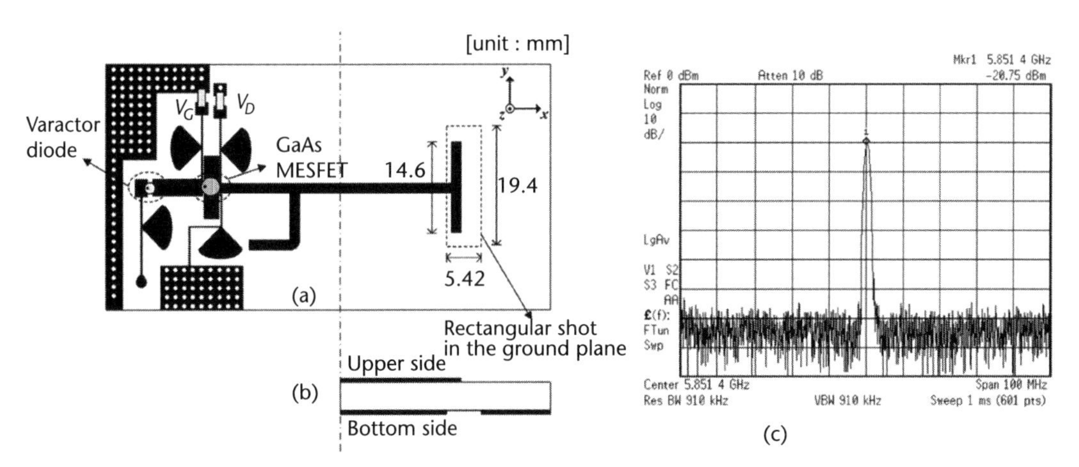

Figure 5.9 Oscillator AIA using a slot radiator: (a) top view of the geometry of the AIA, (b) side view, and (c) measured spectrum. (*From:* [16]. © 2005 IEEE. Reproduced with permission.)

Table 5.2 Antenna Types Used in Amplifier-Based AIAs

Patch	*Slot*	*Printed Yagi*	*Printed Dipole/Monopole*	*Others*
[26, 28–31, 33–37, 39, 42, 45, 47]	[27, 32]	[38]	[41, 43, 44, 46]	[40]

5.3.1 Design Outline

Amplifier design was discussed in detail in Chapter 3. The type of amplifier-based AIAs will determine how the amplifier design will be optimized. If we are designing a transmitter-type AIA, we will optimize its operation towards maximum gain; the *NF* comes second. While if we are designing an AIA with an LNA for a receiver path, then we need to optimize for *NF* first and gain second. Usually, it is recommended or preferred to have a system with flat gain and noise performance over the operating frequency range if possible. When considering PA with nonlinear operation, utilizing the load-pull method is generally preferred. The antenna should be designed with good input matching behavior to simplify cascading it with the amplifier. If the matching is poor at the antenna input and amplifier output or the bands do not overlap well, most of the power will be reflected and not delivered to the antenna causing a very poor design and reduced output power since the gain of the amplifier will not be seen in the radiated power.

Figure 5.10 shows the two configurations of an amplifier-based AIA. Two bias voltages (Bias-1 and Bias-2) are required to provide the current and voltage levels required for the amplifier operation. Based on these bias points, the rest of the circuit is designed, that is, the matching networks (or for tightly integrated designs, the antenna will serve as the proper load for the amplifier). The gate-source (V_{GS}) and drain-gate (V_{DS}) voltages (for Bias 1 and Bias 2, respectively) in an FET active element are required to be set before the design starts as indicated in Chapter 3. The integration between the antenna and the OMN in a transmitting-type AIA or the antenna and the IMN of a receiving-type AIA is critical for proper operation (if the design tightly integrates the antenna with the amplifier, sometimes no matching network is required as long as Z_{load} and the input impedance Z of the active circuit

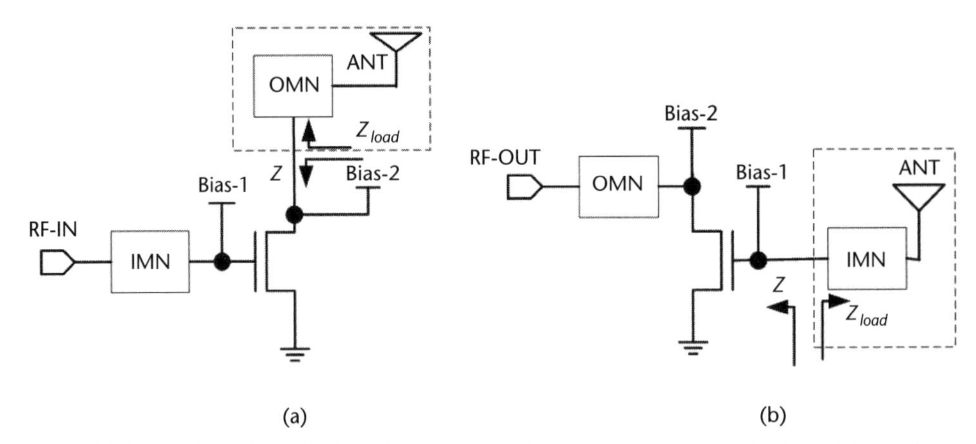

Figure 5.10 Amplifier-based AIA configurations: (a) transmitting side, and (b) receiving side.

are matched and provide the required gain and *NF* within the required bandwidth). The two should be matched over the desired operating bandwidth as was shown in Section 5.1.6.

5.3.2 Examples

The first example considers an active integrated electrically small PIFA-based antenna covering the bands between 1 and 2 GHz [30]. The PIFA antenna is optimized to act as the load at the output of the active transistor for optimum output power and efficiency. The geometry of the AIA is shown in Figure 5.11(a). Two shorting pins at a distance of 7.62 mm were used to have a wider range with a real input impedance. A class-F load provides high efficiency via matching the transistor output impedance (drain impedance) to provide optimum power at the center frequency (f_0). It also provides a short-circuit at the second harmonic ($2f_0$) and an open-circuit condition at the third one ($3f_0$). This class-F load results in a clamped voltage waveform at the drain of the transistor and a current peak when biased near the pinch-off point. For the dimensions given, $Z_{in}(f_0) = 18\Omega$, $Z_{in}(2f_0) = 6 - j4\Omega$ and $Z_{in}(3f_0) = 200 - j77\Omega$, corresponding to the three conditions for the class-F load. A 30 cm × 30 cm ground plane was used below the patch.

An LP750SOT89 pseudomorphic high electron mobility transistor (pHEMT) PA from Filtronics was used. The antenna was optimized to operate over 1 to 2 GHz, and tested at 1.05, 1.55, and 1.8 GHz with a small-signal gain ranging between 13 and 16 dB. The amplifier circuit was implemented on a RT/duroid ($\varepsilon_r = 2.45$) board with 1.58-mm thickness. The measured impedance of the amplifier was approximately 18Ω at f_0. For the bias conditions of V_{DS} = 5V and I_{DS} = 0.25 IDSS, the maximum PAE was 58%, 52%, and 50% at 1.05, 1.55, and 1.8 GHz, respectively. This is illustrated in Figure 5.11(b) where the curves of the PAE versus the input power levels are shown. The PIFA size was altered to cover different frequencies due to the narrow band nature of the method and inability to match over widebands.

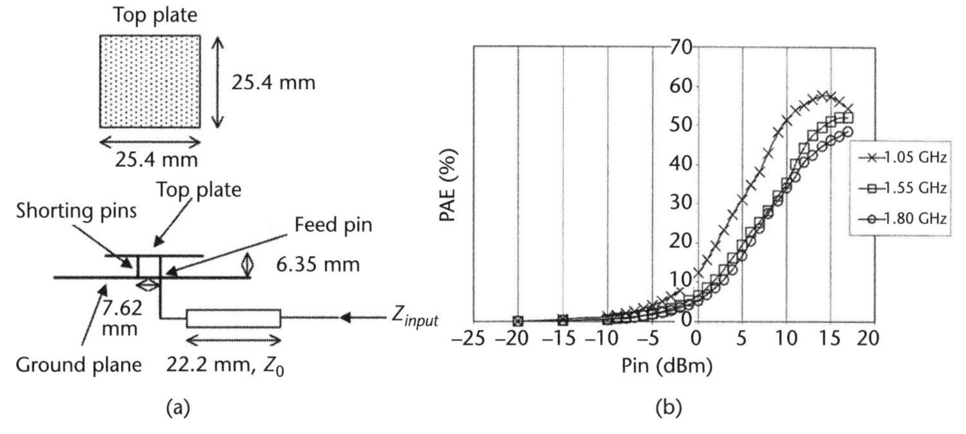

Figure 5.11 A PA AIA: (a) geometry and layout of the PA-AIA, and (b) PAE versus input power (P_{in}) curves at the three frequencies considered. (*From:* [30]. © 2003 IEEE. Reproduced with permission.)

Another class-F AIA-based design using a slot antenna was proposed in [32]. Here, the slot antenna was able to provide acceptable matching (harmonic tuning) at the three frequencies (f_0, $2f_0$, and $3f_0$), via the careful design of its slot shape and open stub length. The slot antenna acted as the OMN of the PA, as well as a harmonic tuned antenna.

The second example showing an LNA-based AIA is shown in Figure 5.12(a) [45]. The antenna is based on a wearable square patch built on flexible foam and fabric substrates with its metallic parts built from copper on polyimide films. The LNA chip was integrated below the patch with two feeding arms to support circular polarization. The AIA can be used in GPS and Iridium satellite communication systems. The rectangular patch had two slots to excite two orthogonal modes for circular polarization generation that is required for GPS and Iridium. While the two feeding arms are of the same length, a hybrid coupler with 90° phase shift between its arms is used to create the phase difference required. The LNA substrate had a thickness of 0.45 mm and made of an aramid textile fabric with a layer of polyimide, that had an $\varepsilon_r = 1.775$, and a tan $\delta = 0.02$. The patch antenna layer had a thickness of 7.25 mm and was made of polyurethane foam with $\varepsilon_r = 1.25$ and tan $\delta = 0.02$. An Anaren XC1400P-03S coupler was used for circular polarization combining. The LNA (MAX2659) is connected to the output of the combiner. A matching network consisting of a 470-pF capacitor and a 6.8-nH inductor is used between the LNA and combiner to match for 50Ω. Note that the LNA chip has a fixed bias voltage, and thus the operating conditions cannot be controlled via altering the bias conditions as in the PA example. Thus, ease of handling and single bias voltage level is traded off with less flexibility in the design.

The fabricated prototype was tested, and the AIA provided a realized gain of 25 dBic with a 1-dB gain bandwidth of at least 150 MHz covering 1.517–1.7 GHz. The

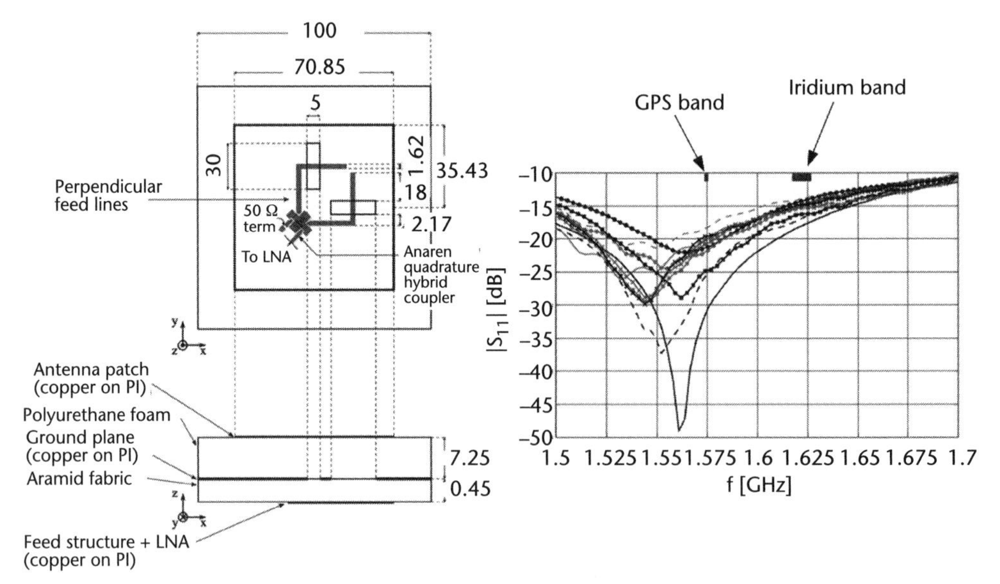

Figure 5.12 GPS/Iridium AIA: (a) geometry of the AIA showing the circularly polarized patch and the LNA (all dimensions are in millimeters), and (b) the reflection coefficient curves of the AIA with different orientations and bends on a human model. (*From:* [45]. © 2013 IEEE. Reproduced with permission.)

3-dB beamwidth was 68°, the axial ratio at 1.6 GHz was 1.866, and the 3-dB axial ratio beamwidth was 106°. The antenna was robust against bending and placement on the body and provided acceptable results in terms of covered bandwidth and gain. Some bending and placement cases provided higher than 5-dB axial ratio values, but these were limited to the case of placing the antenna along *y*-45° with a bent.

5.4 Mixer-Based AIAs

A mixer translates its input frequency to an upper or lower frequency and is consequently denoted as an upconversion or downconversion mixer, respectively. In a transmitter, we need an upconverter, and in a receiver, we usually need a downconverter. Because of its frequency conversion nature, mixer-type AIAs are often called frequency conversion AIAs. The conversion efficiency and gain are the most important performance parameters of mixer-type AIAs. In general, a mixer-type AIA uses the nonlinear characteristic of an active component for frequency conversion. Two-terminal devices such as Gunn diodes, PIN diodes, and three-terminal devices such as HEMT or MESFET are generally used. The main advantage of three-terminal devices is their conversion gain and compatibility with MMIC, but with an extra cost of circuit complexity. In mixer-type AIAs, active devices are biased in the nonlinear region and the device nonlinearity along with a filter provides the frequency conversion required (up or down).

Very few works are reported in this type of AIA, although they have good potential in transceiver modules [48–54]. A mixer-type AIA is presented in [48] using a circular patch and a Gunn diode. This mixer reports a conversion gain over 2 dB from 200–450 MHz. The active antenna VCO operation is also reported by varying the Gunn diode biasing voltage from 7.7V to 10V. Other related works found in the literature such as those in [49, 51, 54] are more or less similar in principle and use two-terminal devices (i.e., diodes) for the nonlinear component. Improved conversion efficiency is achieved using three-terminal devices such as BJT, FET, HEMT, as in [50, 52, 53]. None of these works considered antenna miniaturization through active device integration.

The concept of a self-oscillating mixer AIA has also appeared in [13, 50, 52, 53], where it provides the highest integration level between the antenna and the mixer. In this configuration, one nonlinear device acts as a local oscillator (LO) and another serves as the mixer. To reduce the LO-RF path leakage, a differential topology is always advised at the expense of more active nonlinear devices (i.e., more complexity and area). In the mixing-type AIA, most of the works did not focus on the antenna type itself, rather on the integration. The most integrated types are the self-mixing oscillator types [59]. The antenna types used in the mixer-type AIAs are patch [48, 50, 53, 54], quasi-Yagi [52], and dipole [49, 51] antennas.

5.4.1 Design Outline

While RF mixer design has been extensively addressed in literature, we provide here a brief overview for the completeness of the topic. The reader is advised to refer to [7–10] for more information about RF mixer design fundamentals. A mixer can be obtained via multiplying two incoming signals to yield an upconverted or downconverted output, or by passing the sum of the two signals to a nonlinear device and then collecting the proper frequency term of interest. When an RF signal $v_{RF}(t)$ and an LO signal $v_{LO}(t)$ are passed to a passive nonlinear device (i.e., a diode), the output $i(t)$ will be of the form,

$$i(t) = a_0 + a_1\left(v_{RF} + v_{LO}\right) + a_2\left(v_{RF} + v_{LO}\right)^2 + a_3\left(v_{RF} + v_{LO}\right)^3 + \cdots \tag{5.5}$$

$$v_{RF}(t) = V_{RF}\cos\left(\omega_{RF}t + \varphi\right) \tag{5.6}$$

$$v_{LO}(t) = V_{LO}\cos\left(\omega_{LO}t\right) \tag{5.7}$$

where a_x are the coefficients of the nonlinear model of the diode, ω_{RF} and ω_{LO} are the RF and LO frequencies, respectively, φ is an arbitrary phase shift, and V_{RF} and V_{LO} are the amplitudes of the RF and LO signals, respectively. The even-order nonlinearity, which is the quadratic term, will give the components of $(\omega_{RF} + \omega_{LO})$ and $(\omega_{RF} - \omega_{LO})$, which show the difference IF term (remember that the multiplication of two cosine functions can yield the sum and difference of their arguments). The sum and difference terms are also obtained when the two sinusoids are multiplied using a nonlinear active device. One very common circuit for a double-balanced mixer is the Gilbert cell [10]. Figure 5.13 shows some possible passive and active mixer configurations that one can use. Figure 5.13(a) shows a simple passive diode-based RF mixer. The diode even-order nonlinearity will provide the IF terms. Figure 5.13(b) shows a single-ended active mixer. Note the output matching π-network and the LO injection via an RF transformer. This will isolate the LO from the active transistor source. In Figure 5.13(c), the well-known Gilbert cell is shown. It is a double-balanced mixer. The antenna is connected to the RF terminals of these mixers either via direct integration where the antenna is carefully loaded to match the mixer terminals or via proper matching using appropriate impedance matching networks.

It should be noted that active mixers need lower LO injection levels compared to their passive counterparts. In addition, active mixers require DC biasing, unlike their passive counterparts. The double-balanced Gilbert cell-based mixer has the advantage of no even-order nonlinearity components (they cancel each other), while the odd order ones are the ones responsible for IF generation. It has also the advantage of rejecting common-mode terms and other unwanted components of the incoming signals.

The performance metrics for mixers in general are the conversion loss (L_c), the NF and the image rejection ratio (IRR). L_c is defined as the ratio between the incoming input power (P_{in}) before conversion to the output power (P_{out}) after conversion/mixing at the frequency of interest. This loss is inevitable as the output signal will be one of the multiple terms generated by the nonlinear device; thus, its power will be a small portion of the incoming one (note the coefficients in [5.5]). It is given as

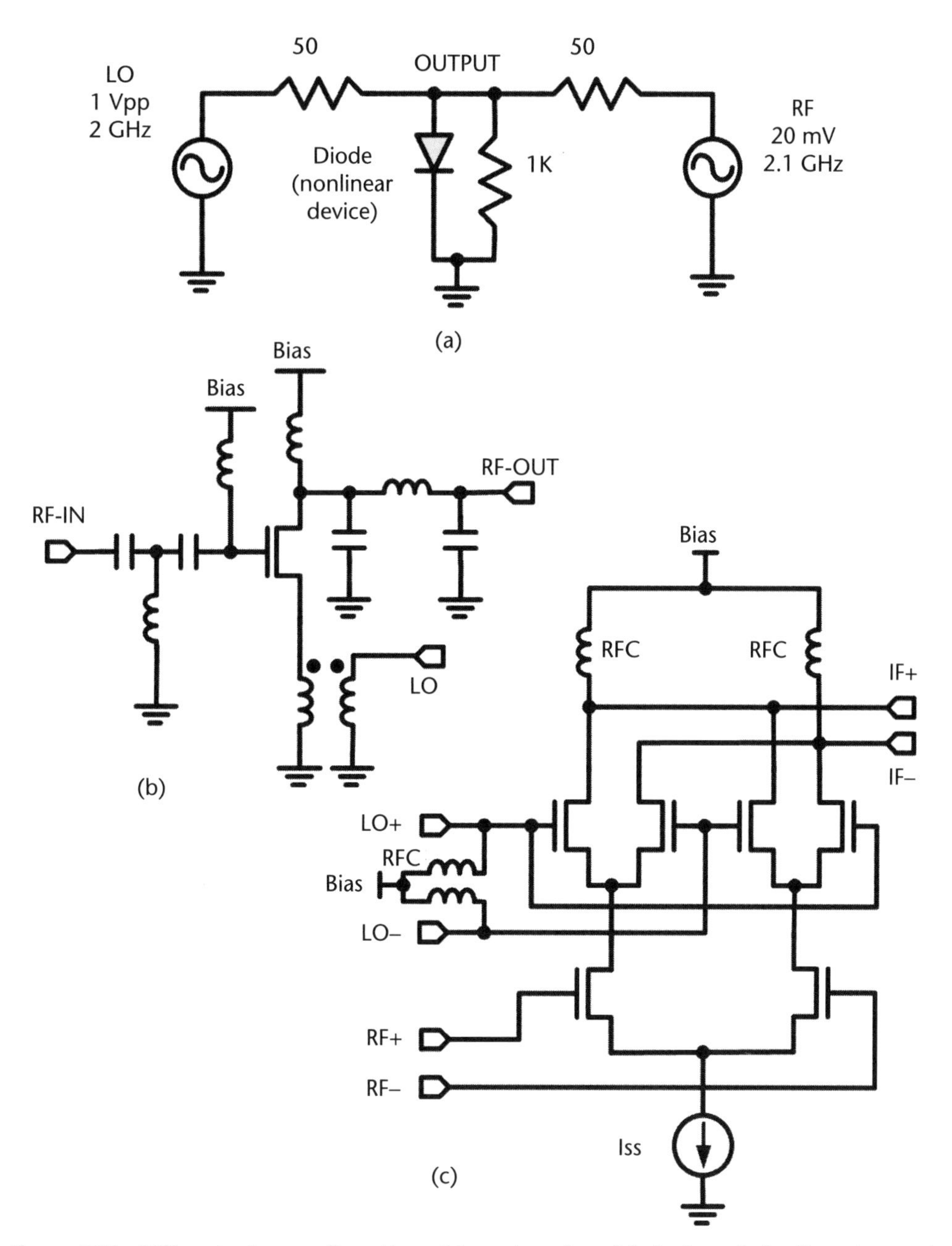

Figure 5.13 Different mixer configurations: (a) passive mixer, (b) single-ended active mixer, and (c) double-balanced Gilbert cell-based mixer.

$$L_c = \frac{P_{\text{in}}}{P_{\text{out}}} \tag{5.8}$$

The conversion gain is the reciprocal of L_c and is less than 1 (i.e., less than 0 dB) for passive mixers, but can be greater than one in a Gilbert cell-based one. This positive conversion gain can relax the amplification requirements in the RF chain.

The NF of the mixer is the decibels form of the noise factor (F), which represents the SNR between the input and output of the mixer (see Section 5.1). It should be noted that the mixer has two inputs, the incoming signal and the LO, and one output. Thus, careful analysis is required when considering single side-band (SSB) or double side-band (DSB) waveforms in terms of the inclusion of the image frequency contributions to the signal and noise factors. Also, the NF of active mixers is usually higher than passive ones.

An image signal is defined as the signal that can provide the same IF as the desired signal due to the difference terms with the LO (the absolute difference terms). This image is not desired as it will distort the original signal if not filtered out before the mixer. Thus, bandpass filters (BPF) are used to remove such image signals. The IRR is defined as the ratio between the mixer desired output power level and that output power due to the image signal, as

$$IRR = \frac{P_{\text{out}}}{P_{\text{out,image}}} \tag{5.9}$$

5.4.2 Examples

Two examples are presented for the mixer-type AIA. The first example presents an active integrated self-oscillating mixer (SOM) using a dual-gate FET. The diagram of the proposed mixer-type AIA is shown in Figure 5.14 [50]. A substrate with 0.5-mm thickness and $\varepsilon_r = 9.6$ was used. One of the two gates (G1) of the FET had the antenna connected to via a $\lambda/4$ impedance transformer to match the antenna to the FET gate. The patch was designed with a center frequency of 3.6 GHz. The second gate (G2) was connected to the LO via a high-quality factor dielectric resonator (DR). The DR will increase the oscillator stability as well. The DR is connected to the gate via a coupling microstrip line to match the impedances. The distance between the DR and the gate was optimized to maintain negative impedance at the FET

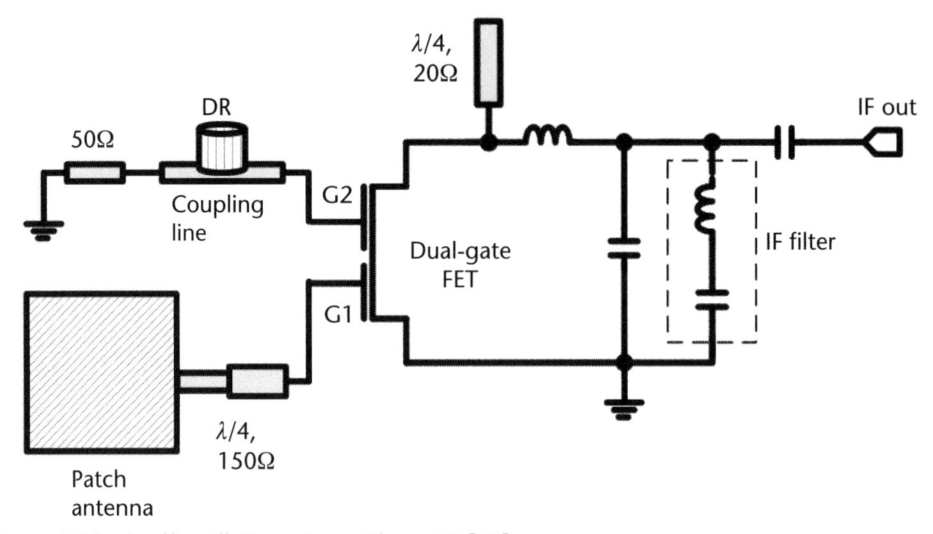

Figure 5.14 A self-oscillating mixer with an AIA [50].

drain terminal. The DR resonance was designed to resonate at a center frequency of 4 GHz. The output IF lowpass filter was designed with a cutoff frequency of 1 GHz to capture the first-order component. The fabricated-mixer AIA provided a 440-MHz IF signal corresponding to the patch and DR frequencies.

The second example is an SOM active integrated antenna operating at millimeter waves and integrated on a microwave monolithic integrated circuit (MMIC). The complete design was made on a 100-μm-thick GaAs substrate with a 0.15-μm pHEMT process. The circuit diagram of the design is shown in Figure 5.15(a) and the microphotograph is shown in Figure 5.15(b). The antenna output is taken at two points of equal amplitude and opposite phase that generate the proper current distribution on the patch and combine in total yielding a balanced patch operation. The radiation efficiency of this antenna on the GaAs substrate was 70%. If the two signals are in-phase, the center of the antenna becomes virtually open circuit with no current alignment in a single direction, the efficiency drops to 0.8%, and the antenna does not operate properly. The dimensions of the patch are shown in Figure 5.15(a). The antenna is designed to operate at 60 GHz.

The outputs of the antenna are fed to two oscillators via coupled lines to minimize the antenna effect on the oscillator operation and stability. The pair of coupled oscillators see one another as the mixing transistor (i.e., M2 or M3) in its feedback path. The transistors M1 and M2 along with the connecting transmission lines make the appropriate phase condition to maintain the oscillator operation with a loop gain larger than 1 (due to M1). When factoring in the effect of the other combination (M1 and M3), the oscillation condition is maintained. By tuning the line length between the M1 gate and the drain of M2 or M3, one can obtain the optimum load impedance between the source-drain for LO operation. Each pHEMT gate has a length of 0.15-μm and 100-μm width. The IF signals are obtained at the output of the quarter-wavelength lines that are used for M2 and M3 biasing without any external circuits. The LO was tuned to work at 58.46 GHz, and thus the

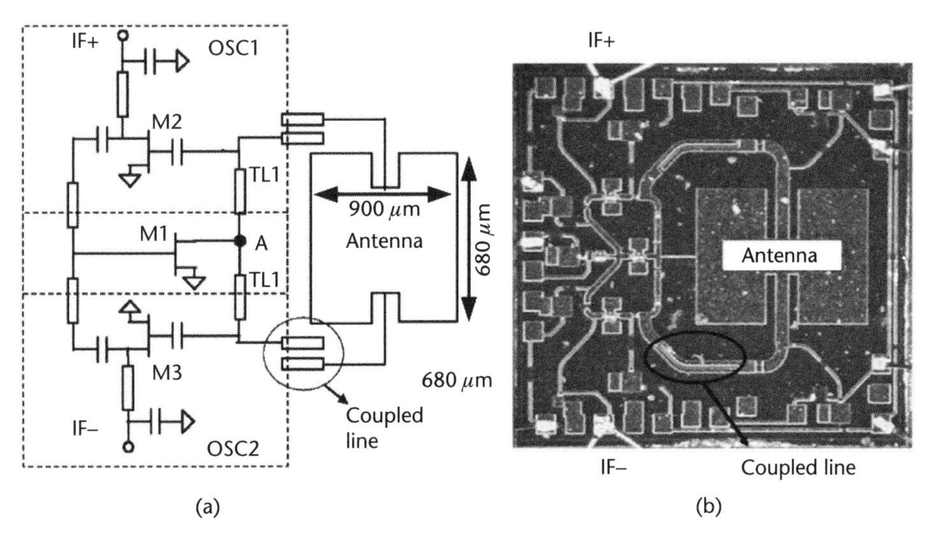

Figure 5.15 An SOM active integrated antenna: (a) geometry of the 60-GHz design, and (b) microphotograph. (*From:* [53]. © 2006 IEEE. Reproduced with permission.)

IF obtained was 1.54 GHz. The obtained maximum effective isotropic conversion gain was −18.9 dBi (this is measured using a standard gain horn of 25 dB at a 0.8-m distance, and then the received signals from a similar horn and the designed SOM AIAs are recorded). The low-conversion gain is due to the not optimum matching conditions adopted. The conversion gain was the highest at 60 GHz due to the high efficiency of the antenna.

5.5 Transceiver-Based AIAs

In current wireless terminals, if the wireless device sends and receives signals, usually the same passive antenna is used for the two-way communications, but the transmit and receive paths are isolated in frequency bands. Thus, they each have their unique bands, and a diplexer is used to separate the frequencies right after the antenna terminal. The antenna is connected to the PA and LNA at the diplexer outputs. An RF switch is sometimes used as well. This is the most common way of connecting the passive antenna to the active parts for transmit (uplink) or receive (downlink) operations.

Diplexer-less transceiver-based AIAs are very few in literature. They can provide simultaneous full-duplex operation if implemented properly. The limited number of designs is due to the isolation issue that arises when designing such integrated systems. Such designs have appeared in [55–58, 60]. All the transceiver-based AIAs without a diplexer were based on patch antennas except for [60] where a meandered monopole was used. Two works [55, 56] relied on using circulators to provide the required isolation between the transmit and receive ports. Self-diplexing operation via the use of open split ring resonators (OSRR), which are metamaterial-based unit cells, was achieved in [57]. This gave the proper isolation between the two paths. In [58], proper isolation was achieved via the use of defected ground structures (DGS), while in [60], PIN diodes were used to switch between the two paths and bands of operation.

5.5.1 Design Outline

Two possible main configurations for transceiver AIAs are shown in Figure 5.16. The first one, which is widely used in current wireless devices and terminals, is based on a passive antenna connected to the active electronics via a diplexer/switch. This will allow the antenna to operate and provide transmit and receive capabilities at different frequencies. Several ready-made off-the-shelf parts can be used in this configuration and usually are optimized and matched for 50Ω. Multiband closed and open-loop matching switches are also used to smoothly switch between frequencies and provide automatic impedance matching. Circulators are sometimes used under this configuration as in [55, 56].

The second configuration which is shown in Figure 5.16(b) shows a diplexer-free design of a transceiver-based AIA. As mentioned in the previous section, several methods are employed to improve the isolation between the transmit and receive ports within the AIA. Two ports on the same passive antenna can be used as one for receive and the other for transmit operations. The isolation is improved via the

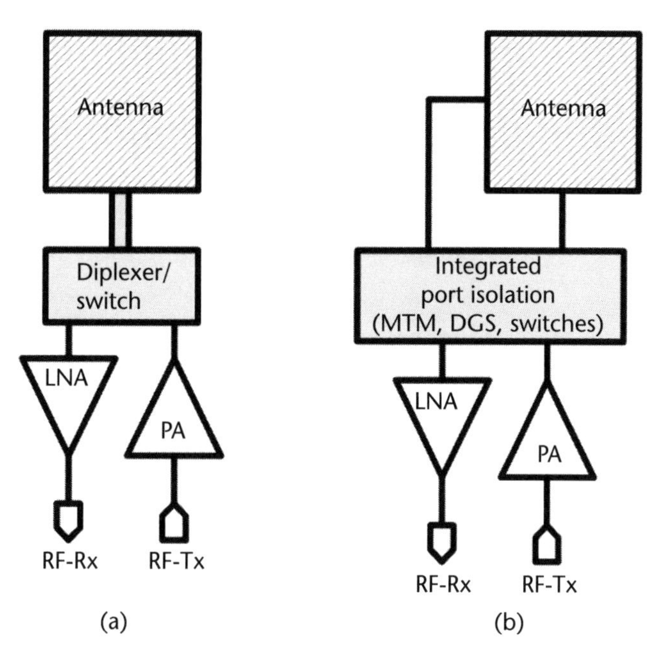

Figure 5.16 Transceiver-based AIAs: (a) diplexer-based configuration, and (b) diplexer free configuration.

use of DGS [58], MTM-based structures within the main radiator to isolate the two modes of operation as in [57] or using switches or PIN diodes as in [60].

5.5.2 Examples

We discuss two examples in this section one for each of the two configurations. An active circulator can be used to isolate the transmit and receive paths as shown in Figure 5.17(a) [55]. Three gain blocks (HP-MGA-86576) are used to connect the transmit port to the antenna and pass the received signal to the output port. The signal can pass from the transmitter to the receiver as well (through a circulator). The path lengths are adjusted to achieve phase cancellation at the receiver. The sections of low and high impedances (thin and thick transmission lines) are used to improve the isolation between the two paths and provide the required matching. The complete planar design is placed on a 50×40 mm^2 board. The isolation was narrow band in this work with 7 MHz of bandwidth with at least 20 dB of isolation between the two paths. Also, the active circulator along with the antenna provided 7 dB of gain on the receive path and 13 dB for the transmit path. The antenna had a center frequency of 3.745 GHz.

The second example utilizes DGS to enhance the isolation between the transmit and receive ports as shown in Figure 5.17(b) [58]. A patch antenna is used, and its two edges are utilized and matched to operate at two frequencies, 1.74 GHz and 1.84 GHz, for the transmit and receive ports, respectively. Bipolar transistors (BFP450, BFT92, NESG2031) were used for the class-E PA and the LNA. Optimum *NF* values were obtained with 78Ω impedances for the LNA, and more than 20-dBm IIP3 was obtained after optimizing its performance. More than 10 dB of

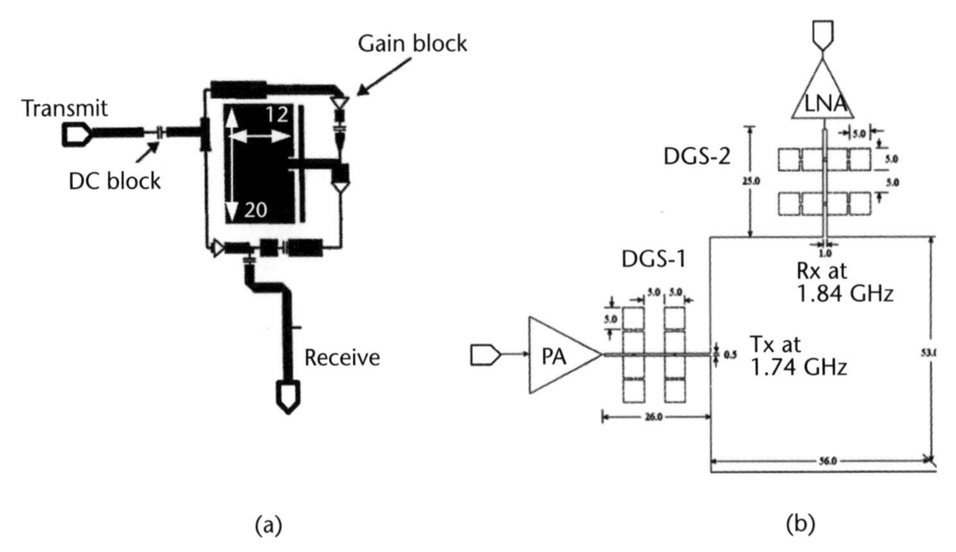

Figure 5.17 Transceiver-based AIAs: (a) with active circulator [55], and (b) using DGS for port isolation improvement [58]; all dimensions are in millimeters. (© 2010 IEEE. Reproduced with permission.)

gain was achieved. The PA provided PAE of approximately 50% in the band of operation. Dumbbell-shaped DGS were used to enhance the isolation with values more than 30 dB.

5.6 Other Types of AIAs

Active antennas, as well as AIAs, are names given to antennas that have active elements within their structure to add to their basic radiation capabilities. Such examples include diode loaded antennas to help in their frequency, pattern, or polarization reconfigurability. Also, on-chip or on-package antennas are considered active when they are closely designed with their RF electronics. Finally, the use of active transistors to build non-Foster circuits and antenna systems for wideband matching and operation can also be considered as AIAs. In this section, we give examples and design features of each of these different AIA configurations.

5.6.1 Frequency, Polarization, and Pattern Reconfigurable Antennas

Frequency reconfigurable AIAs are those antennas that can change their resonance frequency via the use of active elements such as PIN diodes, varactor diodes, or RF switches [61–67]. The basic operating principle lies in the active loading of the antenna. When using PIN diodes, the diode operates as an open-circuit if it is off or a short-circuit if it is on. Thus, it will connect or disconnect different parts of the antenna or create shorting posts to alter the current path and thus change the resonance frequency of the antenna under consideration. A varactor diode can provide a variable capacitance via the change in the reverse bias voltage applied. This will load the antenna actively and will change its resonance frequency based on the applied voltage, that is, capacitance value (remember that any antenna can be represented by a series or a parallel RLC resonant circuit, adding a capacitor will alter the resonance of such a circuit). An RF switch will behave somehow like the PIN diode in principle.

Frequency reconfigurable antennas can be designed using various antenna types such as monopoles [67], patches [61, 63], PIFAs [66], and slots [62]. Other configurations including filtering-antennas (filtennas) [65] also used such active loading. An overview article on reconfigurable antennas is found in [64].

An example of a frequency reconfigurable antenna based on a PIFA element is shown in Figure 5.18 [66]. The frequency reconfigurable antenna was implemented within a MIMO configuration using PIN and varactor diodes together. The PIN diodes (D2) were used to switch between two operating modes, one covering 1,100 MHz and 2,480 MHz, and the second covering 585-MHz, 860-MHz, and 2,410-MHz bands. This was achieved because of the change in the overall length of the antenna when the diode is on/off. A fixed bias of 5V was used to turn the PIN diode on. The varactor diodes (D1) were used to switch the operating bands within the center frequencies of the major bands covered within each mode. The antenna geometry is shown in Figure 5.18(a) and its biasing circuit is shown in Figure 5.18(b). RF chokes with 1-μH values were used to isolate the DC from the AC paths in addition to the decoupling capacitor C1. The obtained band switching as a function of the reverse bias voltage and the PIN diode modes are shown in Figure 5.19(a, b), respectively. Note the main modes obtained from the PIN diode on-off operation and then the fine tuning within the main bands obtained via the change in the reverse bias voltage of the varactor diode. Such reconfigurable features are highly desirable in multiband coverage as well as in cognitive radio (CR) platforms.

The polarization and radiation pattern orientation can be also controlled via loading the antennas with active elements. These are denoted as polarization/pattern reconfigurable active antennas. Several works in literature have investigated active polarization/pattern reconfigurable antennas such as those in [68–77]. The basic idea in the reconfigurability is altering the current path on the antenna to create another polarization or tilt the radiation patterns for various load conditions. The simplest example would be to feed a patch antenna from two adjacent corners to have a 90° phase shift in the generated fields to get two orthogonal polarizations. PIN diodes are widely used in these configurations due to the small size that does not affect the actual antenna characteristics and their ease of biasing. Amplifiers along with electronically controlled phase shifters are also used in active integrated reconfigurable designs [68–70]. Several AIA self-mixing oscillator configurations with pattern reconfigurability were also proposed such as those in [71, 73, 75, 77].

A frequency and polarization reconfigurable antenna array based on circular patch antennas was proposed in [76] and is shown in Figure 5.20. The frequency reconfigurability was achieved via the use of four varactor diodes to cover the bands between 1.5 and 2.4 GHz. Polarization reconfigurability was achieved via feeding each antenna at two points and switching between the two feeds and providing the appropriate phase shifts to obtain four polarizations (vertical, horizontal, right-hand-circular, and left-hand circular). Figure 5.20(a) shows the geometry of the circular patch with the four varactor diodes and two feeding points. Figure 5.20(b) shows the feeding network and the switches along with the phase shifters. One path will

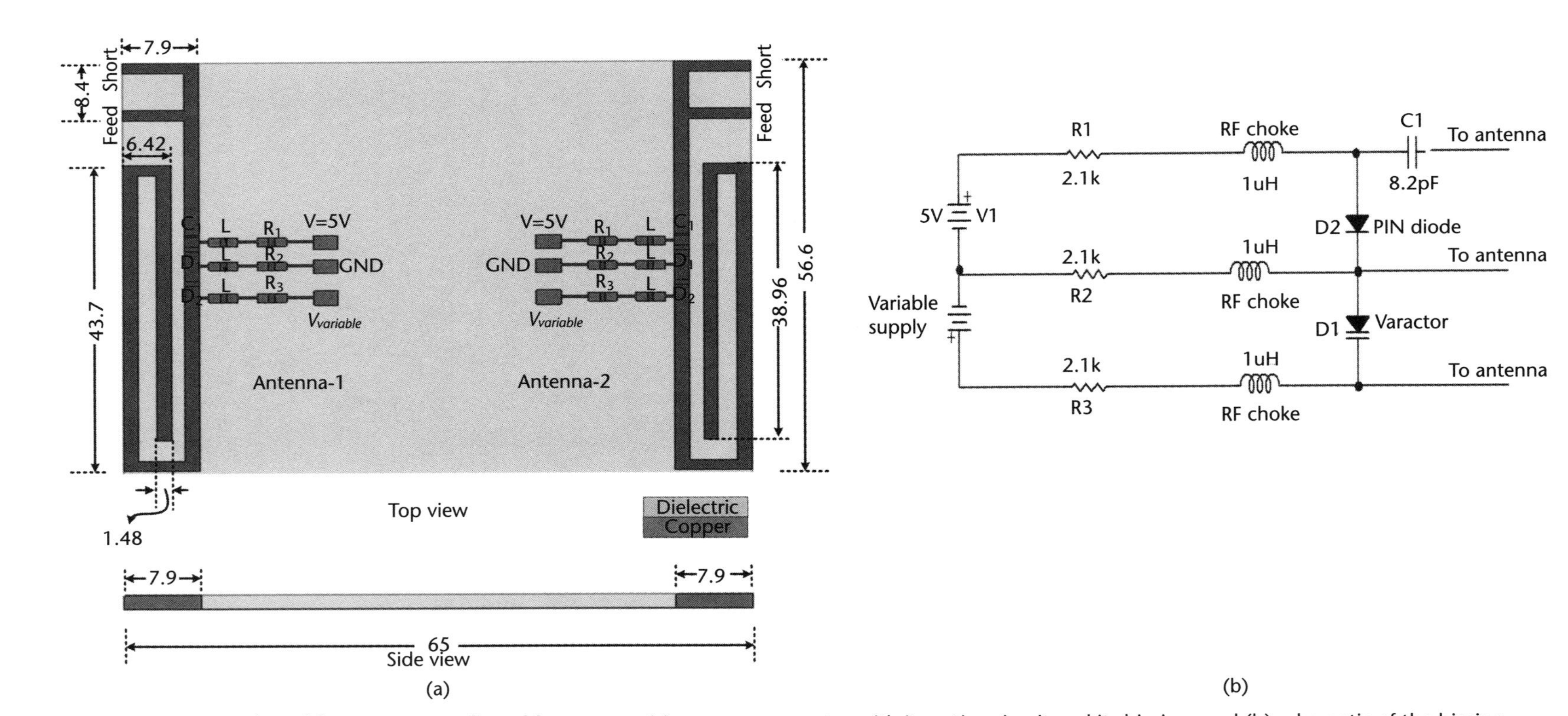

Figure 5.18 A PIFA-based frequency reconfigurable antenna: (a) antenna geometry with its active circuit and its biasing, and (b) schematic of the biasing circuit used for the two diodes [66].

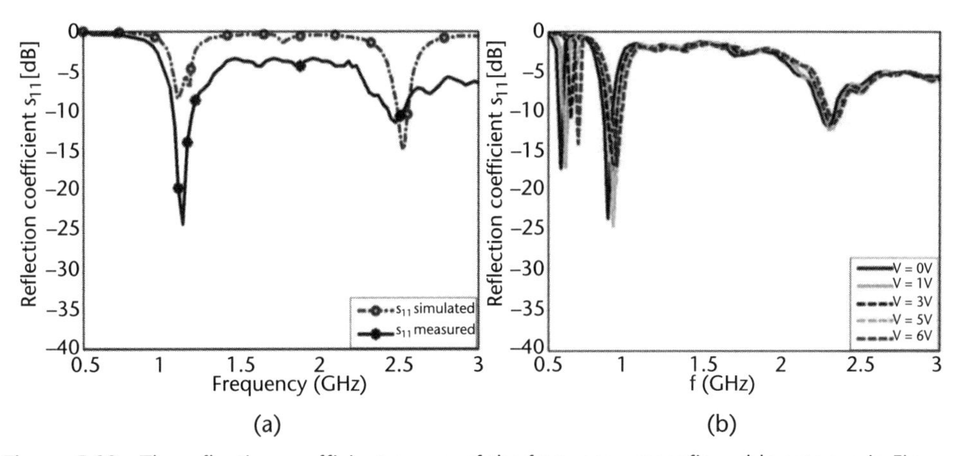

Figure 5.19 The reflection coefficient curves of the frequency reconfigurable antenna in Figure 5.18: (a) mode 1 response (PIN diode OFF), and (b) mode 2 response and the effect of the reverse bias voltage of the varactor diodes (PIN diode ON). (*From:* [66]. © 2015 IEEE. Reproduced with permission.)

be responsible for the linear polarizations (two of them via phase control), and the other (middle two lines) is responsible for the circular polarizations (left-hand and right-hand) via a two-point feed. The obtained frequency and polarization agility responses of this antenna configuration are shown in Figure 5.21. Figure 5.21(a) shows the different bands covered for the different varactor diode capacitance values. Figure 5.21(b) shows the maximum gain curves for the circular polarization cases at different covered bands.

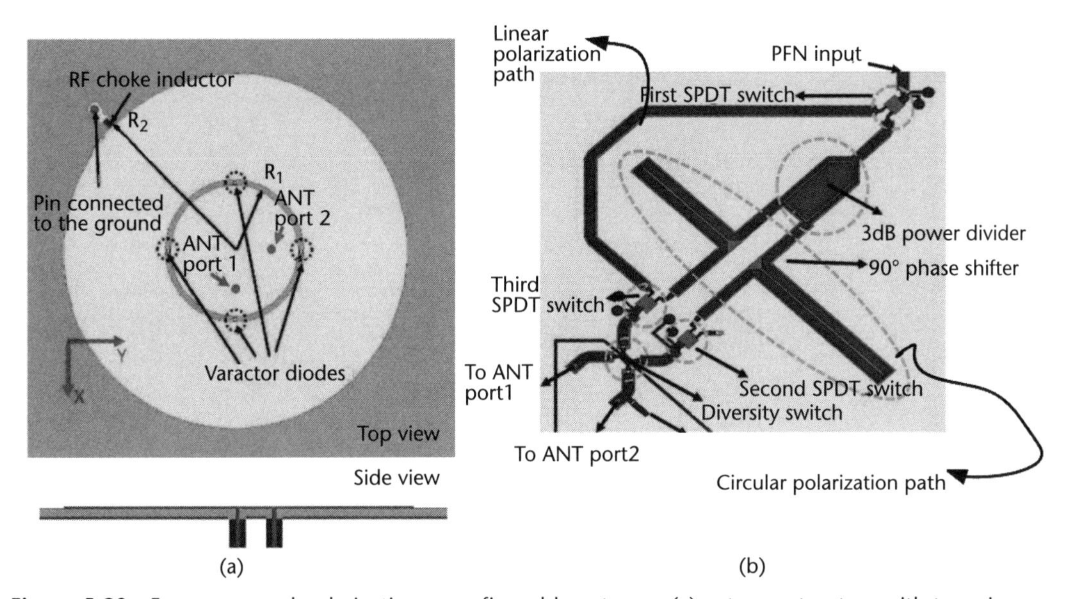

Figure 5.20 Frequency and polarization reconfigurable antenna: (a) antenna structure with two views, and (b) feed network with its switches for providing four-polarization configurations. (*From:* [76]. © 2016 IEEE. Reproduced with permission.)

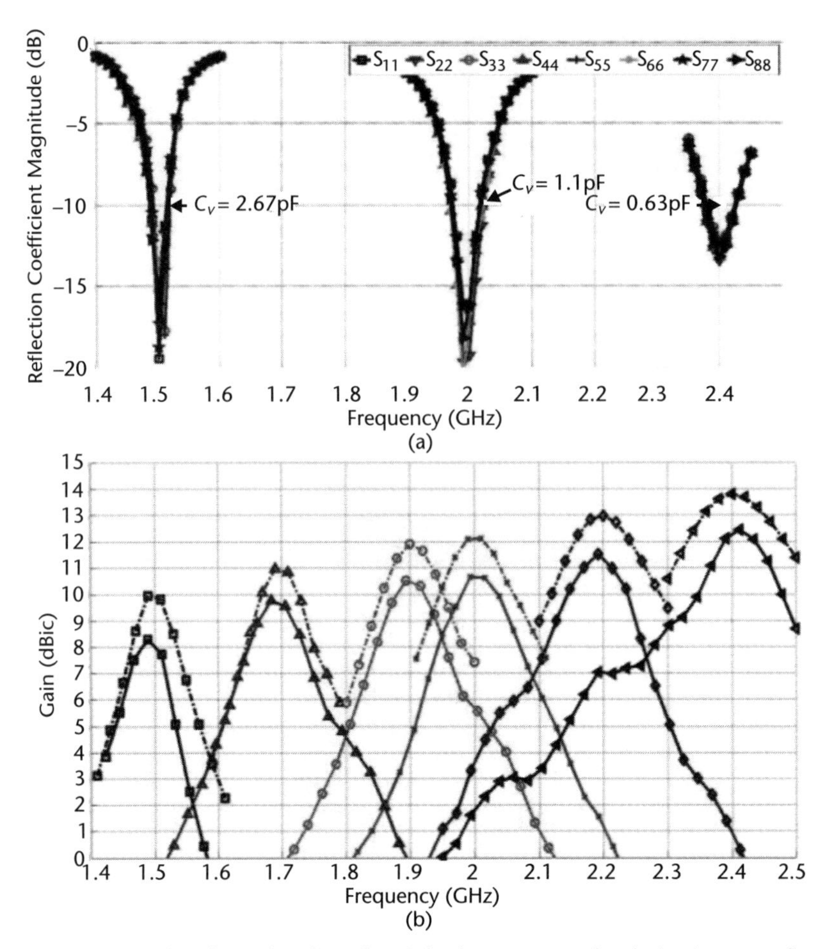

Figure 5.21 Measured and simulated results of the frequency and polarization reconfigurable antenna in Figure 5.20: (a) reflection coefficient curves for different capacitance values, and (b) maximum gain curves for circular polarization cases. (*From:* [76]. © 2016 IEEE. Reproduced with permission.)

5.6.2 On-Chip/On-Package Antennas

On/in chip active antennas have also found their applications in complete RF systems on a chip or in a package. Due to the small size of the chip die or package, millimeter-wave-based designs are the ones that have been considered thus far for such applications due to the small size of the antennas at such high frequencies and the ability to place them along with the RF electronics on the same die or within the same package [78–86]. Very few designs of on-chip AIAs operating below 10-GHz bands were proposed such as the one in [80]. An overview of on-chip AIAs can be found in [83]. Various antenna types have been considered in on-chip/on-package antennas such as dipoles [78, 82, 84, 85], monopoles [80], patch [79, 81, 86], loop/ring [81], among others such as Yagi and PIFA.

On-chip antennas suffer from low efficiency due to the back radiation into the substrate that is suited for active circuits due to its low resistance, but not suitable for antenna operation [83]. In addition, the high dielectric constant of the silicon substrate confines the energy within it. The existence of surface waves due to the thick substrate and its high dielectric constant values also degrades the efficiency of the antenna. Several innovative techniques have been implemented in recent years to overcome such losses such as implementing the antennas on semi-shielded layers (with SiO_2 and metal layers) [79–82], utilizing artificial magnetic conductor (AMC) boundaries below the antennas and on-top of the active electronics portion [84], or having a separate antenna block within the same package [86].

One example of an on-chip antenna implementation is the one in [81]. The antennas were implemented on the top layer on a quartz superstrate of 100-μm thickness and with $\varepsilon_r \sim 3.8$ and $\tan \delta = 0.001$ to provide good radiation efficiency and suppress surface waves. A ground layer isolates the lossy silicon substrate from the antenna for better radiation efficiency. The active electronics were built on the silicon substrate using a 0.13-μm IBM8SF CMOS process. Metal squares were added to satisfy the metal density of the process. Two antennas were implemented using this process, a patch antenna and a slot-ring antenna. Both were resonating at 94 GHz with about 8 GHz of operating bandwidth and almost 50% radiation efficiency. A gain of 2–3 dBi was obtained within 91–94 GHz. Figure 5.22 shows this on-chip antenna. Figure 5.22(a) shows the overall structure of the on-chip antennas considered (i.e., the patch and the slot/ring). Figure 5.22(b) shows the chip stack-up showing the various layers used to isolate the antenna from the silicon substrate. Figure 5.22(c) shows the measured and simulated reflection coefficient curves, and the fabricated prototype of the patch on-chip antenna is shown in Figure 5.22(d).

Another example of an in-chip/in-package is the one shown in Figure 5.23 [86]. Here a different method is used to minimize substrate-based effects on the

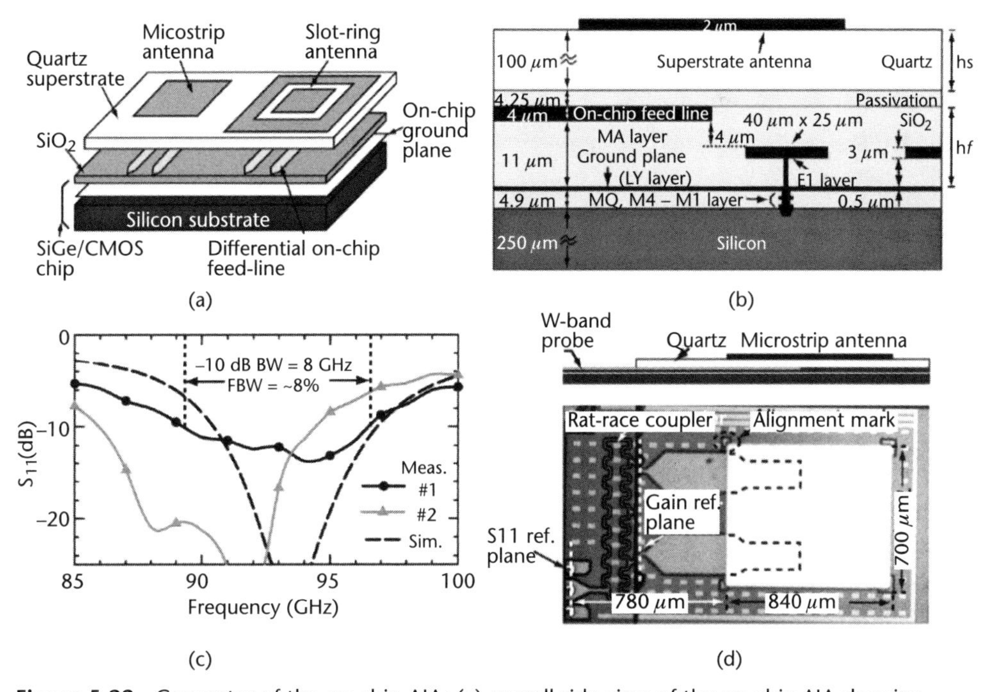

Figure 5.22 Geometry of the on-chip AIA: (a) overall side view of the on-chip AIA showing the two antennas, (b) layer stack-up of the chip layers, (c) reflection coefficient curves, and (d) microphotograph of the fabricated prototype. (*From:* [81]. © 2012 IEEE. Reprinted with permission.)

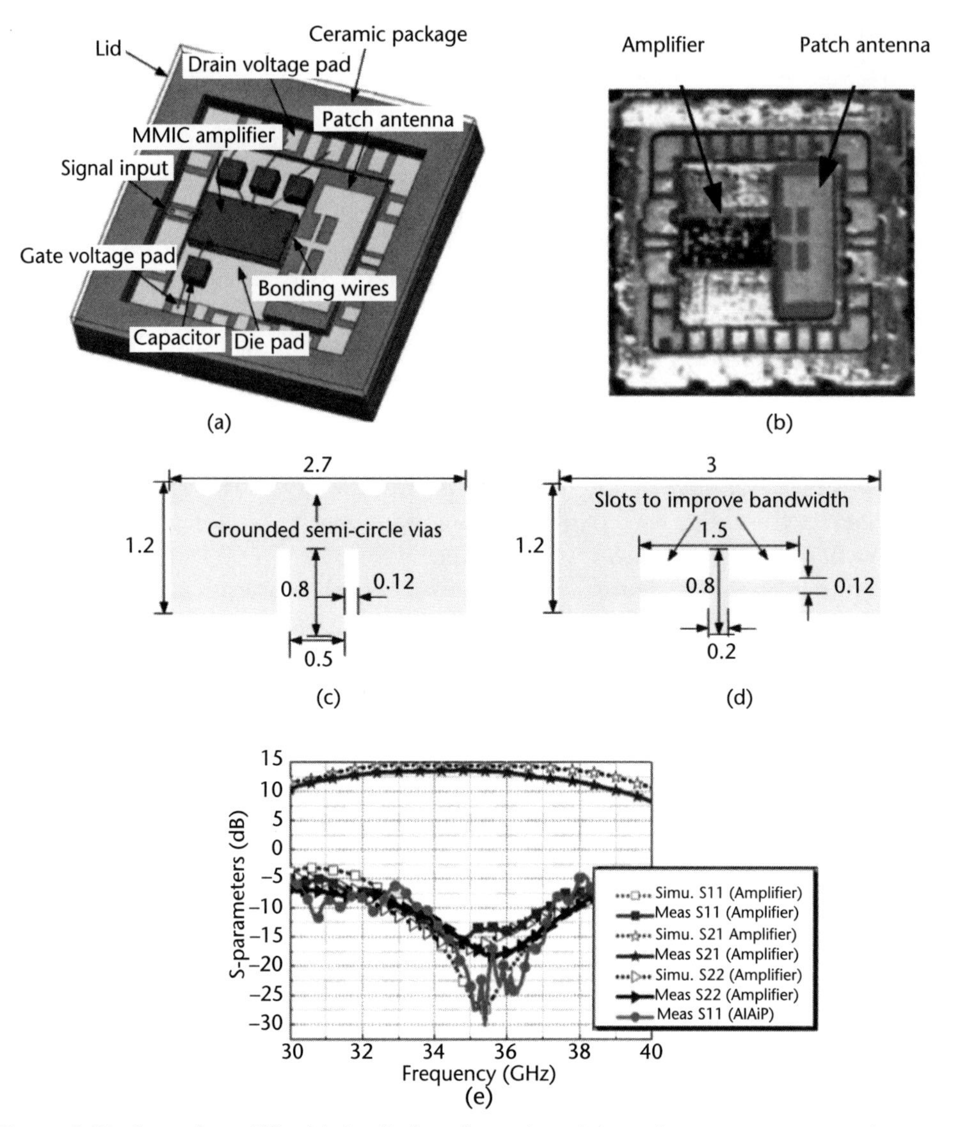

Figure 5.23 In-package AIAs: (a) detailed configuration of the various components, active and antenna parts, (b) microphotograph of the fabricated prototype, (c) quarter-wave patch geometry, (d) half-wave patch geometry, and (e) measured amplifier gain and antenna reflection coefficient curves. (*From:* [86]. © 2017 IEEE. Reprinted with permission.)

integrated in-chip antenna separating the antenna substrate from the amplifier/active electronics one and connecting them via bonding wires as shown in Figure 5.23(a) and its fabricated prototype is shown in Figure 5.23(b). The amplifier is implemented using a 0.15-μm GaAs pHEMT process. It occupies a 1.2×2.2 mm^2 substrate. The amplifier gain with the bonding wire was 14.5 dB. The bonding wire is AIAiP with a diameter of 25 μm and length of 400 μm. The effect of the wire should be incorporated in the design and optimization of the amplifier. The complex output impedances of the amplifier at various frequencies are recorded to match the antenna. Two antennas were investigated, a quarter-wave and a half-wave patch, with the former built on an RO4350B substrate with $\varepsilon_r = 3.66$ and tan δ = 0.0037 and height of 0.338 mm, and the later built on RT6010LM, with ε_r = 10.2 and tan δ = 0.0023 and height of 0.254 mm. Figure 5.23(c, d) shows the two patch configurations with proper impedance and bandwidth optimization via the slot addition. The half-wave patch was used in the final design as it provided 2.1-dB higher gain at the expense of 2% less bandwidth. A ceramic package (Kyocera A473) with the size of 7 mm × 7 mm and ε_r = 8.5 and tan δ = 0.0021 is used to host the AIA in a package. The available cavity within it is 4 × 4 × 0.74 mm^3. Figure 5.23(a) shows the overall structure of the in-package integrated antenna. The complete AIA in package covered 33–37.1 GHz, with amplifier gain values 12.3–13.6 dBi and maximum radiated gain of 18.9 dBi at 35 GHz. The measured response is shown in Figure 5.23(e).

5.6.3 Non-Foster Antennas

One of the fundamental limits of electrically small passive antennas (ESA) in terms of impedance matching is their high quality factor (Q), which affects their bandwidth of operation (i.e., they are very narrowband). Antennas possess larger reactance as their size decreases. Thus, miniaturized antennas show narrow impedance bandwidth as their size gets reduced. This Foster reactance of the miniaturized antennas cannot be compensated for over a wideband by traditional lossless matching networks. This is mainly because of the opposite slope of the positive inductance and capacitance with respect to frequency. As can be observed from Figure 5.24, the impedance due to the capacitance is decreased while it is increased when the frequency is increased. Thus, for a highly reactive load, compensation cannot be achieved over a wideband.

On the contrary, the Foster reactance can be perfectly compensated for by a negative reactance and theoretically providing infinite bandwidth as illustrated in Figure 5.24 [87, 88]. The reactance due to a capacitance is shown to be perfectly balanced by a negative capacitance obtained from a negative impedance converter

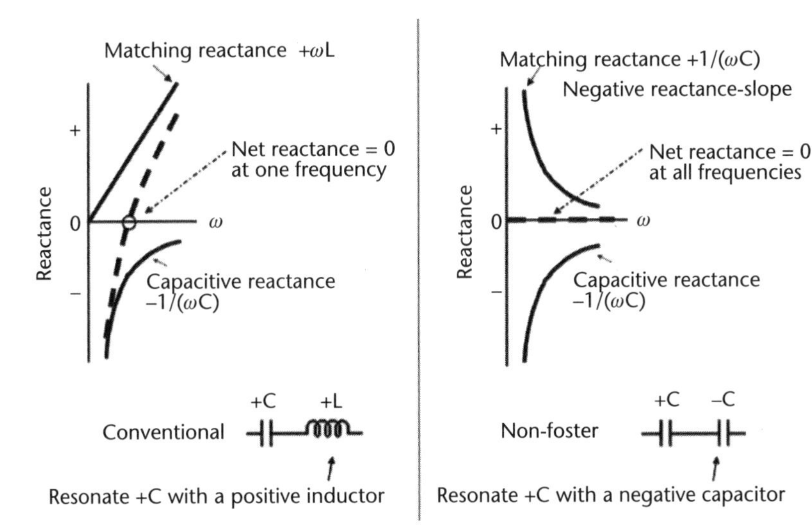

Figure 5.24 Conventional and non-Foster-based impedance matching. (*From:* [88]. © 2009 IEEE. Reproduced with permission.)

(NIC) circuit that is based on active elements with various circuit topologies as shown in Figure 5.25. Thus, it is obvious that the bandwidth limit of small antennas can be successfully overcome by negative reactances via the use of NIC circuits. This was proven in [87]. Although, theoretically speaking, NIC can provide infinite bandwidth at higher frequencies and due to the noticeable effect of parasitic capacitances that increase the positive feedback between the output to the input of such circuits, the potential of unstable behavior increases. Thus, care should be taken when designing NICs at higher frequencies.

Several examples of AIAs utilizing NIC to enhance the operating bandwidth of ESA are shown in [89–94]. Monopole antennas were considered in [88, 89, 92, 94], dipole antennas were considered in [90, 91], and PIFA antennas were considered in [93]. A very high frequency (VHF) monopole antenna operating between 30 and 200 MHz with an NIC circuit for bandwidth improvement was proposed in [89], and an improved gain of at least 10 dB was achieved compared to the conventional narrowband matched case. The antenna length was only 15 cm. Another printed patch antenna operating at 377 MHz with a 5 times improved bandwidth achieved via an NIC circuit is presented in Figure 5.26 [92]. The non-Foster circuit was embedded within the patch to save space and it had a direct effect on the input reactance of the antenna. The design was built on an FR-4 substrate with 0.79-mm thickness and $\varepsilon_r = 4.35$. The overall board size was 100 × 100 mm^2.

5.7 Conclusions

AIAs are of great importance due to their attractive features of smaller size due to tight integration with the active circuits, higher gain, frequency and polarization reconfigurability, the possibility of having miniaturized antenna sizes, and the possibility of achieving very high bandwidth when non-Foster active circuits are used. In this chapter, we provide a complete coverage of all types of AIAs as well as their design guidelines with several examples on each type to make it easier for the reader to have a complete design. Specifically, oscillator, amplifier, mixer, and transceiver-based AIAs were covered in full details. In addition, frequency-reconfigurable,

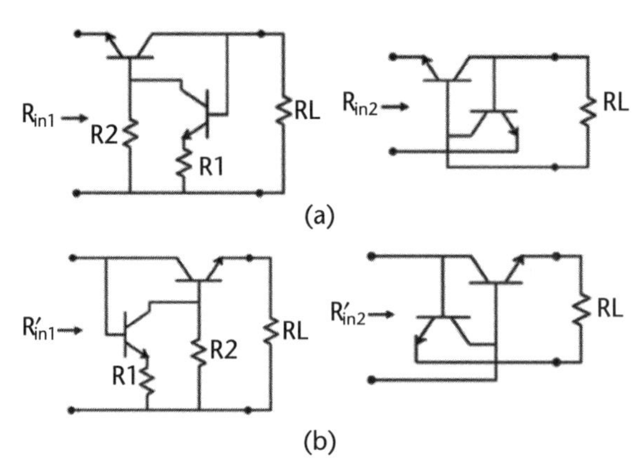

Figure 5.25 Linvill's ideal NIC: (a) open-circuit stable (OCS), and (b) short-circuit stable (SCS). (*From:* [88]. © 2009 IEEE. Reproduced with permission.)

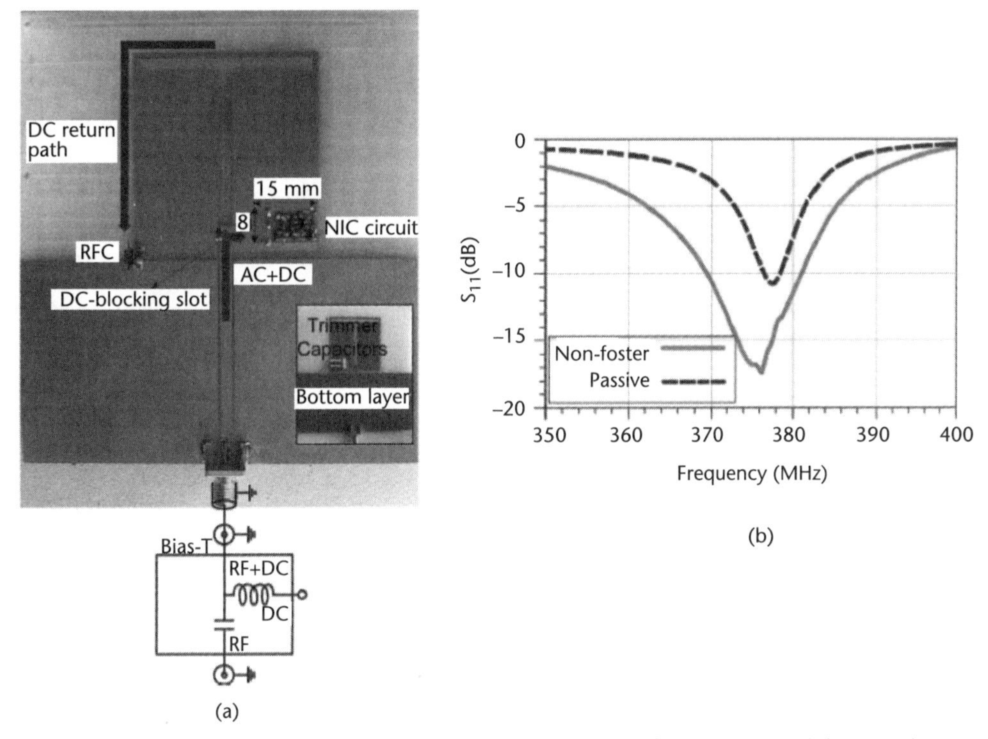

Figure 5.26 A non-Foster AIA monopole antenna: (a) fabricated prototype, and (b) impedance matching comparison of the non-Foster with the conventional passive methods. (*From:* [92]. © 2013 IEEE. Reproduced with permission.)

pattern, and polarization reconfigurable antennas were discussed with several examples showing the biasing circuits for the active parts. Finally, we considered on-chip and in-package AIAs and non-Foster-based ones with several practical examples showing their importance in very small footprint applications and highly miniaturized antenna structures.

References

[1] Shaw, R. W., et al., "Integrated Active Antenna Module for Space Station Multiple Access Communication," *IEEE Microwave Theory and Techniques Digest*, May 1990, pp. 801–804.

[2] Roscoe, D. J., A. Ittipiboon and L. Shafai, "The Development of an Active Integrated Microstrip Antenna," *Antennas and Propagation Society Symposium Digest*, 1991, pp. 48–51.

[3] Lin, J., and T. Itoh, "Active Integrated Antennas," *IEEE Transactions on Microwave Theory and Techniques*, Vol. 42, No. 12, December 1994, pp. 2186–2194.

[4] Pobanz, C. W., and T. Itoh, "Active Integrated Antenna," *IEEE Potentials*, 1997, pp. 6–10.

[5] Qian, Y., and T. Itoh, "Progress in Active Integrated Antennas and Their Applications," *IEEE Transactions on Microwave Theory and Techniques*, Vol, 46, No. 11, November 1998, pp. 1891–1900.

[6] Chang, K., et al., "Active Integrated Antennas," *IEEE Transactions on Microwave Theory and Techniques*, Vol. 50, No. 3, March 2002, pp. 937–944.

[7] Pozar, D., *Microwave Engineering*, Third Edition, New York: Wiley, 2005.

[8] Razavi, B., *RF Microelectronics*, 1st ed., Upper Saddle River, NJ: Prentice Hall, 1998.

[9] Lee, R., *RF Circuit Design*, 2nd ed., New York: Wiley, 2012.

[10] Steers, M., *Microwave and RF Design: A Systems Approach*, SciTech, 2010.

[11] Dhar, S. K., M. S. Sharawi, and F. M. Ghannouchi, "On Microwave Connector De-Embedding and Antenna Characterization," *IEEE Antenna and Propagation Magazine*, 2017.

[12] Chang, K., K. A. Hummer, and G. K. Gopalakrishnan, "Active Radiating Element Using FET Source Integrated with Micro Strip Patch Antenna," *Electronics Letters*, Vol. 24, No. 21, October 1988, pp. 1347–1348.

[13] Liu, J. C., et al., "Double-Ring Active Micro Strip Antenna and Self-Mixing Oscillator in C. Band," *IEE Proceedings on Microwave Antennas Propagation*, Vol. 147, No. 6, December 2000, pp. 479–482.

[14] Bonefacic, D., J. Bartolic, and Z. Mustic, "Circular Active Integrated Antenna with Push-Pull Oscillator," *Electronic Letters*, Vol. 38. No. 21, October 2002, pp. 1238–1240.

[15] Urbani, F., et al., "VCO Active Integrated Antenna with Reactive Impedance Surfaces," *Microwave and Optical Technology Letters*, Vol. 47, No. 1, October 2005, pp. 82–86.

[16] Choi, D. H., and S. O. Park, "A Varactor-Tuned Active-Integrated Antenna Using Slot Antenna," *IEEE Antennas and Wireless Propagation Letters*, Vol. 4, 2005, pp. 191–193.

[17] Yun, G., "Compact Oscillator-Type Active Antenna for UHF RFID Reader," *Electronic Letters*, Vol. 43, No. 6, March 2007.

[18] Ponchak, G. E., M. C. Scardelletti, and J. L. Jordan, "270 C, 1GHz Oscillator-Type Antenna," *Electronic Letters*, Vol. 45, No. 8, April 2009.

[19] Lin, Y. Y., C. H. Wu, and T. G. Ma, "Miniaturized Self-Oscillating Annular Ring Active Integrated Antennas," *IEEE Transactions on Antennas and Propagation*, Vol. 59, No. 10, October 2011, pp. 3597–3606.

[20] Lee, K. J., K. I. Jeon, and Y. S. Kim, "Aperture-Coupled Active Antenna with Push-Pull Oscillator," *Electronic Letters*, Vol. 47, No. 19, September 2011.

[21] Baek, S., et al., "Multi-Band Active Integrated Antenna Using Clapp Oscillator Circuit," *6th International Conference on Telecommunication Systems, Services and Applications*, 2011, pp. 69–72.

[22] Liu, Y. C., and H. Y. Chang, "Design of a V-Band Active Integrated Antenna (AIA) with Voltage Controlled Oscillator," *Proceedings of the 2012 IEEE International Symposium on Antennas and Propagation*, 2012, pp. 1–2.

[23] Wu, C. H., and T. G. Ma, "Self-Oscillating Semi-Ring Active Integrated Antenna with Frequency Configurability and Voltage-Controllability," *IEEE Transactions on Antennas and Propagation*, Vol. 61, No. 7, July 2013, pp. 3880–3885.

[24] Wu, C. H., and T.G. Ma, "Miniaturized Self-Oscillating Active Integrated Antenna with Quasi-Isotropic Radiation," *IEEE Transactions on Antennas and Propagation*, Vol. 62, No. 2, February 2014, pp. 933–936.

[25] Liu, Z. H., Y. W. Chang and T. G. Ma, "Frequency Reconfigurable Self-oscillating Active Integrated Antenna using Metamaterial Resonators," *IEEE 5th Asia-Pacific Conference on Antennas and Propagation (APCAP)*, 2016, pp. 427–428.

[26] Wu, X. D., K. Leverich, and K. Chang, "Novel FET Active Patch Antenna," *Electronic Letters*, Vol. 28, No. 20, September 1992.

[27] Wu, X. D., and K. Chang, "Compact Wideband Integrated Active Slot Antenna Amplifier," *Electronic Letters*, Vol. 29, No. 5, March 1993, pp. 496–497.

[28] Wu, X. D., and K. Chang, "Novel Active FET Circular Patch Antenna Arrays for Quasi-Optical Power Combining," *IEEE Transactions on Microwave Theory and Techniques*, Vol. 42, No. 5, May 1994, pp. 766–771.

[29] Erturk, V. B., R. R. Rojas, and P. Robin, "Hybrid Analysis/Design Method for Active Integrated Antennas," *IEE Proceedings on Microwave Antennas Propagation*, Vol. 146, No. 2, April 1999, pp. 131–137.

[30] Ellis, G. A., and S. Liw, "Active Planar Inverted-F Antennas for Wireless Applications," *IEEE Transactions on Antennas and Propagation*, Vol. 51, No. 10, October 2003, pp. 2899–2906.

[31] Vargas, D. S., et al., "An Active Broadband-Transmitting Patch Antenna for GSM-1800 and UMTS," *Microwave and Optical Technology Letters*, Vol. 41, No. 5, June 2004, pp. 350–354.

[32] Kim, H., I. J. Yoon, and Y. J. Joong, "A Novel Fully Integrated Transmitter Front-End with High Power-Added Efficiency," *IEEE Transactions on Microwave Theory and Techniques*, Vol. 53, No. 10, October 2005, pp. 3206–3214.

[33] Qin, Y., S. Gao, and A. Sambell, "Broadband High-Efficiency Circularly Polarized Active Antenna and Array for RF Front-End Application," *IEEE Transactions on Microwave Theory and Techniques*, Vol. 54, No. 7, July 2006, pp. 2910–2916.

[34] Genc, A., et al., "Active Integrated Meshed Patch Antennas for Small Satellites," *Microwave and Optical Technology Letters*, Vol. 54, No. 7, July 2012, pp. 1593–1595.

[35] Khoshniat, A., and R. Baktur, "A Linearly Polarized Active Integrated Square Micro Strip Patch Antenna," *2011 IEEE International Symposium on Antennas and Propagation (APSURSI)*, 2011, pp. 3082–3084.

[36] Upadhayay, M. D., et al., "Active Integrated Antenna Using BJT with Floating Base," *IEEE Microwave and Wireless Components Letters*, Vol. 23, No. 4, April 2013, pp. 202–204.

[37] Ismail, W., and P. Gardner, "Low Noise Integrated Active Antenna as Image Reject Mixer (IRM)," *6th IEEE High Frequency Postgraduate Colloquium*, 2001, pp. 125–129.

[38] Lin, S., Y. Qian, and T. Itoh, "A Low Noise Active Integrated Antenna Receiver for Monopulse Radar Applications," *6th IEEE High Frequency Postgraduate Colloquium*, 2001, pp. 1395–1398.

[39] Peter, T., et al., "Design of Low Noise Amplifier with Active Integrated Antenna at 5 GHz," *Asia Pacific Conference on Applied Electromagnetics Proceedings*, 2007, pp. 1–5.

[40] Dehbashi, R., K. Forooraghi, and Z. Atlasbaf, "Active Integrated Antenna Based Rectenna Using a New Probe-Fed U-Slot Antenna with Harmonic Rejection," *2006 IEEE Antennas and Propagation Society International Symposium*, 2006, pp. 2225–2228.

[41] Charrier, D., et al., "Design of a Low Noise, Wide Band, Active Dipole Antenna for a Cosmic Ray Radio Detection Experiment," *Antennas and Propagation Society International Symposium*, 2007, pp. 4485–4488.

[42] Aloi, D. N., and M. S. Sharawi, "An Active Tri-Band (AMPS/PCS/GPS) Antenna with Enhanced Cellular-to-GPS Isolation for Automotive Applications," *Microwave and Optical Technology Letters*, Vol. 53, No. 8, August 2011, pp. 1764–1767.

[43] Taachoucher, Y., F. Colombel, and M. Himdi, "Very Compact and Broadband Active Antenna for VHF Band Applications," *International Journal of Antennas and Propagation*, Vol. 2012, 2012, pp. 1–5.

[44] Fan, S. Z., and E. L. Tan, "A Low Cost Omnidirectional High Gain Active Integrated Antenna for WLAN Applications," *IEEE Asia-Pacific Conference on Antennas and Propagation (APCAP)*, August 2012, pp. 124–125.

[45] Dierck, A., H. Rogier, and F. Declercq, "A Wearable Active Antenna for Global Positioning System and Satellite Phone," *IEEE Transactions on Antennas and Propagation*, Vol. 61, No. 2, February 2013, pp. 532–538.

[46] Eichler, J., et al., "Active Low Noise Differentially Fed Dipole Antenna," *10th International Symposium on Antennas, Propagation & EM Theory*, September 2011, pp. 219–223.

[47] Lin, H. N., et al., "Receiving Performance Enhancement of Active GPS Antenna with Periodic Structure," *Progress in Electromagnetics Research Symposium Proceedings*, 2001, pp. 125–129.

[48] Montiel, C. M., L. Fan, and K. Chang, "A Novel Active Antenna with Self-Mixing and Wideband Varactor-Tuning Capabilities for Communication and Vehicle Identification Applications," *IEEE Transactions on Microwave Theory and Techniques*, Vol. 44, No. 12, December 1996, pp. 2421–2430.

[49] Nesic, A., et al., "Active Antenna Integrated with Up Converter for 5 to 24 GHz Band," *4th International Conference on Telecommunications in Modern Satellite, Cable and Broadcasting Services*, October 1999, pp. 194–197.

[50] Zhang, J., Y. Wang, and Z. Chen, "Integration of a Self-Oscillating Mixer and an Active Antenna," *IEEE Microwave and Guided Wave Letters*, Vol. 9, No. 3, March 1999, pp. 117–119.

[51] Nesic, A., et al., "Active Antenna Integrated with Down Converter for 24 to 5 GHz Bands," *4th International Conference on Telecommunications in Modern Satellite, Cable and Broadcasting Services*, October 1999, pp. 198–201.

[52] Sironen, M., Y. Qian, and T. Itoh, "A Subharmonic Self-Oscillating Mixer with Integrated Antenna for 60-GHz Wireless Applications," *IEEE Transactions on Microwave Theory and Techniques*, Vol. 49, No. 3, March 2001, pp. 442–450.

[53] Choi, W., C. Cheon, and Y. Kwon, "A V-Band MMIC Self Oscillating Mixer Active Integrated Antenna Using a Push-Pull Patch Antenna," *IEEE MTT-S International Microwave Symposium Digest*, 2006, pp. 630–633.

[54] Rahim, M. K. A., W. K. Chong, and T. Masri, "Active Integrated Antenna with Image Reject Mixer Design," *The Second European Conference on Antennas and Propagation, (EuCAP)*, 2002, pp. 1–4.

[55] Cryan, M. J., et al., "Integrated Active Antennas with Simultaneous Transmit-Receive Operation," *26th European Microwave Conference*, September 1996, pp. 565–568.

[56] Rahim, M. K. A., et al., "Simultaneous Transmit and Receive Circular Polarized Active Integrated Antenna," *2nd European Conference on Antennas and Propagation (EuCAP)*, 2007, pp. 1–4.

[57] Martinez, F. J. H., et al., "Self-Diplexed Patch Antennas Based on Metamaterials for Active RFID Systems," *IEEE Transactions on Microwave Theory and Techniques*, Vol. 57, No. 5, May 2009, pp. 1330–1340.

[58] Mendez, J. A. T., et al., "Improving Performance of Non-Duplexer Active Transceiver Antenna with Defected Structures," *IET Microwaves, Antennas & Propagation*, Vol. 4, No. 3, November 2008, pp. 342–352.

[59] Saavedra, C. E., B. R. Jackson, and S. S. K. Ho, "Self-Oscillating Mixers," *IEEE Microwave Magazine*, September/October 2013, pp. 40–49.

[60] Valizade, A., P. Rezaei, and A. A. Orouji, "Design of Reconfigurable Active Integrated Micro Strip Antenna with Switchable Low-Noise Amplifier/Power Amplifier Performances for Wireless Local Area Network and WiMAX Applications," *IET Microwaves, Antennas & Propagation*, Vol. 9, No. 9, 2015, pp. 872–881.

[61] Luxey, C., et al., "Dual-Frequency Operation of CPW-Fed Antenna Controlled by Pin Diodes," *Electronic Letters*, Vol. 36, No. 1, January 2000, pp. 2–3.

[62] Peroulis, D., K. Sarabandi and L. P. B. Katehi, "Design of Reconfigurable Slot Antennas," *IEEE Transactions on Antennas and Propagation*, Vol. 53, No. 2, February 2005, pp. 645–654.

[63] Sheta, A. F., and S. F. Mahmoud, "A Widely Tunable Compact Patch Antenna," *IEEE Antennas and Wireless Propagation Letters*, Vol. 7, 2008, pp. 40–42.

[64] Costantine, J., et al., "Reconfigurable Antennas: Design and Applications," *Proceedings of the IEEE*, Vol. 103, No. 3, March 2015, pp. 424–437.

[65] Tawk, Y., J. Costantine, and C. Christodoulou, "Reconfigurable Filtennas and MIMO in Cognitive Radio Applications," *IEEE Transactions on Antennas and Propagation*, Vol. 62, No. 3, March 2014, pp. 1074–1083.

[66] Hussain, R., and M. S. Sharawi, "A Cognitive Radio Reconfigurable MIMO and Sensing Antenna System," *IEEE Antennas and Wireless Propagation Letters*, Vol. 14, 2015, pp. 257–260.

[67] Hussain, R., and M. S. Sharawi, "Wide-Band Frequency Agile MIMO Antenna System with Wide Tunability Range," *Microwave and Optical Technology Letters*, Vol. 58, No. 9, September 2016, pp. 2276–2280.

[68] Yang, X. X., S. S. Zhong, and S. C. Gao, "A Novel Polarization-Agile Active Micro Strip Antenna Array with LNA," *5th International Symposium on Antennas, Propagation, and EM Theory*, 2000, pp. 94–97.

[69] Zhong, S. S., X. X. Yang, and S. C. Gao, "Polarization-Agile Micro Strip Antenna Array Using a Single Phase-Shift Circuit," *IEEE Transactions on Antennas and Propagation*, Vol. 52, No. 1, January 2004, pp. 84–87.

[70] Gao, S., A. Sambell, and S. S. Zhong, "Polarization-Agile Antennas," *IEEE Antennas and Propagation Magazine*, Vol. 48, No. 3, June 2006, pp. 28–37.

[71] Leon, G., et al., "Novel Polarization Agile Micro strip Antenna," *IEEE Antennas and Propagation Society International Symposium*, 2008, pp. 1–4.

[72] Ferrero, F., et al., "A Novel Quad-Polarization Agile Patch Antenna," *IEEE Transactions on Antennas and Propagation*, Vol. 57, No. 5, May 2009, pp. 1562–1566.

[73] Vazquez, C., et al., "Receiving Polarization Agile Active Antenna Based on Injection Locked Harmonic Self Oscillating Mixers," *IEEE Transactions on Antennas and Propagation*, Vol. 58, No. 3, March 2010, pp. 683–689.

[74] Petosa, A., "An Overview of Tuning Techniques for Frequency Agile Antennas," *IEEE Antennas and Propagation Magazine*, Vol. 54, No. 5, October 2012, pp. 272–296.

[75] Wu, C. H., and T. G. Ma, "Pattern-Reconfigurable Self-Oscillating Active Integrated Antenna with Frequency Agility," *IEEE Transactions on Antennas and Propagation*, Vol. 62, No. 12, December 2014, pp. 5992–5999.

[76] Babakhani, B., S. K. Sharma, and N. R. Labadie, "A Frequency Agile Micro Strip Patch Phased Array Antenna with Polarization Reconfiguration," *IEEE Transactions on Antennas and Propagation*, Vol. 64, No. 10, October 2016, pp. 4316–4327.

[77] Singh, R. K., A. Basu, and S. K. Koul, "Asymmetric Coupled Polarization Switchable Oscillating Active Integrated Antenna," *Proceedings of the Asia-Pacific Microwave Conference*, 2016, pp. 1–4.

[78] Biebl, E. M., "Integrated Active Antennas on Silicon," *Microwave and Optoelectronics Conference*, 1997, pp. 279–284.

[79] Huang, K. K., and D. D. Wentzloff, "60GHz On-Chip Patch Antenna Integrated in a 0.13μm CMOS Technology," *Proceedings of 2010 IEEE International Conference on Ultra-Wideband*, 2010, pp. 1–4.

[80] Marnat, L., et al., "On-Chip Implantable Antennas for Wireless Power and Data Transfer in a Glaucoma-Monitoring SoC," *IEEE Antennas and Wireless Propagation Letters*, Vol. 11, 2012, pp. 1671–1674.

[81] Ou, Y. C., and G. M. Rebeiz, "Differential Microstrip and Slot-Ring Antennas for Millimeter-Wave Silicon Systems," *IEEE Transactions on Antennas and Propagation*, Vol. 60, No. 6, June 2012, pp. 2611–2619.

[82] Amadjikpe, A. L., et al., "Integrated 60-GHz Antenna on Multilayer Organic Package with Broadside and End-Fire Radiation," *IEEE Transactions on Microwave Theory and Techniques*, Vol. 61, No. 1, January 2013, pp. 303–315.

[83] Cheema, H. M., and A. Shamim, "The Last Barrier," *IEEE Microwaves Magazine*, January/February 2013, pp. 79–91.

[84] Jiang, L., J. F. Mao, and K. W. Leung, "A CMOS UWB On-Chip Antenna with a MIM Capacitor Loading AMC," *IEEE Transactions on Antennas and Propagation*, Vol. 59, No. 6, June 2012, pp. 1757–1764.

[85] Zihir, S., et al., "60GHz 64 and 256-Elements Wafer-Scale Phase-Array Transmitters Using Full-Reticle and Subreticle Stitching Techniques," *IEEE Transactions on Microwave Theory and Techniques*, Vol. 64, No. 12, December 2016, pp. 4701–4719.

[86] Song, Y., et al., "A Compact Ka-Band Active Integrated Antenna with a GaAs Amplifier in a Ceramic Package," *IEEE Antennas and Wireless Propagation Letters*, Vol. 16, 2017, pp. 2416–2419.

[87] Linvill, J. G., "Transistor Negative-Impedance Converters," *Proceedings of the I.R.E.*, 1953, pp. 725–729.

[88] Sussman-Fort, S. E., and R. M. Rudish, "Non-Foster Impedance Matching of Electrically-Small Antennas," *IEEE Transactions on Antennas and Propagation*, Vol. 57, No. 8, August 2009, pp. 2230–2241.

[89] White, C. R., J. S. Colburn, and R. G. Nagele, "A Non-Foster VHF Monopole Antenna," *IEEE Antennas and Wireless Propagation Letters*, Vol. 11, 2012, pp. 584–587.

[90] Zhu, N., and R. W. Ziolkowski, "Broad-Bandwidth, Electrically Small Antenna Augmented with an Internal Non-Foster Element," *IEEE Antennas and Wireless Propagation Letters*, Vol. 11, 2012, pp. 1116–1120.

[91] Tang, M. C., N. Zhu, and R. W. Ziolkowski, "Augmenting a Modified Egyptian Axe Dipole Antenna with Non-Foster Elements to Enlarge Its Directivity Bandwidth," *IEEE Antennas and Wireless Propagation Letters*, Vol. 12, 2013, pp. 421–424.

[92] Mirzaei, H., and G. V. Eleftheriades, "A Resonant Printed Monopole Antenna with an Embedded Non-Foster Matching Network," *IEEE Transactions on Antennas and Propagation*, Vol. 61, No. 11, November 2013, pp. 5363–5371.

[93] Alam, M. N., R. A. Dougal, and M. Ali, "Electrically Small Broadband VHF/UHF Planar Antenna Matched Using a Non-Foster Circuit," *Microwave and Optical Technology Letters*, Vol. 55, No. 10, October 2013, pp. 2494–2497.

[94] Elfrgani, A. M., and R. G. Rojas, "Biomimetic Antenna Array Using Non-Foster Network to Enhance Directional Sensitivity over Broad Frequency Band," *IEEE Transactions on Antennas and Propagation*, Vol. 64, No. 10, October 2016, pp. 4297–4305.

CHAPTER 6

A Codesign Approach for Designing AIAs

In the previous chapter, we saw the advantages and increasing importance of having AIAs in various applications. In the majority of these previous designs, the AIA was done with the 50Ω matching requirement between the active part and the antenna part. This requirement is not always a necessity, and optimized designs in terms of efficiency and lower complexity can be obtained for narrowband and wideband applications without satisfying this 50Ω matching condition. Making sure that complex conjugate matching of the impedances between the two stages is satisfied to ensure maximum power transfer is what a designer should aim for to obtain more efficient designs [1–6].

The dawn of the fifth generation (5G) of wireless communications dictates the use of highly dense antenna arrays that require optimized placement and integration within the massive multiple-input-multiple-output (mMIMO) technology. The use of integrated radio frequency (RF) front-ends and AIAs becomes mandatory [7, 8]. That gives the codesign methodology of AIAs presented in this chapter a high value for adaptation and use in future antenna systems.

In this chapter, we will present a general design procedure for this codesign approach, where the amplifier and the antenna are designed together (and not independently as is usually the case), and the effect of each of the two on the other is taken early in the design stages to obtain a design that is integrated with excellent matching and power transfer features. We will illustrate that this codesign approach can yield improved performance on single as well as multi-antenna designs, and is applicable to narrowband, wideband and ultrawideband (UWB) configurations of AIAs.

6.1 Detailed AIA Codesign Procedure

The codesign approach between the active and passive parts of an AIA has appeared some time back, but not in a unified and detailed design procedure that can be applied seamlessly. In [9], a codesign approach was provided by first extracting the *S*-parameter matrix from a passive structure and then feeding that into a harmonic balance-based circuit simulator to account for the effect of the passive component into the active amplifier design. The antenna was assumed to be perfectly matched (i.e., 50Ω) in this process.

Several works that focused on the AIA concept without limiting their focus on providing 50Ω matching between the different stages appeared in [1–3, 6] targeting rectifying antennas (rectennas) for energy harvesting applications. In these works, the impedance matrix of the antenna part was extracted using the field simulator, and then it was passed to the harmonic balance-based simulator for the active (rectifier) circuit design. This provided better efficiency values as compared to the standard 50Ω real matching techniques. Since these works targeted rectenna designs for energy harvesting applications, they did not consider having amplifiers within their active parts. Also, none of them considered any added antenna miniaturization due to active integration, thus the sizes of the antennas and designs were rather large.

The codesign procedure of AIAs has been applied to several antenna designs of various types and bandwidths with a unified approach in [4, 5]. Figure 6.1 shows the block diagram summarizing the design process. The process utilizes two computer-aided design (CAD) packages, one for EM simulations (i.e., HFSS, CST, IE3D) and the other for microwave circuit simulations (ADS, Microwave Office [MWO]). The antennas are modeled within the EM simulation package, and then the frequency dependent *S*-parameters are extracted and passed to the circuit simulation tool. Several optimization and analysis steps are performed, after which the antenna can be optimized. The antenna is then integrated within a cosimulation model for an overall assessment of the system behavior consisting of the AIA. The design procedure is as follows:

1. Model the desired antenna type according to the design specifications using the EM field solver package/software. After making sure it is suitable at the desired frequency band(s), extract its *S*-parameter data using an appropriate

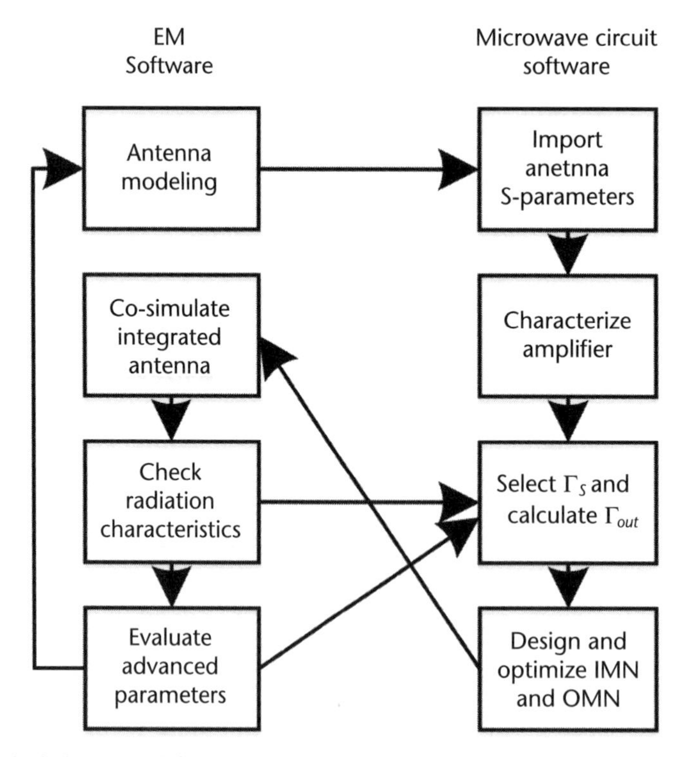

Figure 6.1 Block diagram of the AIA codesign procedure.

format compatible with your microwave circuit simulator tool/software. However, the suitability of the antenna at the desired frequency band can be observed by its bandwidth potential, radiation efficiency, and radiation gain. The single-port S-parameter curve obtained represents the complex input reflection coefficient of your antenna (Γ_{ant}) as a function of the swept frequency.

2. Pick a suitable transistor/amplifier and optimize its performance in terms of gain or NF (see Chapter 3 for additional information on this). A trade-off is always chosen between these two requirements, and based on the selected point, the source reflection coefficient (Γ_s) is determined. The amplifier might need to be designed to cover more than the required antenna bandwidth. Using the S-parameters of the unmatched transistor/amplifier (S_{11}, S_{21}, S_{12}, and S_{22}) that are obtained from the microwave circuit simulation package, we can find the amplifier output reflection coefficient (Γ_{out}) using

$$\Gamma_{\text{out}} = S_{22} + \frac{S_{12}S_{21}\Gamma_s}{1 - S_{11}\Gamma_s} \tag{6.1}$$

3. Upon determination of Γ_s and Γ_{out}, input and output matching networks can be designed and optimized. Chapter 2 provides the details for designing such a matching network. The input matching network (IMN) should match the source (i.e., 50Ω) to the amplifier input (Γ_s), and the output matching network (OMN) should match the antenna input (Γ_{ant}) to the output of the amplifier (i.e., $\Gamma_L = \Gamma^*_{\text{out}}$) as shown in Figure 6.2.
4. In the integrated antenna system, the antenna input port is not directly accessible, and thus its reflection coefficient Γ_{ant} (integrated) cannot be measured, but we can calculate it from the impedance of the matching network (MN) (Z_{MN}) and its own impedance (Z_{ant}) according to

$$\Gamma_{\text{ant}}(\text{integrated}) = \frac{Z_{\text{ant}} - Z^*_{MN}}{Z_{\text{ant}} + Z_{MN}} \tag{6.2}$$

5. To obtain the gain and NF of the integrated system shown in Figure 6.2 (i.e., a transmitting AIA), a similar process can be adopted for a receiving AIA; most microwave circuit simulators require a two-port model for the various

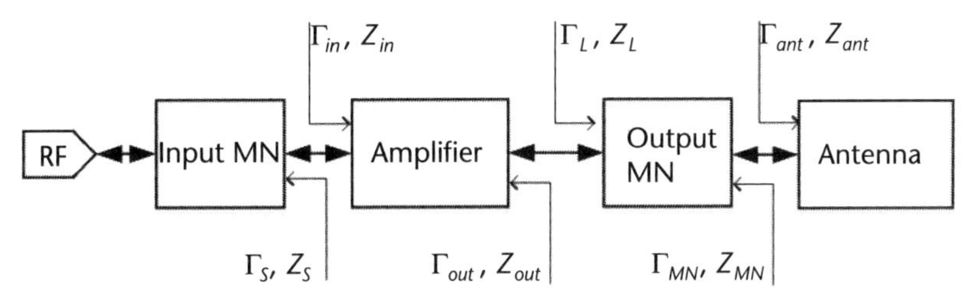

Figure 6.2 Transmitting AIA block diagram.

parts of the circuit. This can be done at a single frequency if we replace the antenna with its equivalent impedance at that frequency. However, this will allow the analysis at a single frequency point only. To be able to sweep the response and have it over a range of frequencies, some techniques exist to convert the single-port antenna model into a two-port one as outlined in [10]. Another option is to use modern tools that have the integration capability within them such as using CST and ADS. ADS will provide the *NF* analysis and network gain, while CST will assess the total efficiency and radiated fields.

6. The last step is evaluating the radiation behavior of the integrated system in terms of total efficiency, radiated patterns, gain, and the like, and if further optimization is required, the matching network design is repeated starting from step 2 by selecting another Γ_s as in step 1. Otherwise, if the performance requires a modified antenna design, then you should start from step 1 again.

The generic procedure above can be applied on any AIA design irrespective of its antenna type, operating frequency, or application. As an illustration of the versatility of this technique, we will apply this procedure on narrowband monopole and patch antennas, wideband monopole and patch antennas, and UWB monopole antennas in the sections that follow. In addition, for the UWB case, we will consider a multi-antenna (MIMO) system and see how this method can be extended and used to optimize an integrated MIMO antenna system by taking the coupling between adjacent elements into the analysis.

6.2 Narrowband AIA Codesign Examples

A narrowband patch based AIA was introduced in Section 5.1.6 that was used to demonstrate the performance metrics of such integrated systems. In this section, we will present another AIA design of a receiving antenna system consisting of a monopole antenna and an LNA that are designed using the codesign approach highlighted in Section 6.1. We will show that active integration can sometimes be useful when considering antenna miniaturization if the antenna efficiency is not degraded much at lower bands. This is a function of the antenna structure and type, and should be checked by the designer to ensure that miniaturization will not severely affect the radiation efficiency as in electrically small antennas (ESAs) this is an inevitable consequence [11]. An example that is based on a narrowband patch antenna was designed using the codesign approach in [5] for the readers to check as well.

Let us consider the antenna geometry shown in Figure 6.3. A monopole antenna is designed and optimized to resonate within 3.2–3.8 GHz. An RO4350B substrate is used with 0.76 mm thickness and ε_r = 3.48 with tan δ = 0.004. Figure 6.3(a) shows the simulation model in CST, and the fabricated prototype is presented in Figure 6.3(b). All dimensions are in given in millimeters. The antenna was fed using an SMA connector from the board edge. The simulated and measured reflection coefficient curves are compared in Figure 6.3(c), and very good agreement between simulations and measurements is observed. The simulated radiation efficiency as well as the Q-factor are shown in the same figure. It should be noted that for this

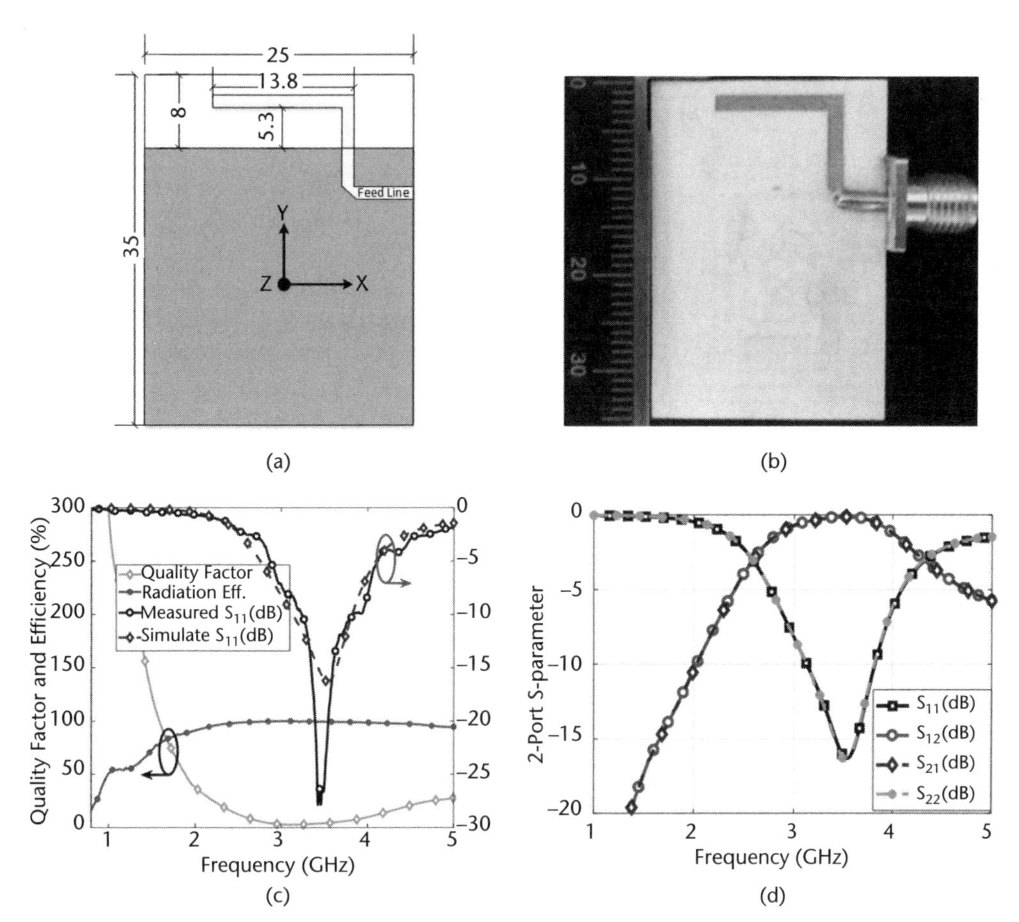

Figure 6.3 Monopole antenna geometry and response: (a) detailed geometry, (b) fabricated prototype, (c) simulated and measured responses, and (d) two-port model.

particular antenna, the radiation efficiency has a wideband response, and thus this antenna can be miniaturized with minimum effect on its radiation efficiency using the codesign approach, but at the expense of the narrower operating bandwidth (i.e., higher Q-factor). To conduct the codesign approach highlighted in Section 6.1, we need to convert the single-port antenna into a two-port network to be able to include its effect in the microwave circuit simulations with the amplifier design using ADS or MWO. The process highlighted in [10] is used to extract the two-port *S*-parameters of the antenna, and its two-port parameters are shown in Figure 6.3(d). After obtaining the two-port network representation of the antenna, we can move on to the amplifier design.

We are interested to design an LNA using the ATF 53189 transistor model [12]. Following the recommended data sheet, Figure 6.4(a) shows the schematic of biasing network of the amplifier with stability network. The design is optimized to operate between 1 and 3 GHz. The following parameter values are used: $R_{\text{stab}} = 100\Omega$, $C_{\text{stab}} = 4.7$ pF, $C_1 = C_2 = 1$ nF, $L_1 = L_2 = 25$ nH, $V_{DS} = 3.5$V, $V_{GS} = 0.7$V, and $I_{DS} = 110$ mA. The obtained response curves are shown in Figure 6.4(b) for the gain, *NF* and stability metrics (K and $|\Delta|$).

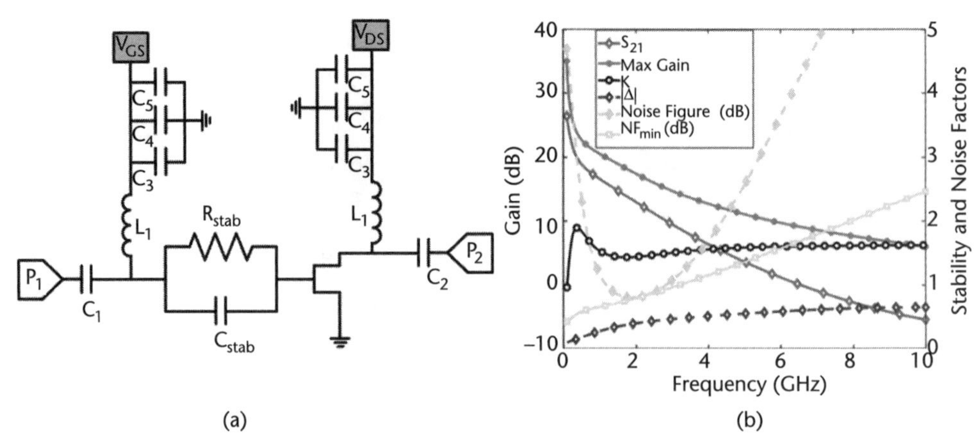

Figure 6.4 Amplifier design using the ATF53189 transistor covering 1–3 GHz: (a) biasing and stability networks, (b) response curves.

The codesign approach starts by setting the band of interest, which is in this example 1.85–1.91 GHz, covering several LTE subbands in the current 4G standard. The two-port antenna model is to be integrated with the LNA design via an IMN, for tracking the complex impedances without any 50Ω restriction, while the output of the LNA should be matched to 50Ω to be connected to the transmission line for testing via an OMN. This is conducted in the microwave circuit simulation tool (ADS). This is shown in Figure 6.5(a) as a block diagram that also highlights the various parameters that are considered in the design process. Next, the gain and noise circles of the LNA are plotted (as in Figure 6.5[b]) and a Γ_s is selected based on the balance between the *NF* and gain. The corresponding Γ_{out} is calculated using (6.1). The IMN will match the antenna impedance to the complex input impedance of the LNA, and the OMN will match the complex impedance of the LNA to 50Ω. The complete circuit is shown in Figure 6.6(c). The OMN used is of a distributed topology (i.e., transmission line sections) for better accuracy (as lumped components have tolerances), while the IMN was of a mixed topology utilizing capacitors and transmission lines to reduce its size (other solutions using lumped components can be done as well as presented in Chapter 2).

The widths of all the transmission lines were 1.7 mm (depending on the substrate thickness and material properties to have 50Ω), and the lengths were: TL_1 = 2.42 mm, TL_2 = 4.57 mm, TL_3 = 1.34 mm, TL_4 = 1.56 mm, TL_5 = 9.72 mm, TL_6 = 10 mm, TL_7 = 13.22 mm, and TL_8 = 7.15 mm. C_1 to C_5 capacitor values were like the ones in Section 5.1.6, while C_6 = 20 pF and C_7 = 2.2 pF for this design.

The results of the codesigned LNA-based AIA are shown in Figure 6.6. The gain and *NF* results are taken from ADS and are shown in Figure 6.6(a). It is clear that the design does provide low *NF* in the band of interest with gain levels around 15 dB. The antenna response can also be deduced using (6.2) and the result of this is plotted in Figure 6.6(b). It is clear that it operates within a narrow bandwidth of around 50 MHz (a consequence of miniaturization), and the transmission coefficient (available due to the two-port model of the antenna) shows little loss within it (i.e., good radiation efficiency). The ADS circuit response is then integrated within

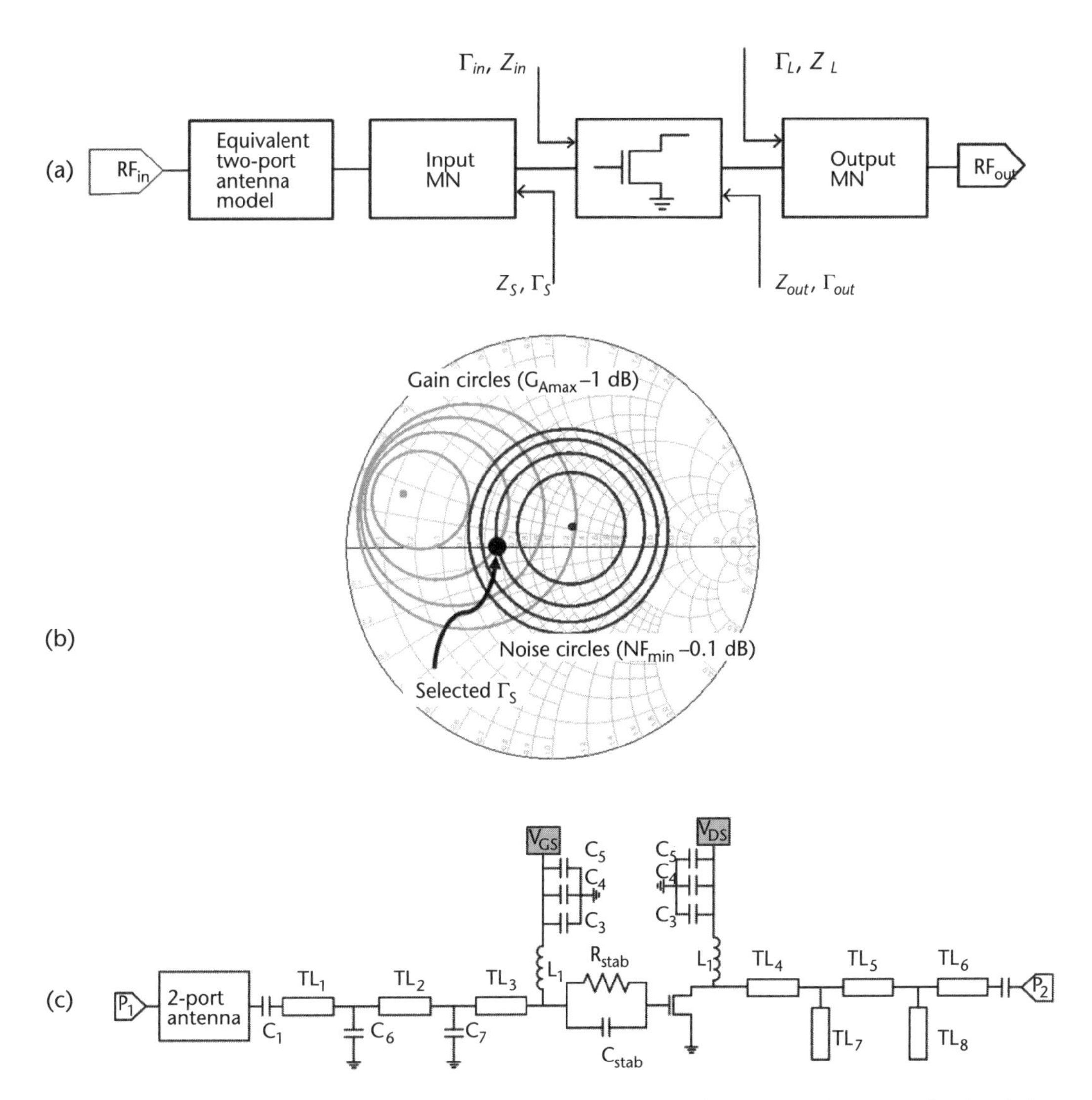

Figure 6.5 LNA-miniaturized monopole-based AIA design: (a) block diagram, (b) gain and noise circles of amplifier, and (c) final design of the AIA circuit.

the full-wave simulation tool (i.e., CST), and the effect of the complete system on the radiated fields is obtained. The elevation cut at $\varphi = 0°$ is shown in Figure 6.6(c) while the elevation cut at $\varphi = 90°$ is shown in Figure 6.6(d) (refer to Figure 6.3 for antenna orientation). It is clear how the AIA response provides a significant boost in gain as compared to the passive stand-alone antenna.

The reader should note that the design presented is based on the linear operation region (small-signal model) of the amplifier. When power amplifiers (PAs) with nonlinear region operations are to be considered, load-pull analysis and the design of the PA should be first conducted to optimize its power response as mentioned in Chapter 3 considering the antenna behavior. Only few works touched upon this area such as [8], but a complete codesign approach applied to nonlinear PA is expected to be more involved and is a topic of further investigation.

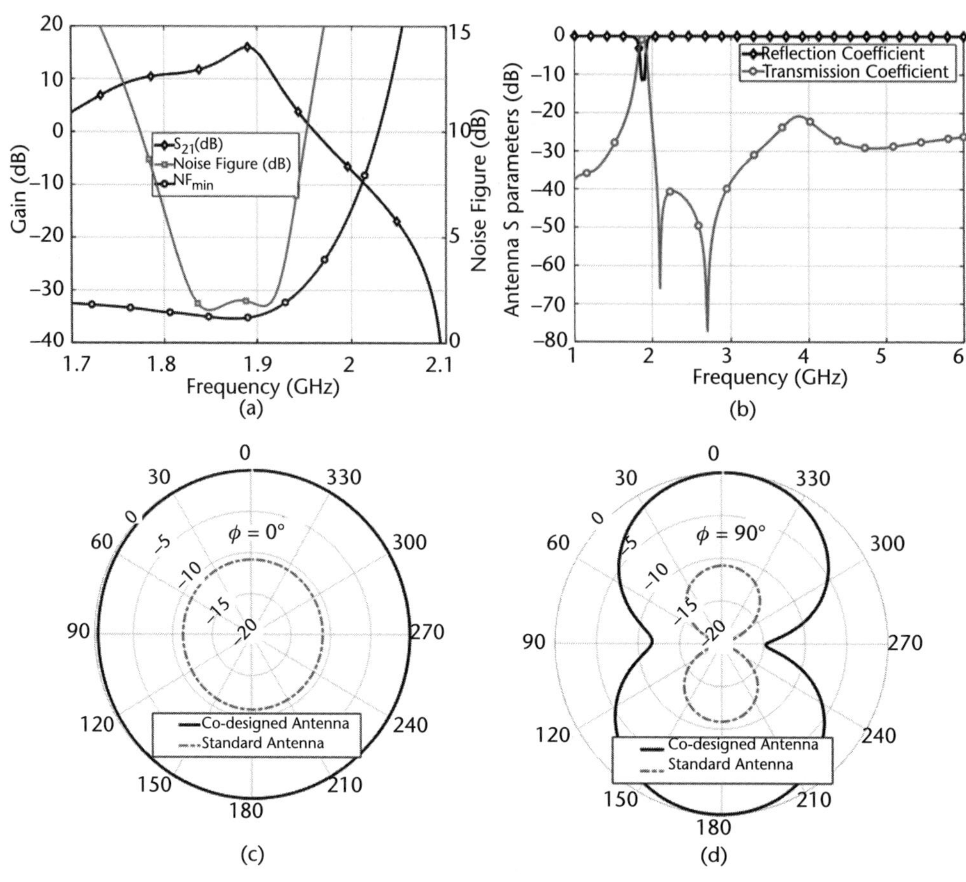

Figure 6.6 Final results of the designed LNA monopole AIA: (a) gain and *NF*, (b) reflection coefficient, (c) normalized realized gain elevation cut at $\varphi = 0°$, and (d) normalized realized gain elevation cut at $\varphi = 90°$.

6.3 Wideband AIA Codesign Examples

In this example, we will redesign the same monopole antenna presented in Section 6.2, but at a higher frequency than the one used before, in order to gain some more bandwidth. The configuration now is a transmitting AIA. The overall AIA bandwidth is being controlled by the antenna bandwidth (the narrower one) as we saw in Chapter 5. The target band in this design example is 2.3–2.6 GHz, with a center frequency around 2.45 GHz.

Since we are using the same antenna as well as the sample amplifier model as in Section 6.2, and Figures 6.3 and 6.4 show the antenna as well as the amplifier structures and responses. We start after obtaining the antenna two-port model, and importing it within ADS. The block diagram we are proposing is the one shown in Figure 6.7(a) for a transmitting AIA. The gain and noise circles are obtained from ADS and shown in Figure 6.7(b). For a transmitting AIA, the system gain is of primary interest. However, it is still important to see the system noise performance. To do so, a Γ_L point is selected for the desired gain and related noise performance based on the Γ_s value according to (6.3) and noise circles. With this method, the optimum gain of the system can be selected while observing the variation the system noise performance.

$$\Gamma_s^* = \Gamma_{\text{in}} = S_{11} + \frac{S_{12}S_{21}\Gamma_L}{1 - S_{11}\Gamma_L} \tag{6.3}$$

The trace of the antenna input impedance is also shown on the Smith chart in Figure 6.7(b). Once the Γ_L and consequently, the Γ_s are decided, the OMN is used to match the antenna impedance to the Γ_L value selected and the IMN is found to satisfy $\Gamma_{\text{in}} = \Gamma_S^*$. This is performed at the center frequency of 2.45 GHz and optimized in the range of 2.3–2.6 GHz.

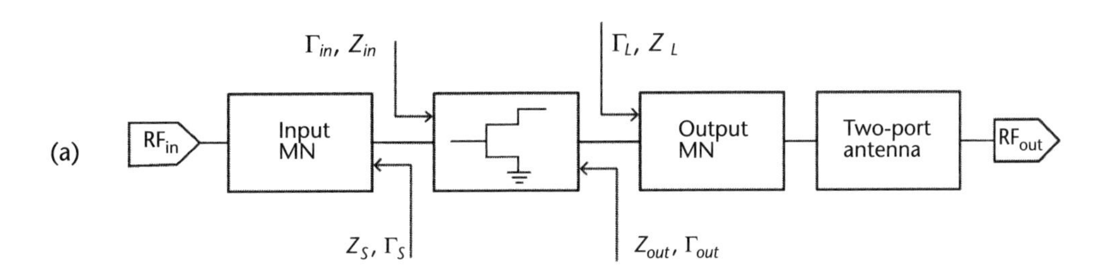

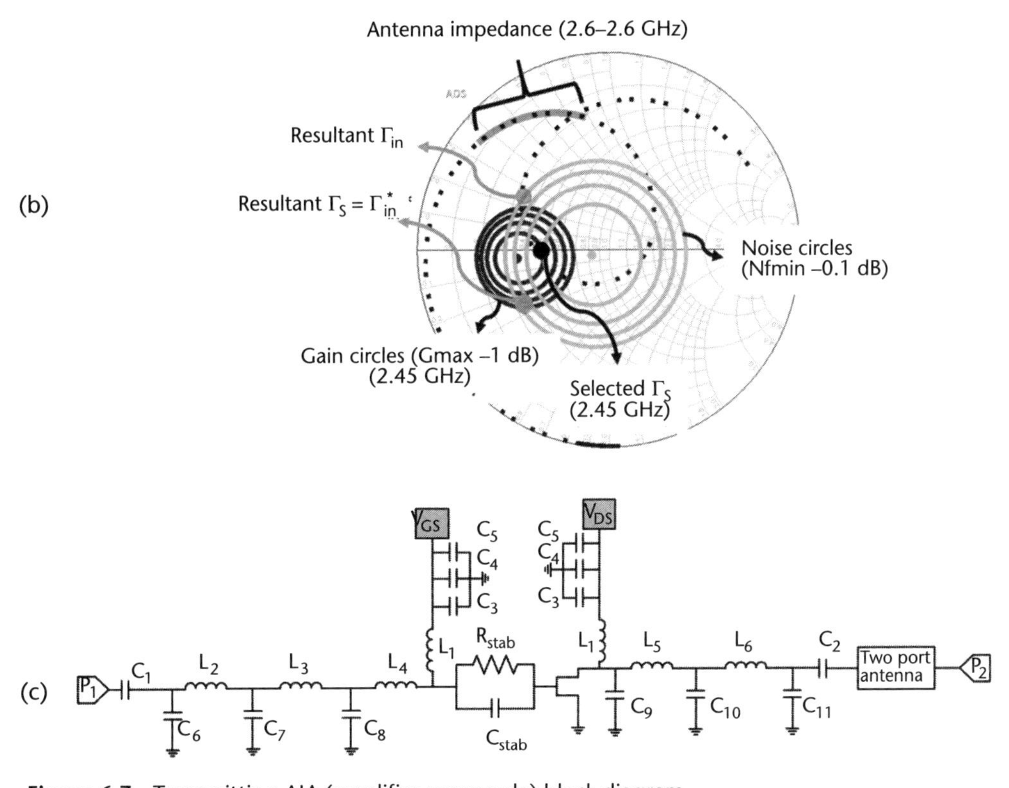

Figure 6.7 Transmitting AIA (amplifier-monopole) block diagram.

In ADS, the IMN and OMN (Figure 6.7[c]) are optimized to provide a flat gain in the desired operating band of 2.3–2.6 GHz, that the antenna can support. This will also result in antenna miniaturization as the original antenna operates between 3.2 and 3.8 GHz as was shown in Section 6.2. The circuit can also be optimized for harmonic suppression as the antenna effect is also integrated within the codesign approach (via the two-port model). A complete model in ADS is shown in Figure 6.8 with all the parameters shown. Several iterations of the optimization process can take place within ADS at this step to optimize the various parameters with all components placed and their effect is taken into consideration.

The obtained response curves for the transmitting AIA designed are shown in Figure 6.9. It is clear that we have a flat gain response from 2.3–2.6 GHz, with 15.6 dB of gain. The *NF* is below 2 dB, and the matching is acceptable (i.e., below −10 dB) with good margin at the input and output of the AIA. To assess the radiation behavior, the optimized amplifier and matching network designs are imported into CST as shown in Figure 6.10(a). The radiation patterns obtained are shown in Figures 6.10(b, c) for the elevation ($\varphi = 0°$) and elevation ($\varphi = 90°$) cuts, respectively. The effect of active integration is obvious when compared to the passive antenna response alone. Note that the gain is more than 15 dBi as the matching increases the power delivered in the case of the properly matched AIA, while for the stand-alone passive antenna, extra losses will be encountered due to mismatches.

Another example of a wideband AIA codesign based on a patch antenna is presented in [5]. The AIA response is again limited by the response of the radiating element. The bandwidth of the patch is increased via the introduction of a slot in its structure. The AIA covered the band 2.45–2.6 GHz and was integrated with an LNA instead of a transmitting amplifier (or a linear power amplifier). The obtained efficiency of the AIA based on the wideband patch was approximately 50%, with a miniaturization of almost 50%. This is a great feature of active integration, as the AIA presented in [5] was built on an FR-4 substrate, and thus the efficiency of the patch is not expected to exceed 65% at best. The efficiency was slightly compromised after the miniaturization, and the active integration took care of the complex impedance of the antenna that results from miniaturizing the antenna.

6.4 UWB AIA Codesign Approach

In Chapter 4, we introduced MIMO antenna design and its importance in current 4G and future 5G communication systems. MIMO antennas will be integrated in all future mobile terminals as it is a confirmed technology for all future wireless standards. We addressed the challenges and issues that accompany MIMO antenna designs, and in Chapter 4 we presented several solutions. The major design requirement that should always be satisfied is having high-port isolation between adjacent antenna elements to ensure high power efficiency, and to ensure low ECC values via tilting the beams spatially to have uncorrelated channels (streams) and then achieve high capacities.

In this section, we will apply the codesign approach of AIAs to the design of an UWB two-element MIMO antenna system. We will demonstrate how such integration is advantageous and can provide better performance than the conventional

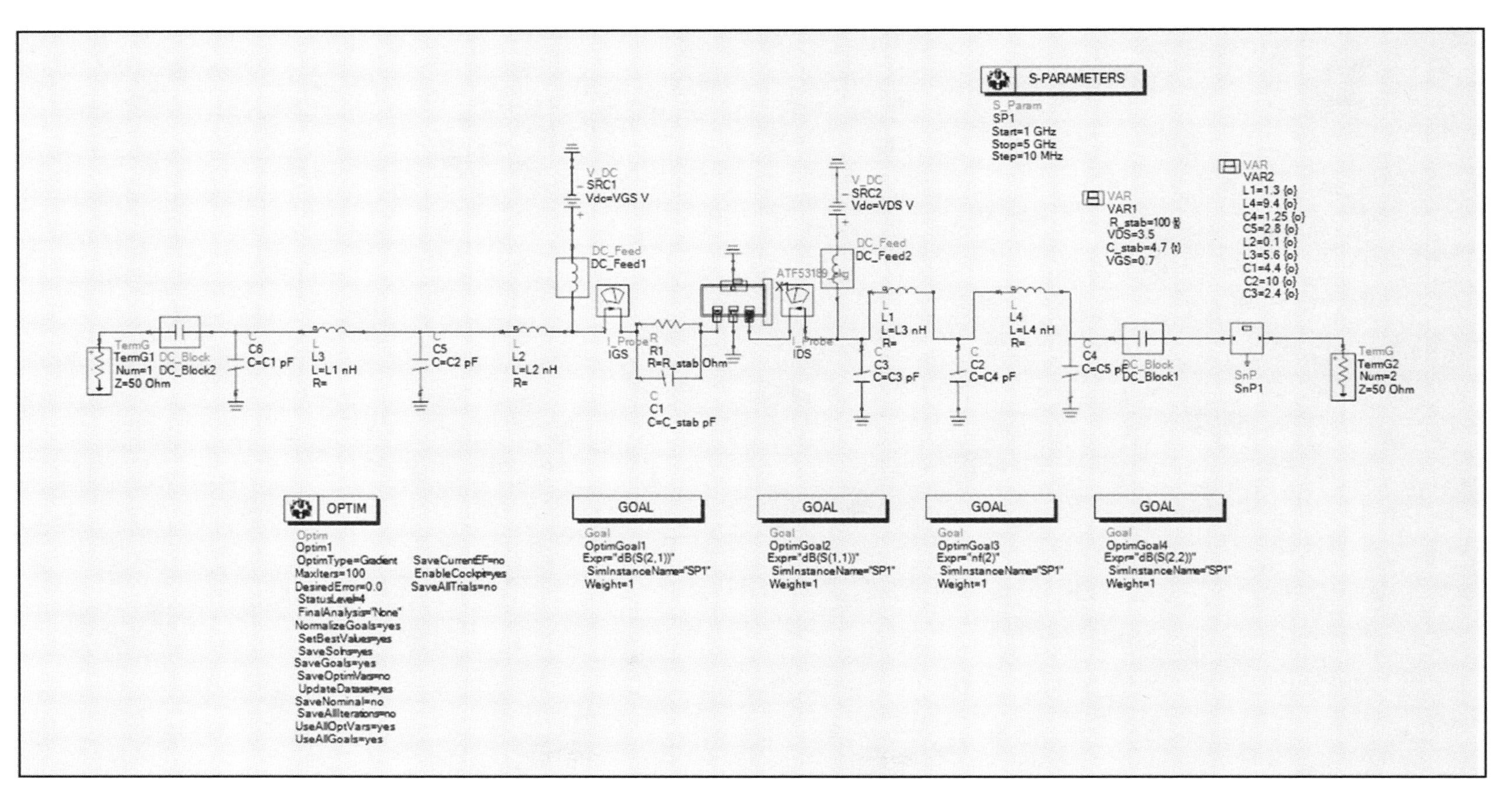

Figure 6.8 ADS schematic for the transmitting AIA simulation/optimization.

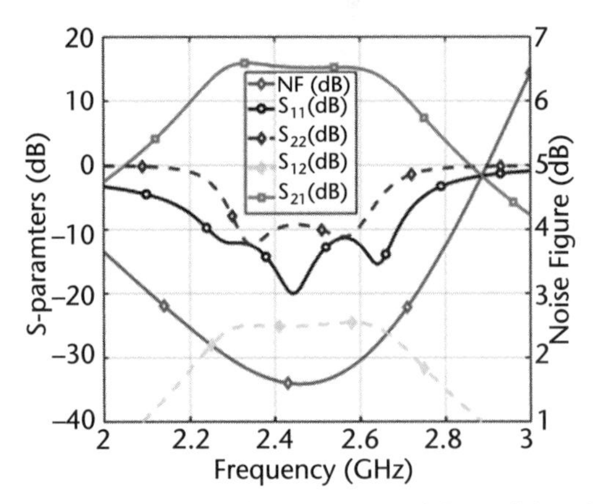

Figure 6.9 Reflection coefficient, gain, and *NF* response of the wideband transmitting AIA.

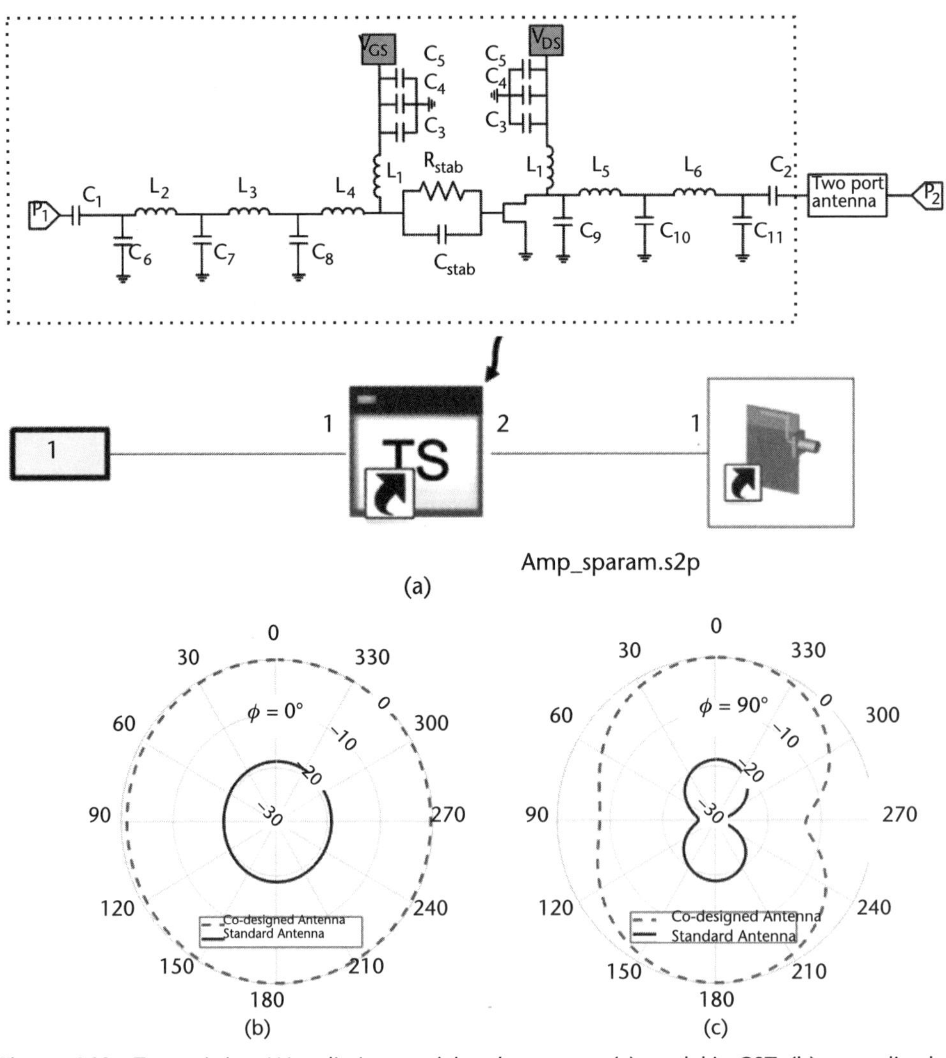

Figure 6.10 Transmitting AIA radiation model and response: (a) model in CST, (b) normalized realized gain elevation cut at $\varphi = 0°$, and (c) normalized realized gain elevation cut at $\varphi = 90°$.

50Ω design procedures when it comes to better isolation between adjacent ports and ECC values of the complete integrated system. The design presented here follows that in [4] presented by the authors.

Figure 6.11(a) shows the geometry of the two-element semi-ring MIMO-based antenna system. The antennas are modeled using CST on an RO4350B dielectric substrate with thickness of 0.76 mm, $\varepsilon_r = 3.5$, and tan $\delta = 0.004$. A single antenna element size occupies 12 × 24 mm^2. A substrate size of 50 × 90 mm^2 is used to represent a small smart phone backplane and to place the two-antenna as close as possible representing a practical scenario. First, the MIMO antenna system is fabricated and its S-parameters are measured in order to obtain the two-port model of the two-antenna element system. Note here that the coupling between the two

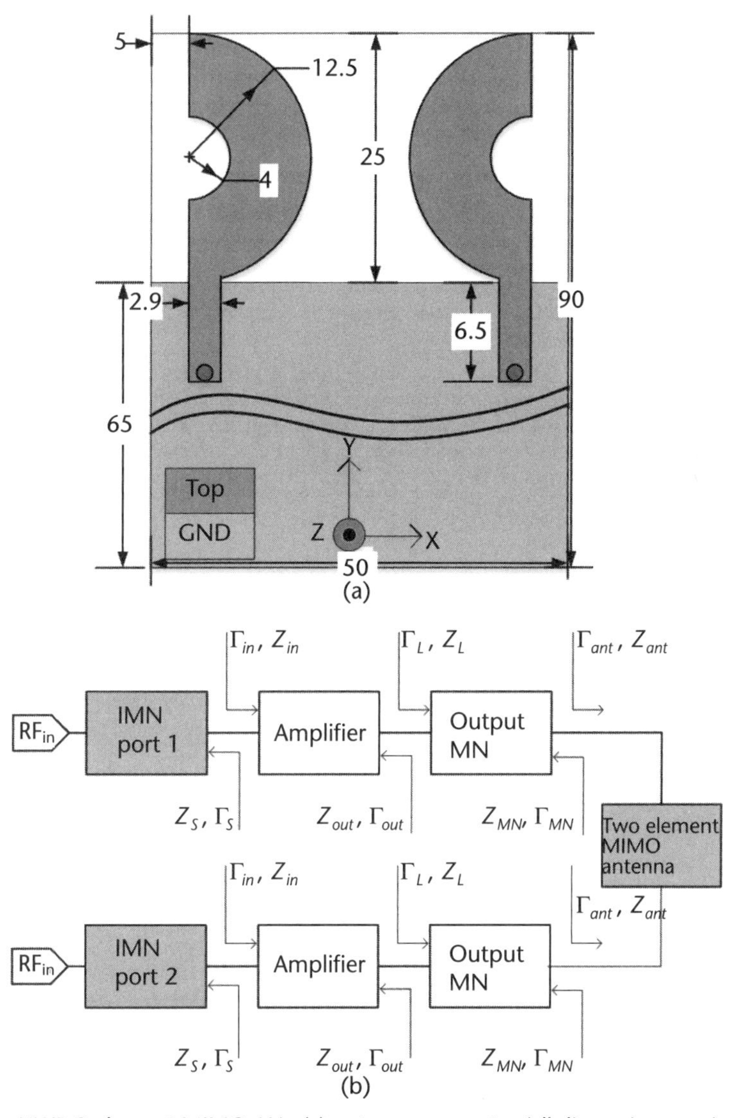

Figure 6.11 UWB 2-element MIMO AIA: (a) antenna geometry (all dimensions are in millimeters), and (b) AIA block diagram.

antennas is considered in the two-port model since currently the two-element MIMO antenna system represents a two-port block as shown in Figure 6.11(b).

The antenna system performance is shown in Figure 6.12. It is observed that the matching of the two antenna elements is not very good between 1.8 and 5.8 GHz (UWB) and the port isolation needs improvement. The discrepancies between measured and simulated models are due to hand soldering of the connectors and the tolerances of the parts used, but this will not affect the codesign method or results much and the advantage will still be clear at the end. The measurements were made using 50Ω connectors. We will proceed with the active integration codesign procedure and see that the matching as well as isolation are improved for the overall system as a direct consequence of the procedure and no need for individual fine tuning and tedious optimization steps.

For the amplifier design part, we use a commercially available wideband (1–6 GHz) RF amplifier (GVA 63+ from mini-circuits) [13]. Initially, we need to check the characteristics of the amplifier and use its measured values in the codesign approach. A test circuit is designed to characterize and measure the amplifier *S*-parameters (Figure 6.13). The test board used was also made on an RO4350B as shown in Figure 6.13(b). High gain was observed, and *NF* and stability calculations followed. The minimum *NF* was in the range of 3.5–4 dB and the amplifier was unconditionally stable. The measured *S*-parameter curves obtained are shown in Figure 6.13(c).

We now codesign the AIA UWB MIMO antenna system via integrating the amplifiers (transmitting mode) and the UWB MIMO antenna system. Note that matching the antenna complex impedance to Γ_L might not yield the optimum matching performance in terms of radiation efficiency and bandwidth. Thus, the optimization process is accomplished via designing the OMN such that the optimum Γ_{ant} (integrated) is achieved by using (6.2). This is different from the conventional procedure by considering Γ_{ant} as the primary design parameter along with the amplifier gain and noise figures. With this in mind, the best (optimum) antenna reflection matching and efficiency are achieved, which is crucial in such AIA designs.

To demonstrate the difference between the codesign approach and that of a standard 50Ω cascaded one, a simple comparison is shown in Figure 6.14. An amplifier was designed based on 50Ω output matching in Figure 6.14(a) following conventional methods. Then, the amplifier was directly connected in cascade with the UWB. The curves show the response of the antenna reflection coefficient (Γ_{ant}, refer to Figure 6.11) of the antenna alone and then when the amplifier is connected.

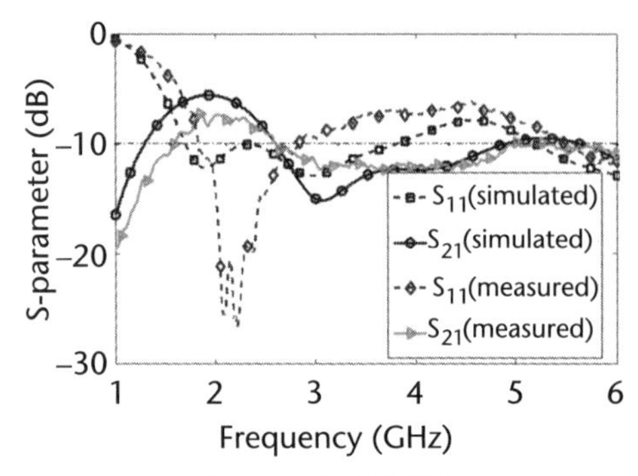

Figure 6.12 *S*-parameters of the two-element UWB MIMO antenna system.

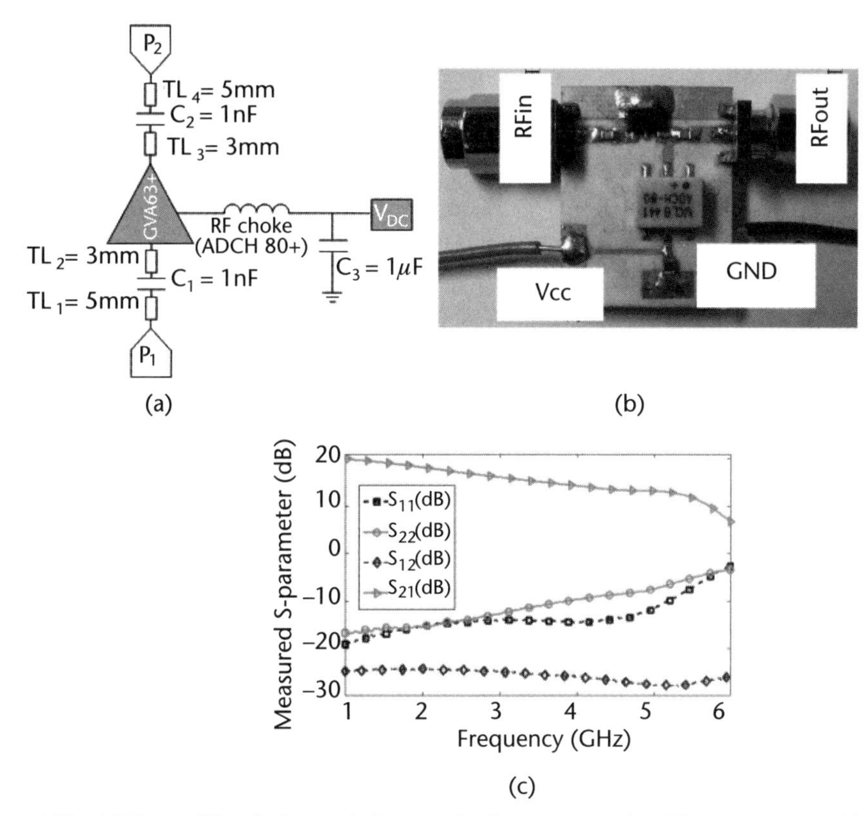

Figure 6.13 UWB amplifier design and characterization connected to 50Ω connectors: (a) test circuit, (b) prototype, and (c) *S*-parameters.

A noticeable degradation in the range between 2.3 and 4 GHz is seen at the antenna input due to the mismatch with the amplifier that was optimized for 50Ω operation. This degradation will affect the overall system efficiency and bandwidth. This mismatch is due to the fact that the two subsystems (i.e., amplifier and antenna) are no longer connected to the reference 50Ω, but rather to each other input impedances that can deviate from this reference value, which will not yield optimal matching (according to the −10-dB reference point) in the band of interest. Some multiband systems relax this to a −6-dB point for matching, but still, the performance can be

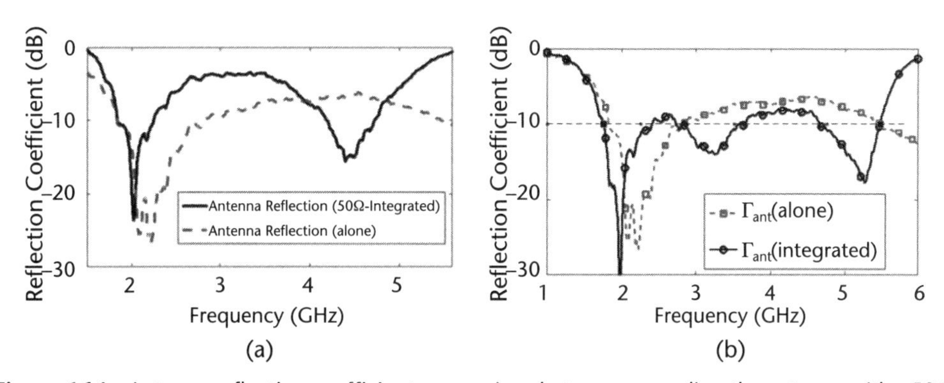

Figure 6.14 Antenna reflection coefficient comparison between cascading the antenna with a 50Ω matched amplifier in (a) versus codesign approach in (b).

noticeably degraded. Figure 6.14(b) shows a much better matching and bandwidth response when the codesign approach is followed targeting the optimization of Γ_{ant} (integrated) as highlighted in Section 6.2.

The final circuit of the UWB AIA MIMO with its IMN and OMN obtained using the codesign procedure is shown in Figure 6.15. The component values on the two paths are similar due to the symmetry of the design. The values are $C_1 = C_4 = 0.1$ pF, $C_2 = 0.4$ pF, $C_3 = 0.7$ pF, $L_1 = 1$ nH, $L_2 = 0.5$ nH, the transmission line lengths are $TL_1 = 10$ mm, $TL_2 = TL_3 = TL_4 = TL_5 = 3$ mm. All line widths are 1.7 mm (RO4350B dielectric substrate with thickness of 0.76 mm, $\varepsilon_r = 3.5$, and $\tan \delta = 0.004$). Although a prematched amplifier is used in this example, the same design process (the generic one highlighted in Section 6.2) is always applicable to synthesize the IMN and OMN within the codesign approach. Now, when considering Figure 6.15(b), the impedance matching via the codesign approach has improved for the overall system. This is one of the advantages. In addition, the complete fabricate prototype of the complete system in Figure 6.15(a) is shown in Figure 6.16(a, b). The measured S-parameters of the complete system through ports

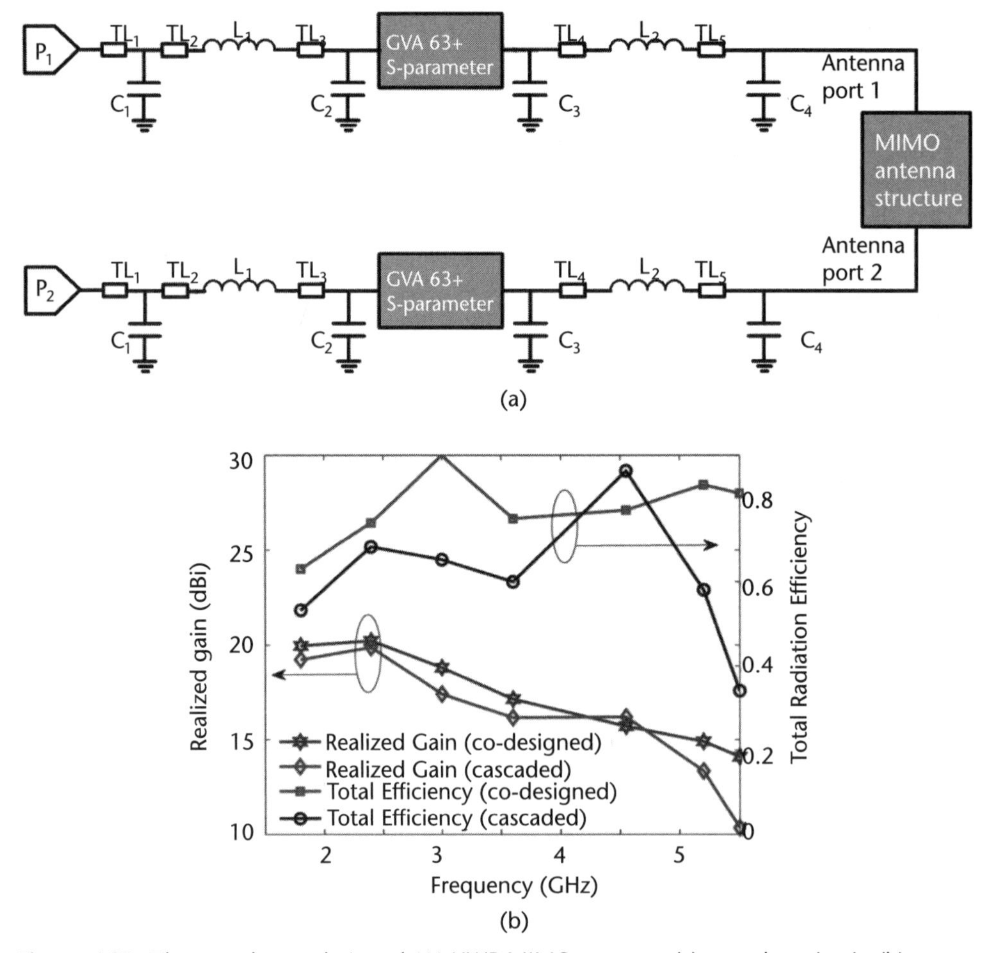

Figure 6.15 The complete codesigned AIA UWB MIMO antenna: (a) complete circuit, (b) realized gain and efficiency compared to a cascaded 50Ω-based design.

1 and 2 are presented in Figure 6.16(c). It can be clearly observed that the system port isolation has now improved by several decibels (around 2 GHz) because of this codesign approach and active integration. The closed-loop gain can be used to verify this improvement. In other words, the port isolation can be defined as in (6.4) where the loss of the matching networks can be neglected for simplification. Since the isolation by the amplifier is greater than the gain, the port isolation is generally greater than the antenna isolation. Such improved isolation has a direct effect of the power transfer efficiency and reduction of coupling between the paths within a MIMO antenna configuration.

$$\begin{aligned} S_{21(P1,2)} &= S_{21(MN1)} + S_{21(AMP1)} + S_{21(OMN1)} + S_{21(\text{ant})} \\ &\quad + S_{12(OMN2)} + S_{12(AMP2)} + S_{12(OMN2)} \\ &= S_{21(AMP1)} + S_{21(\text{ant})} + S_{12(AMP2)} \end{aligned} \tag{6.4}$$

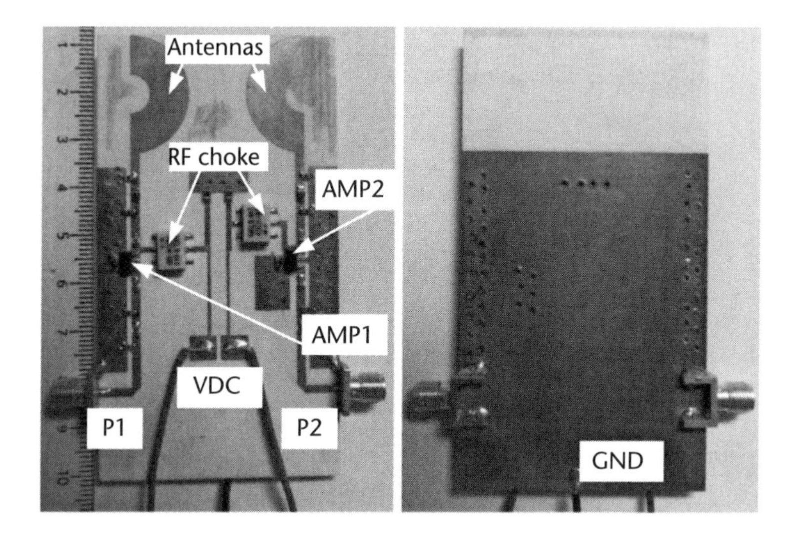

(a)

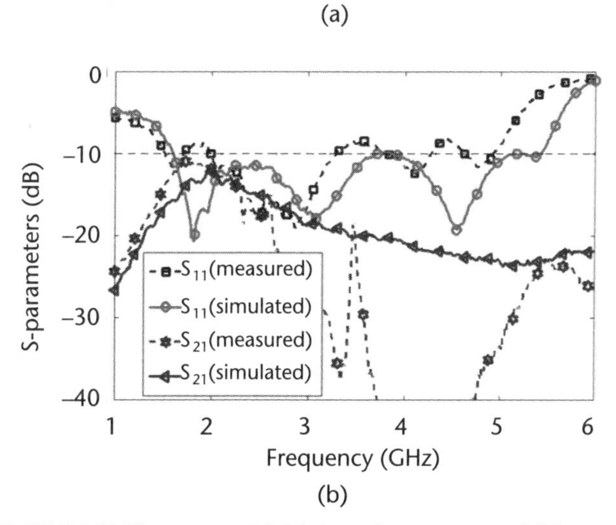

(b)

Figure 6.16 AIA UWB MIMO antenna: (a) fabricated prototype and (b) measured and simulated *S*-parameters.

To be able to assess the amplifier gain within the integrated system, the complete path is simulated in ADS when port 2 of the amplifier is connected to the antenna input port. The integrated antenna gain was found to be more than 13 dB in the entire operating bandwidth between 1.8 and 5.5 GHz. The final design validation step is to check the obtained radiated fields and efficiencies of such an AIA. In addition, since this is a MIMO configuration, the ECC should be evaluated from the radiated fields. Cosimulation using CST was performed and the maximum realized gain (with matching effects) was found to be higher than 14 dBi in the covered bands. This high gain is a direct consequence of the amplifier gain as well as the proper matching between the various parts of the system using the codesign approach. The total radiation efficiency was more than 60% over the entire operating band which is considered of high practical value. The measured normalized gain patterns were obtained from a Satimo-Star lab chamber and are shown in Figure 6.17. The radiated fields show the responses at 2.4 GHz and 5.2

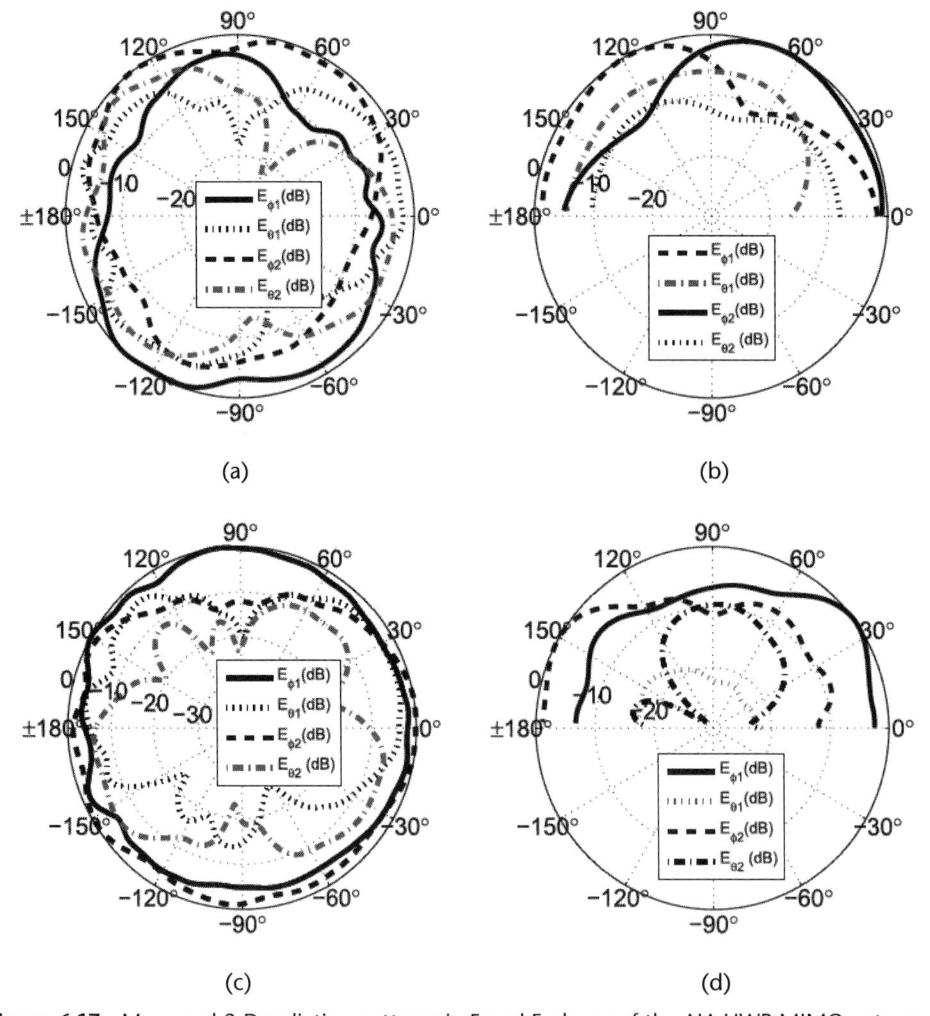

Figure 6.17 Measured 2-D radiation patterns in E and E planes of the AIA UWB MIMO antenna: (a) $\varphi = 0°$ plane cut at 2.4 GHz, (b) $\theta = 90°$ plane cut at 2.4 GHz, (c) $\varphi = 0°$ plane cut at 5.2 GHz, and (d) $\theta = 90°$ plane cut at 5.2 GHz. (*From:* [4]. © 2016 IEEE. Reprinted with permission.)

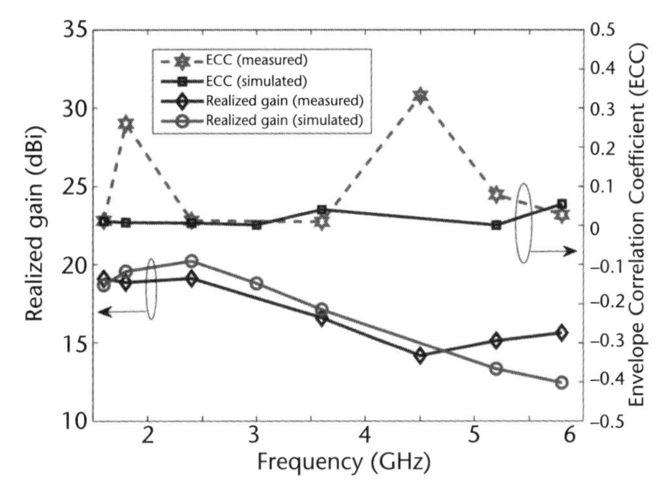

Figure 6.18 Measured and simulated realized gain and ECC curves. (*From:* [4]. © 2016 IEEE. Reprinted with permission.)

GHz, respectively. The two-dimensional (2-D) plane cuts over the elevation angles at $\varphi = 0°$ and azimuth ones at $\theta = 90°$ are shown for the two bands. Note the tilt in the patterns that will yield enhanced ECC performance.

The maximum simulated and measured gains as well as ECC values from the 3-D complex radiation patterns are shown in Figure 6.18. It is clear that the modeled and measured curves are in very good agreement. The discrepancy in the ECC curve values can be attributed to the measurement setup where the cable and the mast holding the antenna are present, and in the simulation model they are not; thus, they might have affected the fields at certain frequencies.

6.5 Conclusions

The design and use of AIAs will be essential in future wireless communication systems due to their advantages on the overall system performance without any increase in the design complexity. In this chapter, we present a detailed codesign approach that can provide noticeable performance improvement as compared to the conventional 50Ω-based interfacing of the active and antenna sections of an AIA. The codesign approach can achieve better efficiency, realized gain, and design miniaturization when properly applied. Several detailed design examples were provided for narrowband, wideband, and UWB antenna systems, with one example covering a MIMO design as well. We have shown monopole and patch based designs, but the procedure can be applied to any antenna type.

References

[1] Georgiadis, A., G. Andia and A. Collado, "Rectenna Design and Optimization Using Reciprocity Theory and Harmonic Balance Analysis for Electromagnetic (EM) Energy Harvesting," *IEEE Antennas and Wireless Propagation Letters*, Vol. 9, 2010, pp. 444–446.

[2] Costanzo, A., et al., "Co-Design of Ultra-Low Power RF Microwave Receivers and Converters for RFID and Energy Harvesting Applications," *IEEE International Microwave Symposium (IMS)*, 2010, pp. 856–859.

[3] Visser, H. D., S. Keyrouz, and A. B. Smolders, "Optimized Rectenna Design," *Wireless Power Transfer*, Vol. 2, No. 1, 2015, pp. 44–50.

[4] Dhar, S. K., et al., "An Active Integrated Ultra-Wideband MIMO Antenna," *IEEE Transactions on Antennas and Propagation*, Vol. 64, No. 4, 2016, pp. 1573–1578.

[5] Sharawi, M. S., et al., "Miniaturised Active Integrated Antennas: A Co-Design Approach," *IET Microwaves, Antennas & Propagation*, Vol. 10, No. 8, 2016, pp. 871–879.

[6] Costanzo, A., et al., "Co-Design Strategies for Energy-Efficient UWB and UHF Wireless Systems," *IEEE Transactions on Microwave Theory and Techniques*, Vol. 65, No. 5, 2017, pp. 1852–1863.

[7] Shinjo, S., et al., "Integrating the Front End," *IEEE Microwave Magazine*, July/August 2017, pp. 31–40.

[8] Hasegawa, N., and N. Shinohara, "C-Band Active-Antenna Design for Effective Integration with a GaN Amplifier," *IEEE Transactions on Microwave Theory and Techniques*, Vol. 65, No. 12, 2017, pp. 4976–4983.

[9] Erturk, V. B., R. J. Rojas and P. Roblin, "Hybrid Analysis/Design Method for Active Integrated Antennas," *IEE Proceedings on Microwaves, Antennas and Propagation*, Vol. 146, No. 2, 1999, pp. 131–137.

[10] Dhar, S. K., M. S. Sharawi, and F. M. Ghannouchi, "On Microwave Connector De-embedding and Antenna Characterization," *IEEE Antennas and Propagation Magazine*, Vol. 60, No. 2, April 2018, pp. 1–10.

[11] Volakis, J., C. -C. Chen, and K. Fujimoto, *Small Antennas: Miniaturization Techniques & Applications*, New York: McGraw-Hill Education, 2010.

[12] ATF-53189 transistor data sheet and application note, http://www.farnell.com/datasheets/1227369.pdf

[13] GVA 63+ transistor data sheet and application note, https://www.minicircuits.com/pdfs/GVA-63+.pdf

APPENDIX A

Using ADS Tutorial

Advanced Design System (ADS) is an electronic design automation (EDA) tool that is used for microwave, radio frequency (RF), and high-speed application designs. It is used to design and analyze passive and active circuits as well as conducts system-level simulations. It is widely used in academia and industry and provides standard based design and verification tools for wireless technologies. It provides standard design libraries as well as the ability to codesign and simulate circuit and EM-based integrated models.

In this tutorial, we will provide step-by-step instructions using ADS to model and simulate two examples. The first one is an RF bandpass filter based on lumped elements, and the second is an RF amplifier example based on its imported *S*-parameters. The tutorial provided is based on ADS2017.

Example A.1: RF Lumped Component-Based BPF

1. Open the ADS software package, and choose a new workspace from File>>new>>New Workspace. Choose a desired workspace name and directory to create in as shown in Figure A.1.

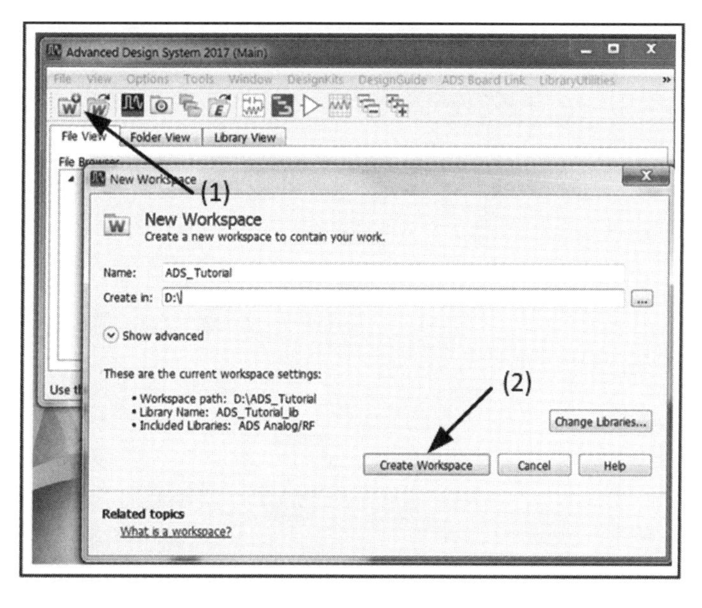

Figure A.1 Creating a new workspace in ADS.

2. We will create an RF filter circuit; thus, we need to open a new schematic. This can be done by clicking the "New Schematic" icon as shown in Figure A.2. Give a name to this circuit, that is, "cell_1."
3. Choose the required circuit elements from the left side menu as shown in Figure A.3 according to the designed RF filter values.

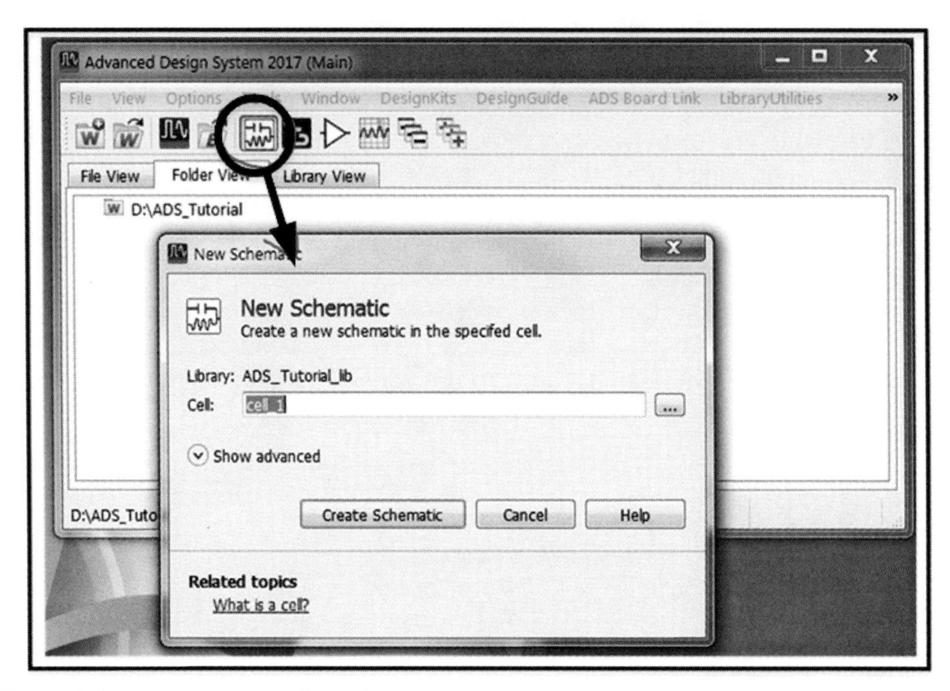

Figure A.2 Creating a new schematic.

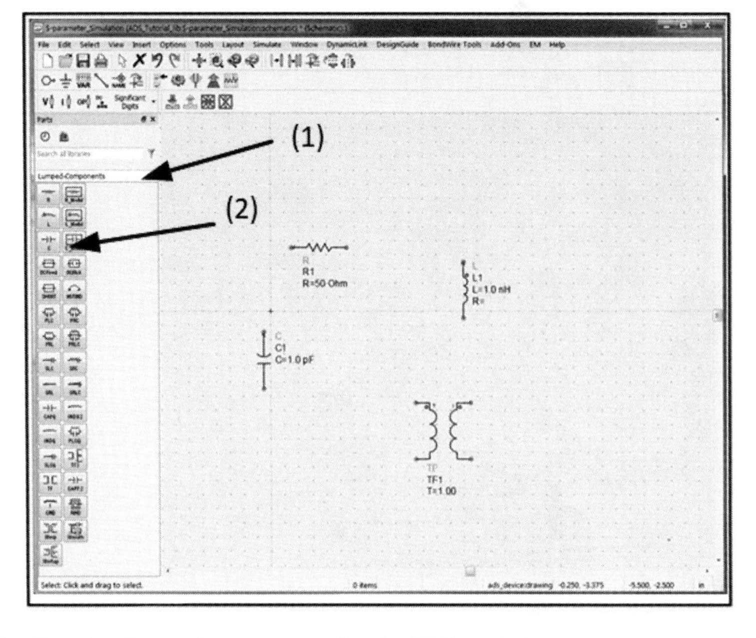

Figure A.3 Choosing lumped components for the RF filter design.

4. Since we will be simulating the response of the RF BPF using the *S*-parameters, we need to insert an *S*-parameter simulation block and set up the frequency ranges. To choose an *S*-parameter simulation, we choose "Simulation *S*-parameters" from the left menu as shown in Figure A.4. Connect the filter parts and add "Termination" elements of 50Ω on both sides of the lumped element BPF as shown in Figure A.5. Then to set up the *S*-parameter gear box, double-click on the gear box and fill in the information shown in the top right corner of Figure A.5.

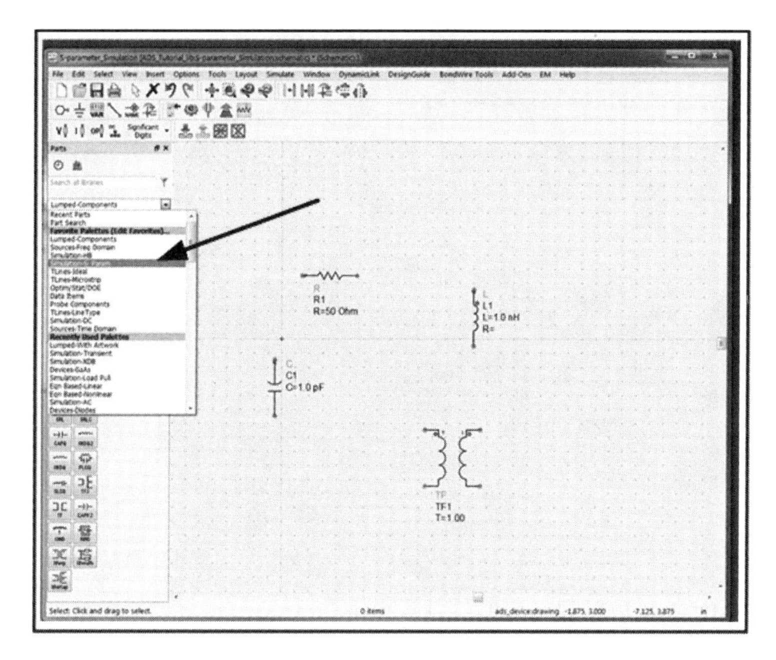

Figure A.4 Choosing the *S*-parameters' simulation block.

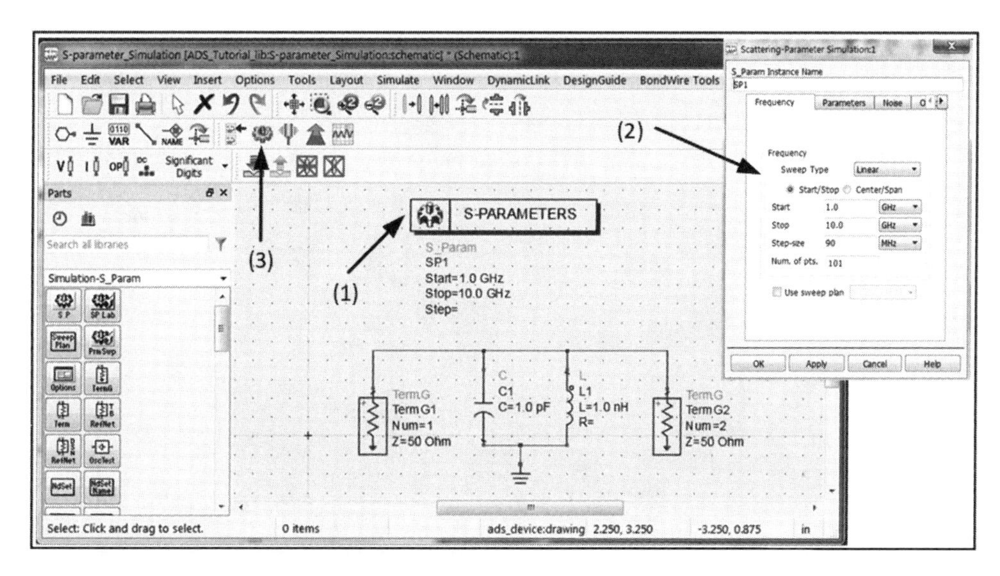

Figure A.5 Setting up the *S*-parameter simulation.

5. To run the simulation, click the gear box in the tool bar as shown in Figure A.5, step 3. After the simulation is completed, a data display window will open.
6. From the Data Display Window, select "rectangular plot," choose the simulation type "*S*-parameters," add "S11" and "S21" to display the return loss and insertion loss curves, respectively, and click OK, as shown in Figure A.6. The obtained response is shown in Figure A.7 for this BPF centered at 5 GHz.

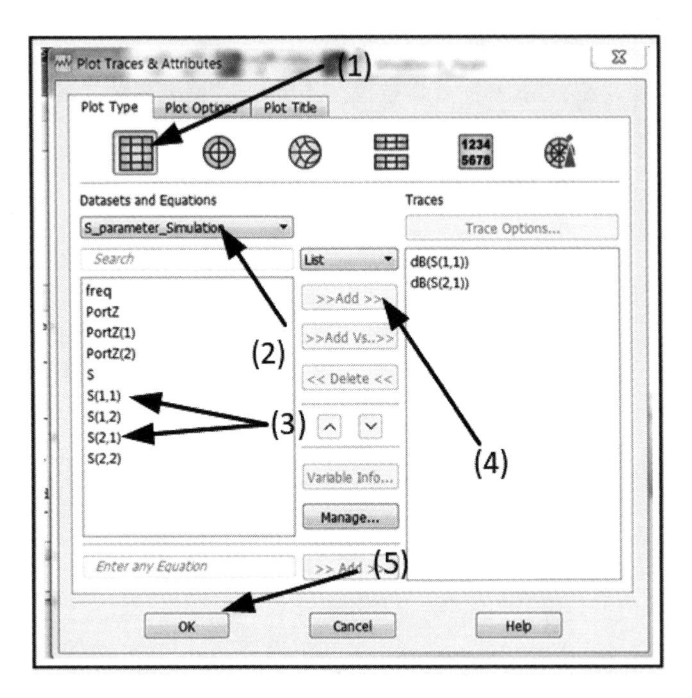

Figure A.6 Displaying the filter response curves.

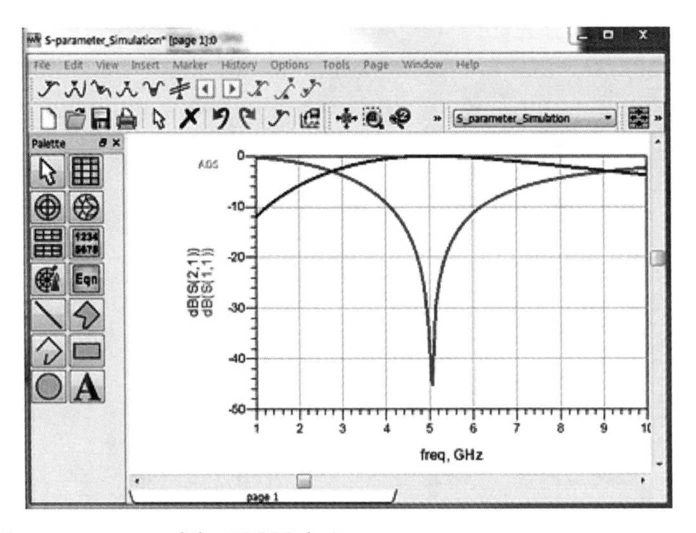

Figure A.7 Response curves of the RF BPF design.

Example A.2: RF Amplifier Characteristics Based on Its *S*-Parameters

1. To import an S2P file (whether is it for an active or a passive component), go to "Data Item" pallet and select the "SnP block" as shown in Figure A.8. Double-click it and set the number of ports as 2. Browse for the desired file in your computer to import and connect with this block and click "apply" and then "OK."
2. To observe the stability parameters of the imported transistor *S*-parameter file, go to the "*S*-parameter" pallet, choose an "*S*-parameter" gear box, set up the frequency ranges to cover 0.1 to 10 GHz, select "noise calculation" from the "noise" tab, and click "OK." This is shown in Figure A.9. Make sure that you connect your S2P block to terminating input and output impedances (i.e., 50Ω in this example). Add also the stability and gain blocks to be able to view the stability and gain circles of the amplifier.
3. Run the simulation and then plot the *S*-parameter curves as well as the stability parameters in a rectangular plot as shown in Figure A.10.
4. To plot the power gain and available gain circles (check Chapter 3), get the "GPCircle" and "GaCircle" simulation option blocks from the "*S*-parameter" pallet and add to the schematic as shown in Figure A.11. Then "Run" the simulation again (press the simulation gear) to plot the circles. In the Data Display Window, plot the circles on a Smith chart as shown in Figure A.12.
5. Any other extra analysis can be conducted now based on the basic steps demonstrated above. It should be noted that in this example we did not optimize the amplifier response at any frequency; rather we wanted to learn how to plot the various parameters and import a specific transistor *S*-parameters and view its general response. Chapter 3 can be used now to design specific examples for maximum gain or minimum *NF* via matching network selection and design.

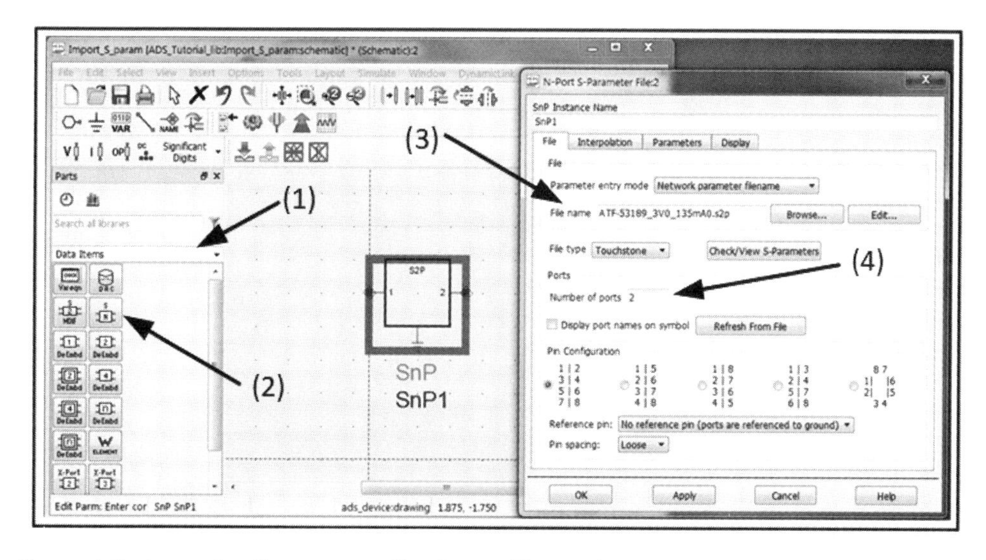

Figure A.8 Importing *S*-parameters file of a specific transistor.

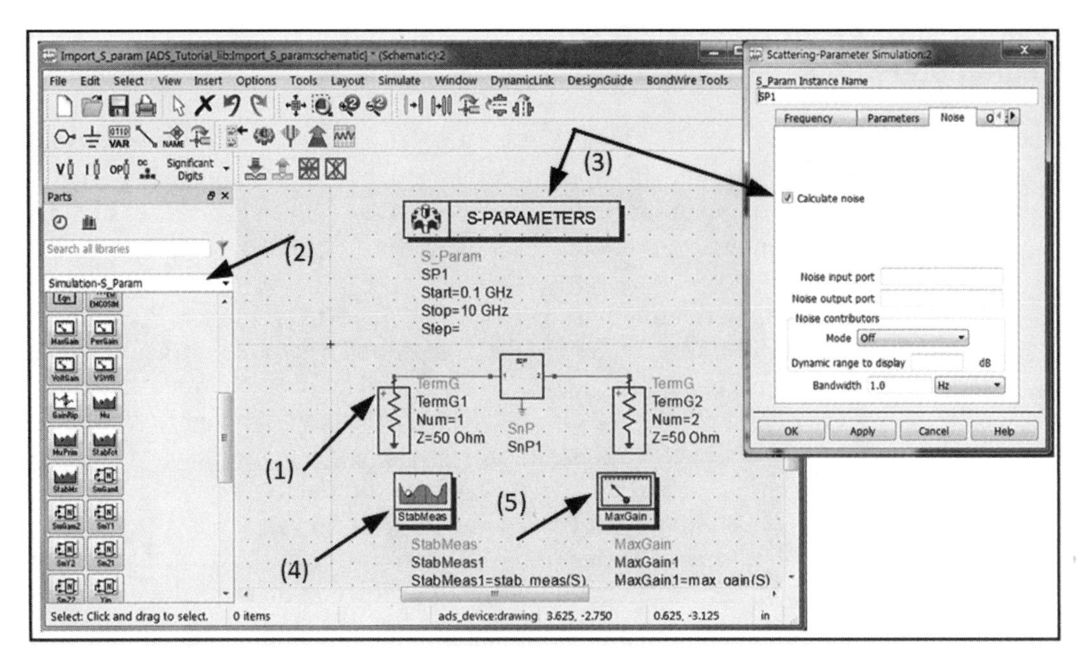

Figure A.9 Setting up the simulation of the RF amplifier block.

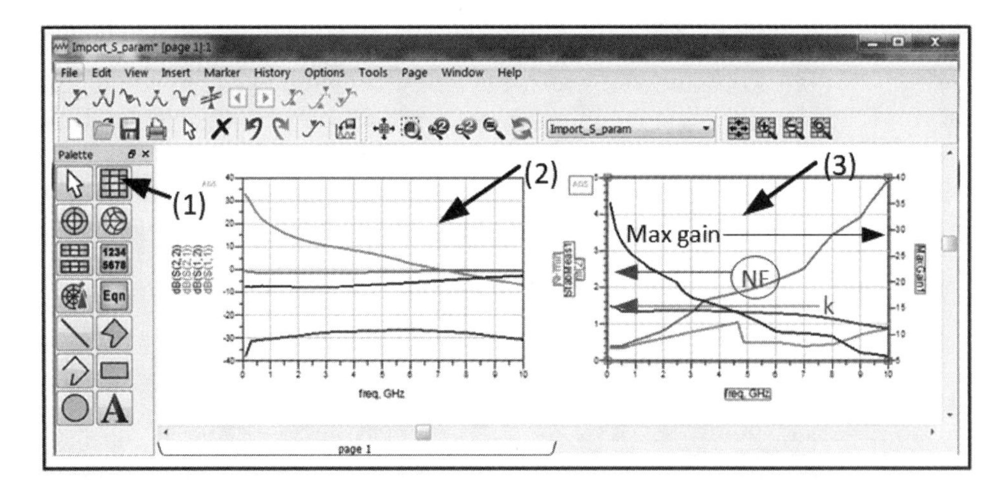

Figure A.10 Plotting the results of simulating the RF amplifier terminated block.

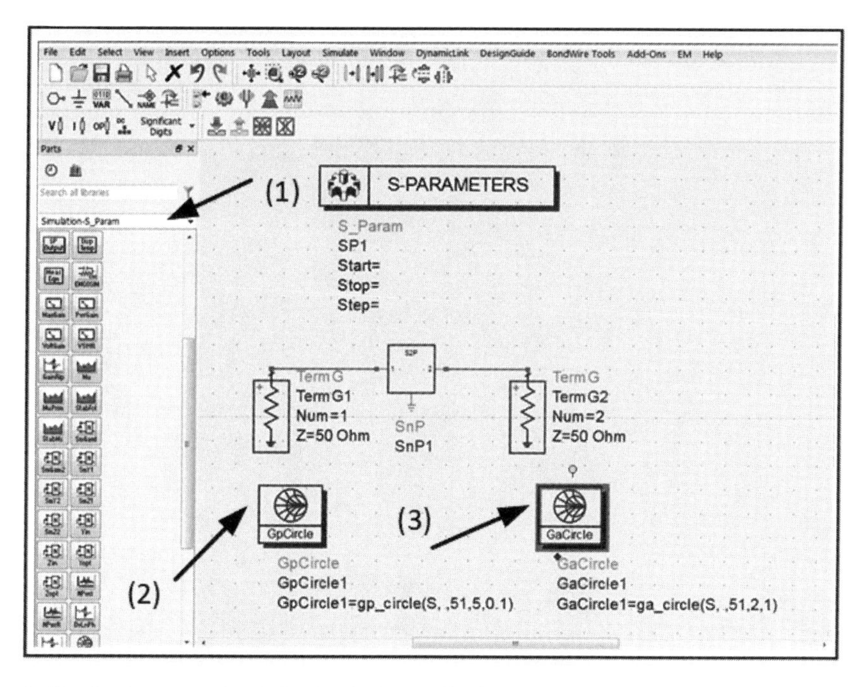

Figure A.11 Setting up gain circle simulation and calculations.

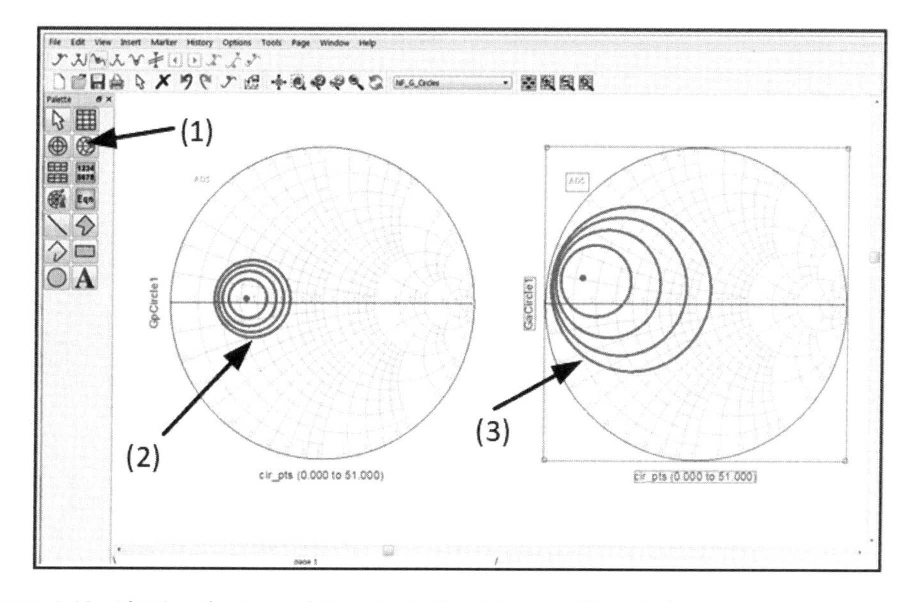

Figure A.12 Plotting the Gp and Ga gain circles of the amplifier block.

APPENDIX B
Using MWO Tutorial

Microwave Office (MWO) is an electronic design automation (EDA) software produced by AWR Inc./National Instruments. The MWO design suite encompasses all the tools essential for high-frequency integrated circuits (IC), PCB, and module design, including: linear and nonlinear circuit simulators, harmonic balance simulations, EM analysis tools, integrated schematic and layout (can be used for schematic entry, analysis, and PCB board layout and extraction), statistical design capabilities, and parametric cell libraries with built-in design-rule check (DRC).

In this tutorial, we will show the detailed steps for obtaining the frequency response of an RF filter. This tutorial covers the 2010 (9.05r build) version of MWO.

Example B.1: Lowpass RF Filter

After invoking MWO, you will get the main window as shown in Figure B.1. On the left side, you will see the side tree where you can add different portions of the design such as circuits and system blocks.

A. Start a New Project

To start a new project, go to File → New Project as shown in Figure B.2. Then save the project with an appropriate name such as "Project1." Do, File → Save Project As, then type in the name and click "save."

B. Set Environment Options

From the "Options" menu, choose "Project Options" as shown in Figure B.3.

Click the "Global Units" tab and set the parameters as in Figure B.4.

C. Create Schematics

To add your circuit schematic, right click on the "Circuit Schematic" block on the left side tree, and enter the name of your circuit (i.e., "Schematic1"). This is shown in Figure B.5.

Draw the circuit using elements from the left side tree, after choosing the "Elements" tab.

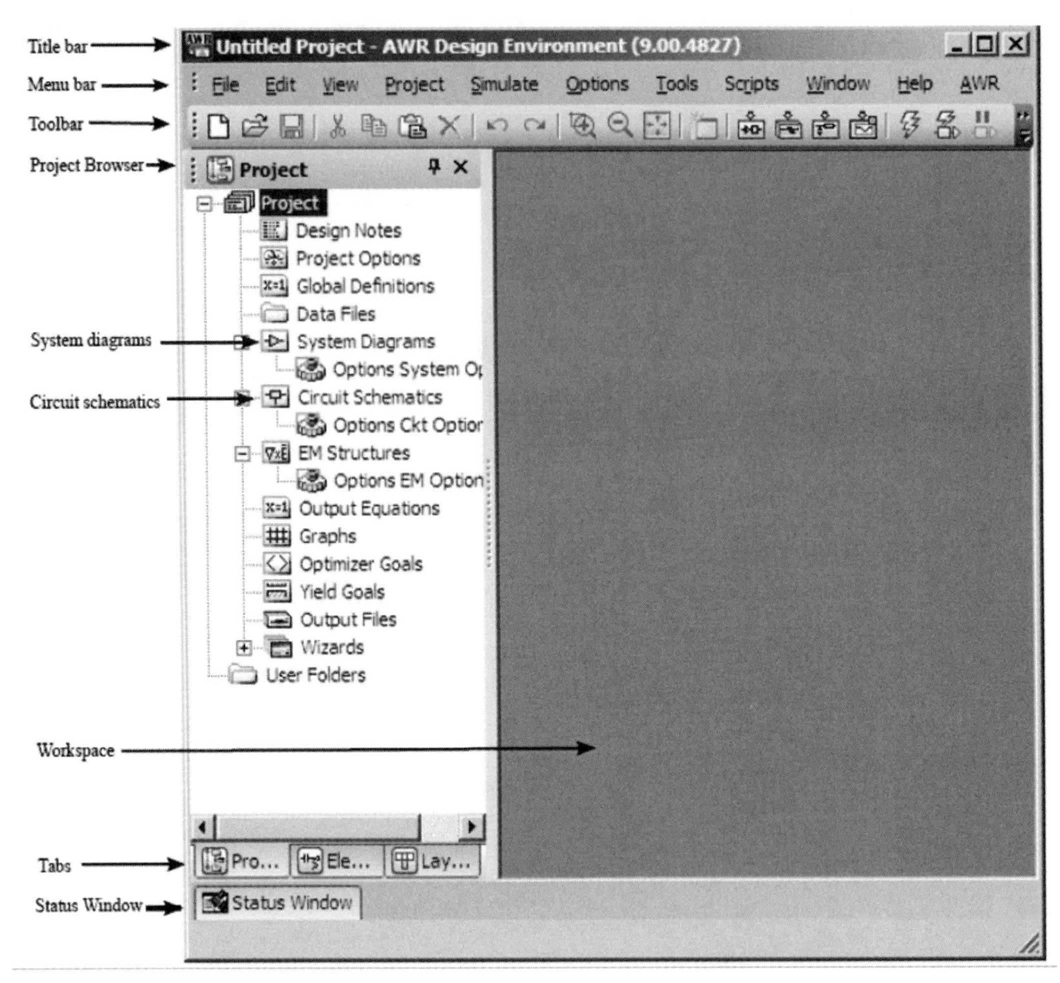

Figure B.1 MWO main PROJECT window.

D. Set the Simulation Parameters

From the "Options" menu, choose "Project Options." Now in the "Frequencies" tab, select the start and end frequencies for your simulation (you need to have an idea about the range you are interested in from the design you are simulating). Fill in the values that are shown in Figure B.6. Click "OK."

Now, after setting up the simulation parameters, click Simulate from the top side bar menu as shown in Figure B.7.

E. Plot the Simulated Results

After the simulation is complete without any errors, create a new graph by clicking on the "Graph" box in the left tree. Then fill in the name of the graph and the type of the graph needed. Figure B.8 shows the graph window where we named it "S21, S11, and S22" graph. This will open the graph window (a rectangular plot).

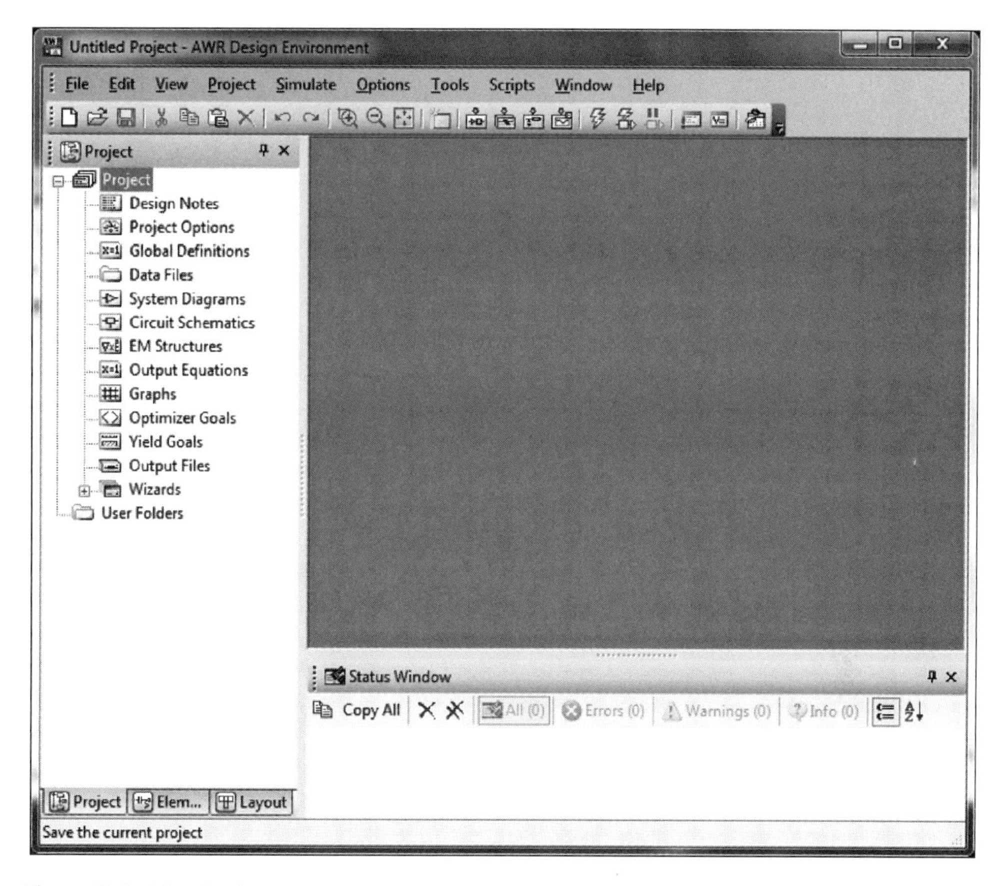

Figure B.2 New Project tree.

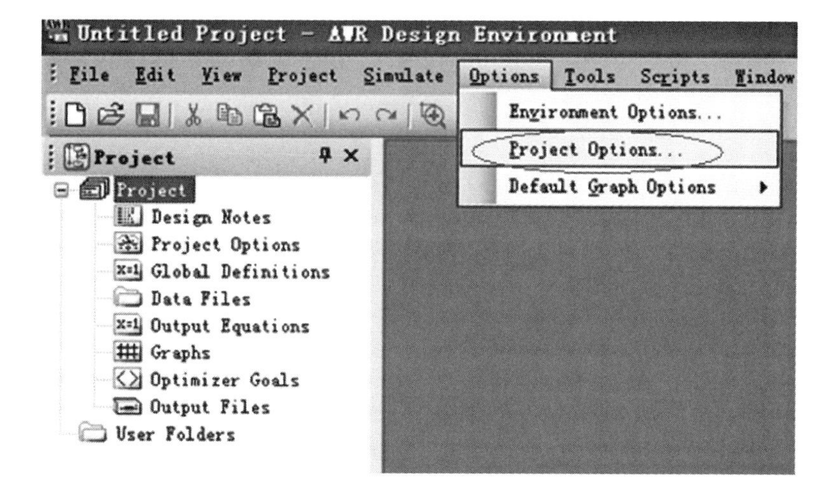

Figure B.3 Setting up Project Options.

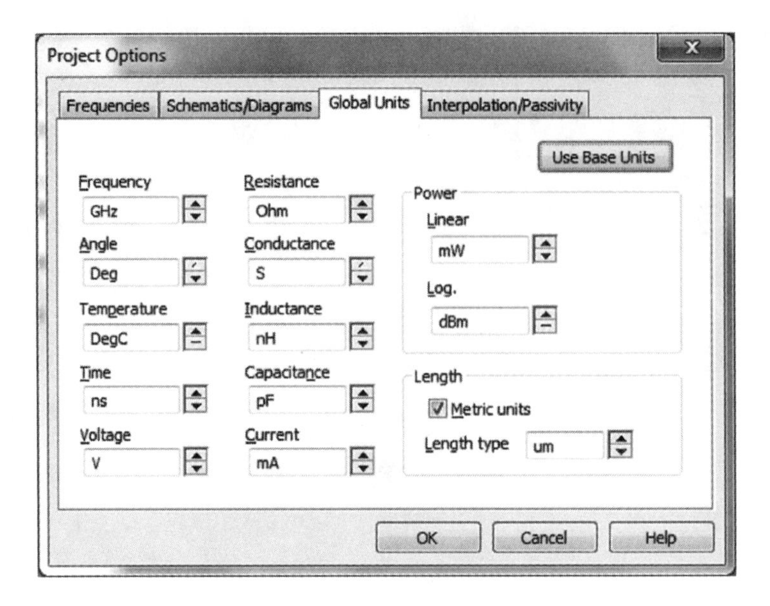

Figure B.4 Global Units settings.

You need to add the curves now. From the “Graph” tree item, right click on the rectangular plot in which you created “S21, S11, and S22” and choose “add Measurement.” In that window choose the *S*-parameters (or other linear/nonlinear parameters) that you need to plot. This is shown in Figure B.9. Make sure that you choose the “Data Source Name” to be the schematic with which you are dealing (the circuit you built).

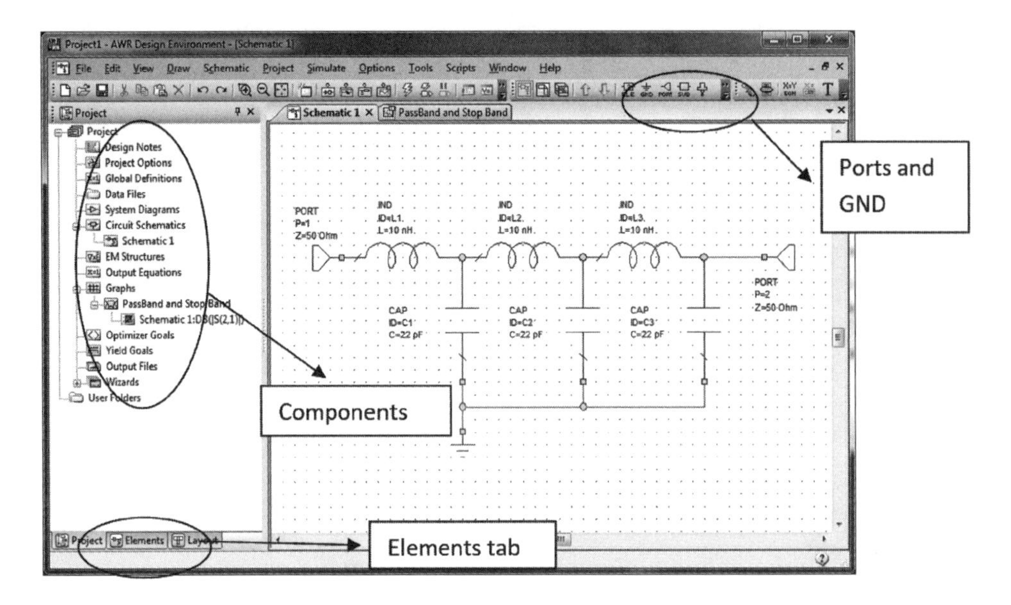

Figure B.5 Circuit schematic creation.

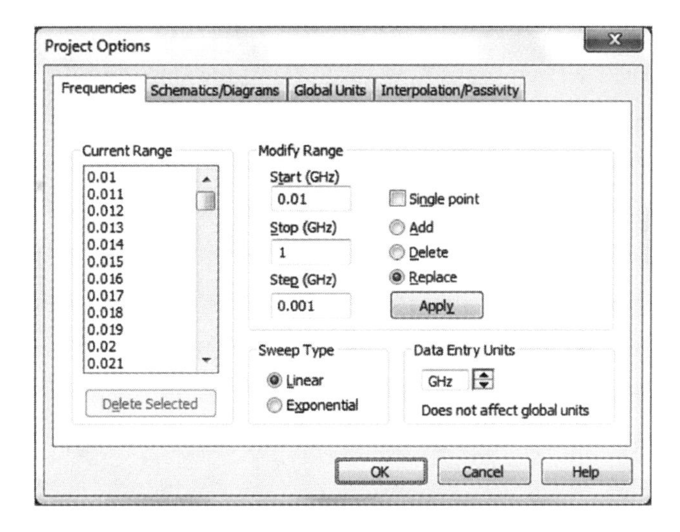

Figure B.6 Setting up simulation frequencies.

The curve will show up on the plot. If it does not, rerun the simulation and it should appear as shown in Figure B.10. You can add another curve by repeating the previous step. Plot "S11" and "S22" magnitudes on the same graph. You will obtain Figure B.11.

You can modify the axis setting by right-clicking on the graph, and selecting "Parameters." There you can change the way you want to display the curves.

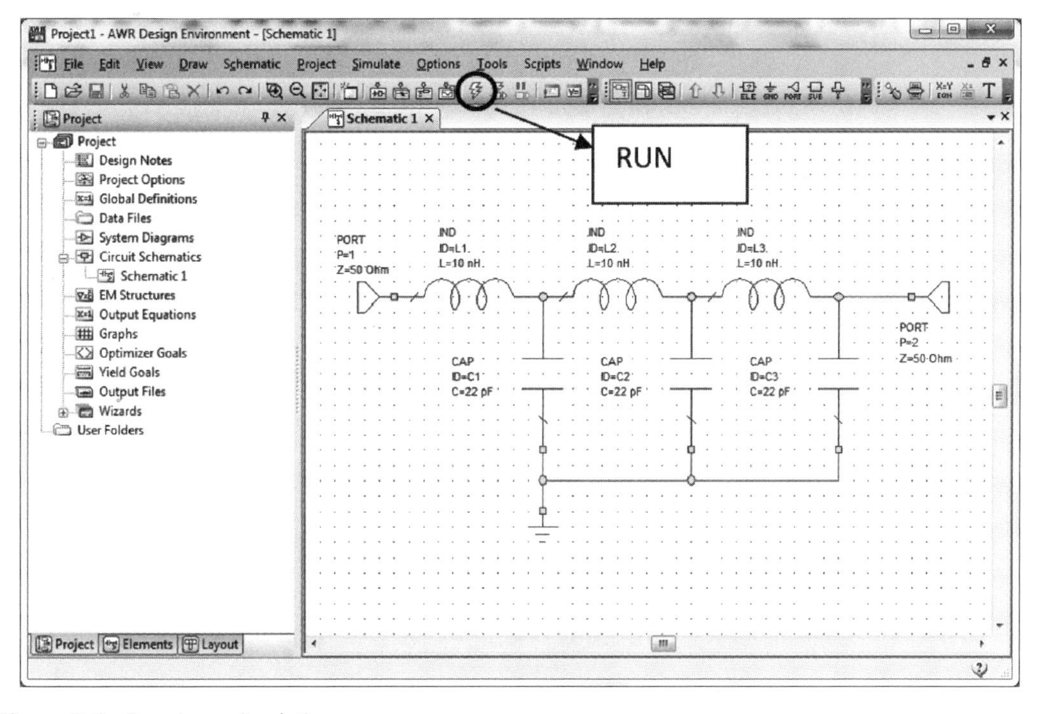

Figure B.7 Running a simulation.

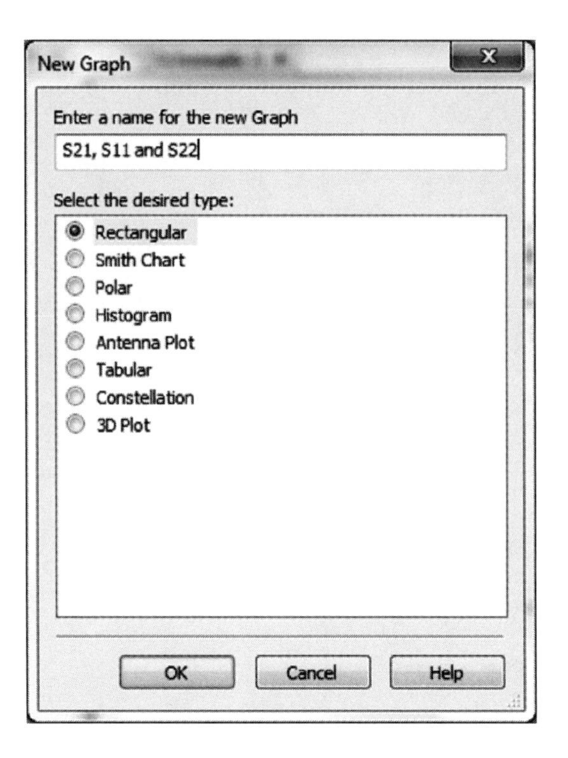

Figure B.8 New Graph window.

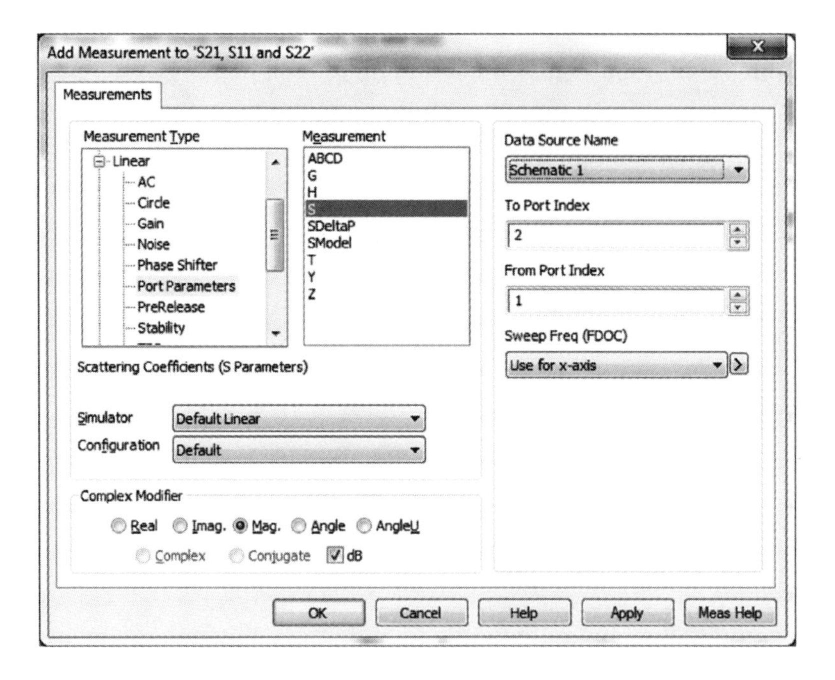

Figure B.9 Adding curves to the results plot.

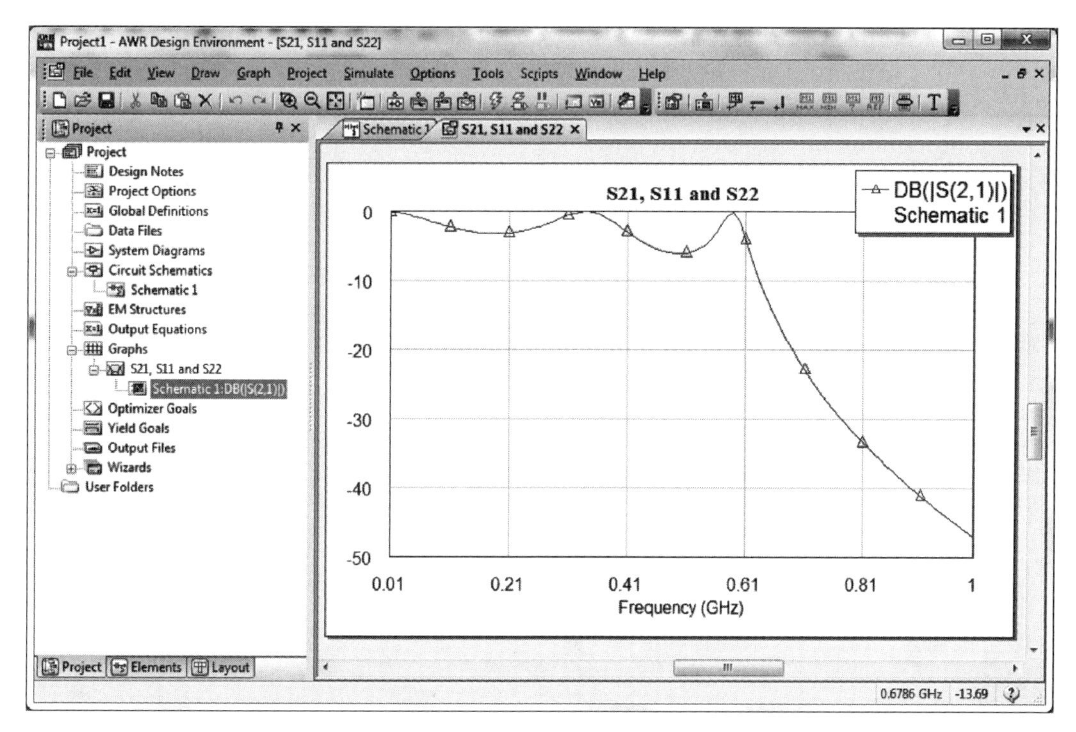

Figure B.10 Adding a simulation curve to the output graph.

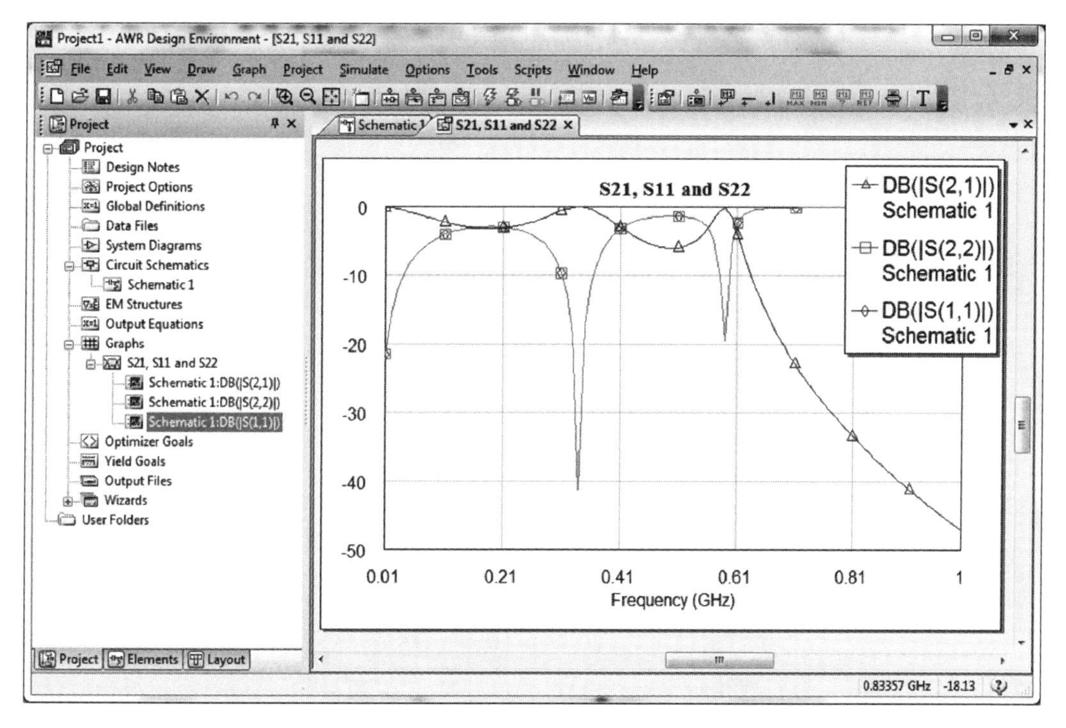

Figure B.11 Final results.

To tune the circuit output via varying the circuit element values, we can use the tuning tool. You can choose this from the main bar menu as shown in Figure B.12. Using the screwdriver of the tuning tool, click on the values of L1, C1, L3, and C3. They will all change color.

Click on "Window" menu and choose "Tile Vertical" to see both the circuit and the graph "S21, S11, and S22." Then choose the "Tune" analyzer bar from the top menu (next to the tuning tool). The window of the tuning tool will appear on your workspace as shown in Figure B.13. Adjust the values of L1, C1, L3, and C3 and see the direct effect on the final LPF performance. As an exercise for the reader,

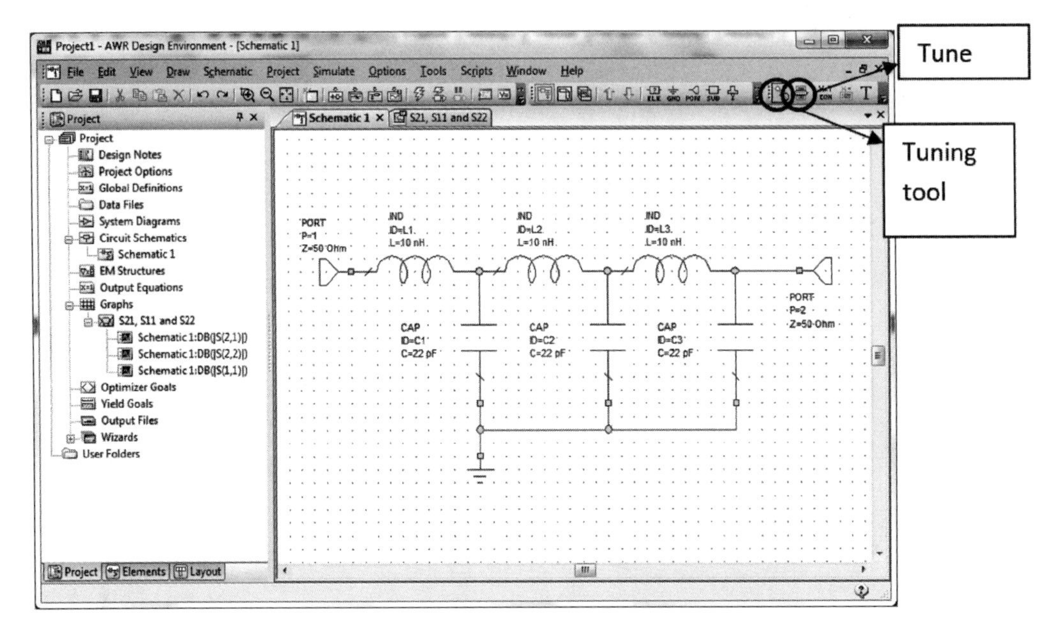

Figure B.12 Tuning element values using the tuning tool.

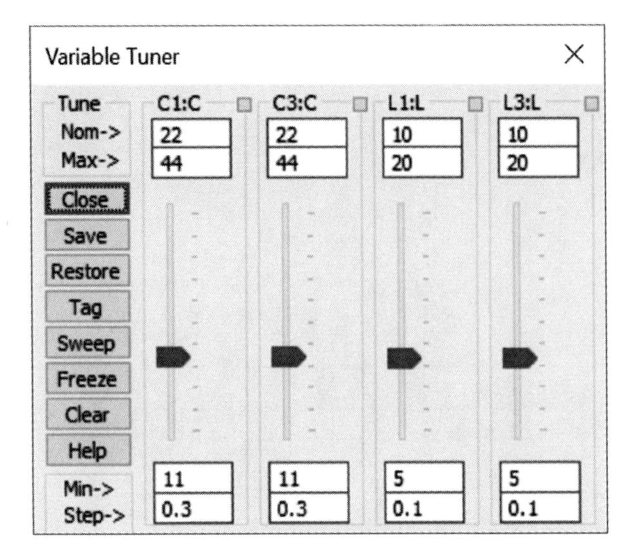

Figure B.13 Variable tuner window (slide indicators to change values instantly).

what combination will give you a flat passband for |S11|, |S22| (i.e., remove the dips in the passband)?

Example B.2: RF Filter Design with PCB Trace Effects

In this example, we will design and simulate another RF filter, but we will incorporate the effect of the connecting transmission lines that occur when you build the circuit on a PCB. We need to calculate the width of the microstrip transmission lines based on the PCB specifications given. Assume that the PCB material is FR4 with dielectric constant of 4.4 and loss tangent of 0.02. The board thickness is 0.8 mm, and we want the transmission line to operate in the passband of the filter; thus, we will design the lines to operate in the 0.5-GHz frequency range (as in Example B.1).

There is a tool within MWO that will help you find the impedance/width of the transmission lines based on the PCB specs that you have. From the tools Menu, choose "TxLine," and you will have the window as in Figure B.14. Use any name for the dielectric as FR4 is not listed there. Fill in the portion in material properties as shown in Figure B.14, as well as the conductor type. Fill in the electrical properties, height, and thickness. Press the arrow going right, and your microstrip line width will be calculated. In this configuration, it will be 1.5 mm.

To define the substrate on which we will lay out the microstrip transmission lines, we need to choose the substrate "MSUB" from the Substrates menu of the Elements tab. Once you place the substrate symbol to the schematic, double click on it to see its configuration window as shown in Figure B.15. Fill in the parameters as shown in the figure for the material properties and geometry. Give the substrate the name "SUB1." You will need this name later on to define the transmission lines.

Place the transmission lines between every two lumped elements in the filter designed in Example B.1. You need to use the transmission line model "MLIN." In

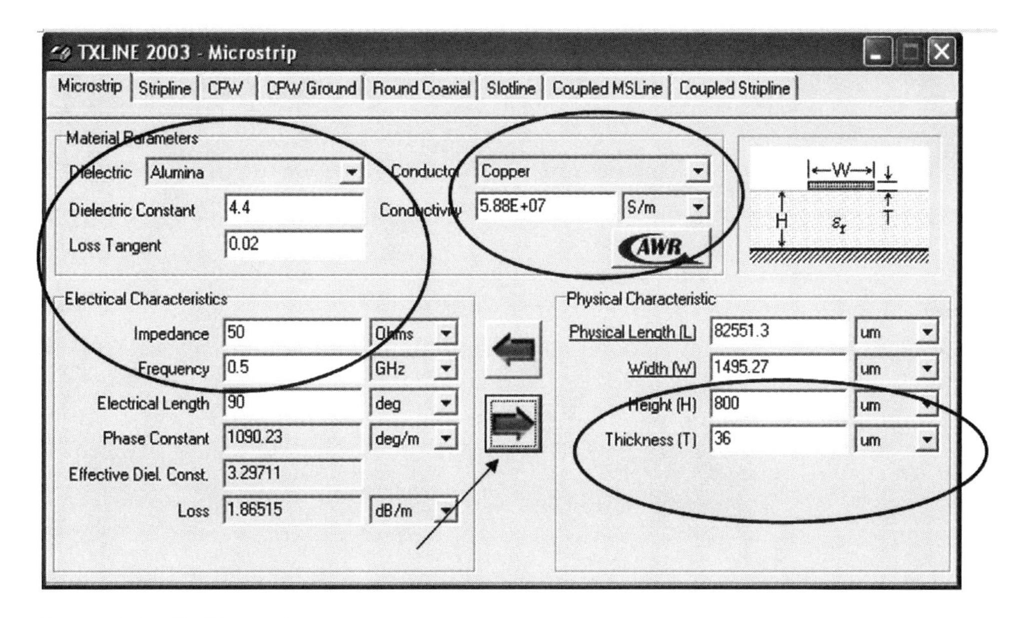

Figure B.14 Tx-Line tool.

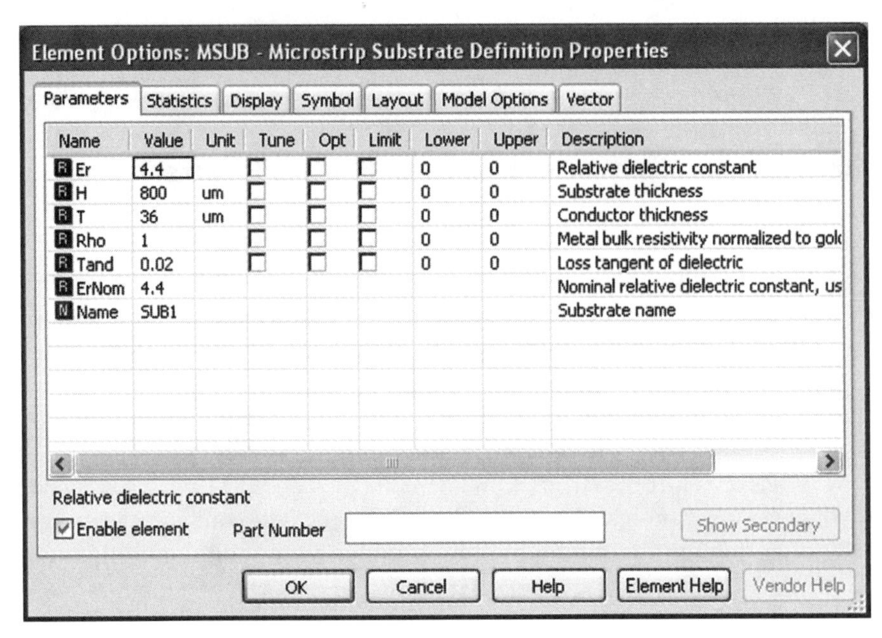

Figure B.15 Substrate parameters.

your first "MLIN" element, fill up its properties as shown in Figure B.16 (W = 800 μm, and MSUB = SUB1). Note that the width is equal to 0.8mm while the length can be determined based on the PCB layout (for this example, you can choose length arbitrarily and see the effect). You will end up with a schematic that looks like Figure B.17.

Again, you need to add the curves now. From the "Graph" tree item, right-click to create a rectangular plot. Add an *S*-parameter magnitude measurement as shown in Figure B.18. Create another graph to plot the phase of the filter as shown in Figure B.19.

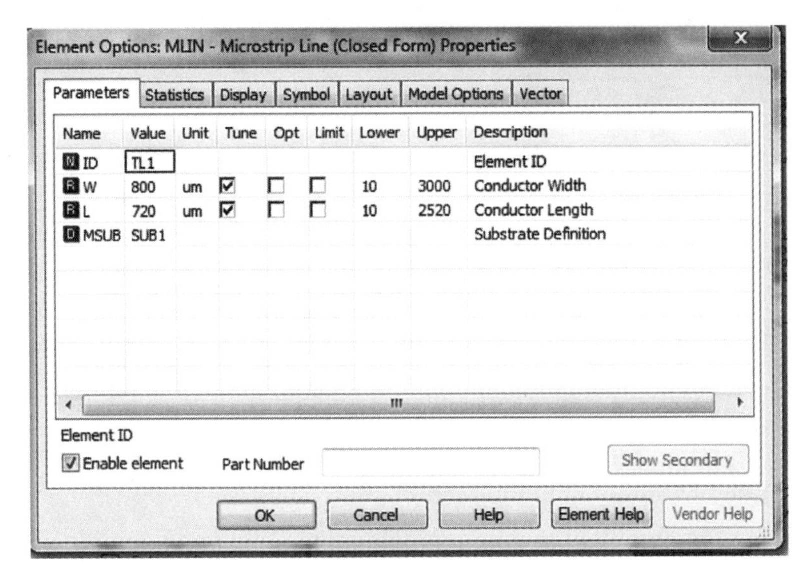

Figure B.16 Microstrip transmission line parameters.

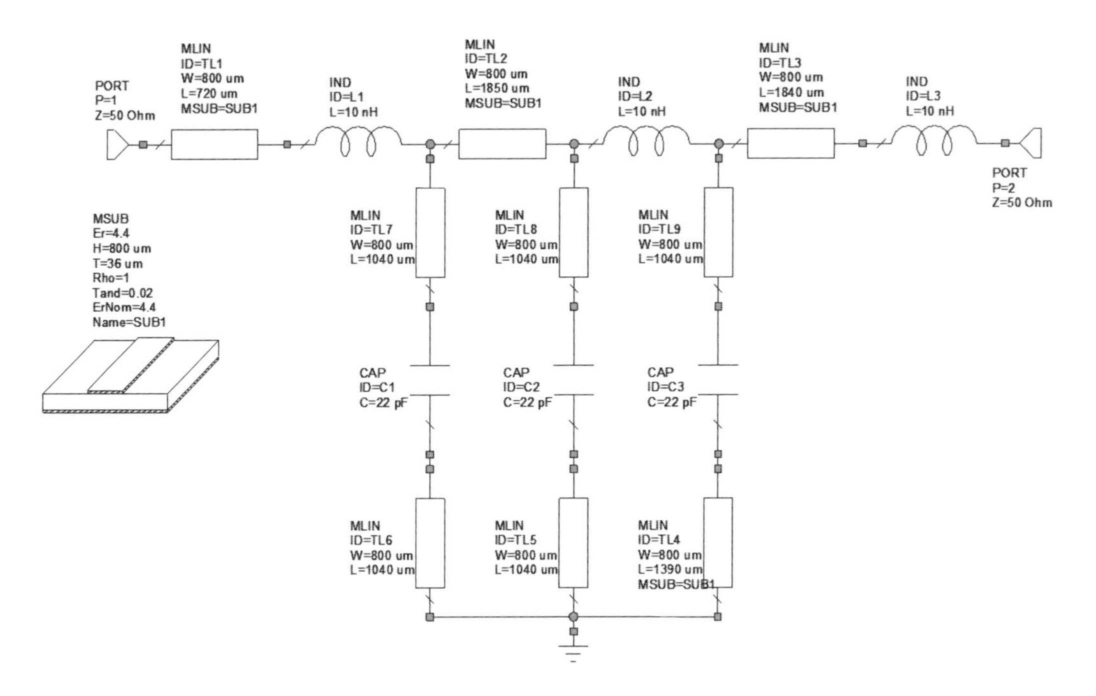

Figure B.17 Final schematic incorporating the PCB traces.

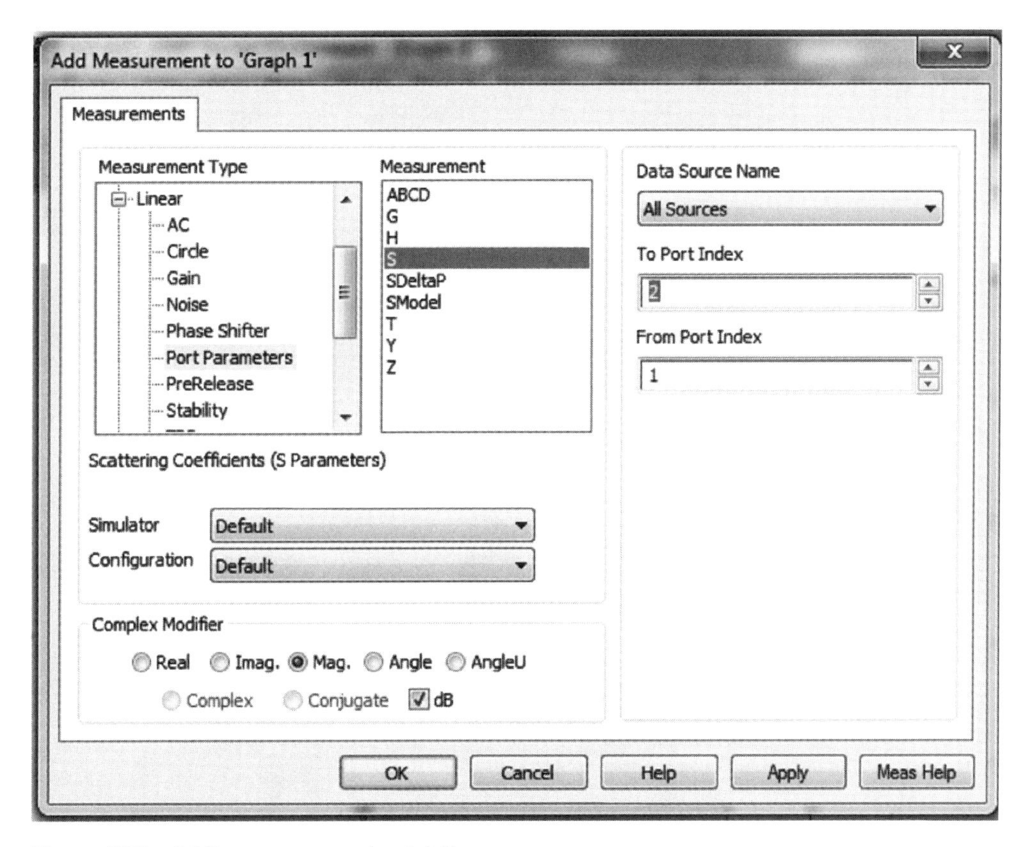

Figure B.18 Adding new curves to plot the response.

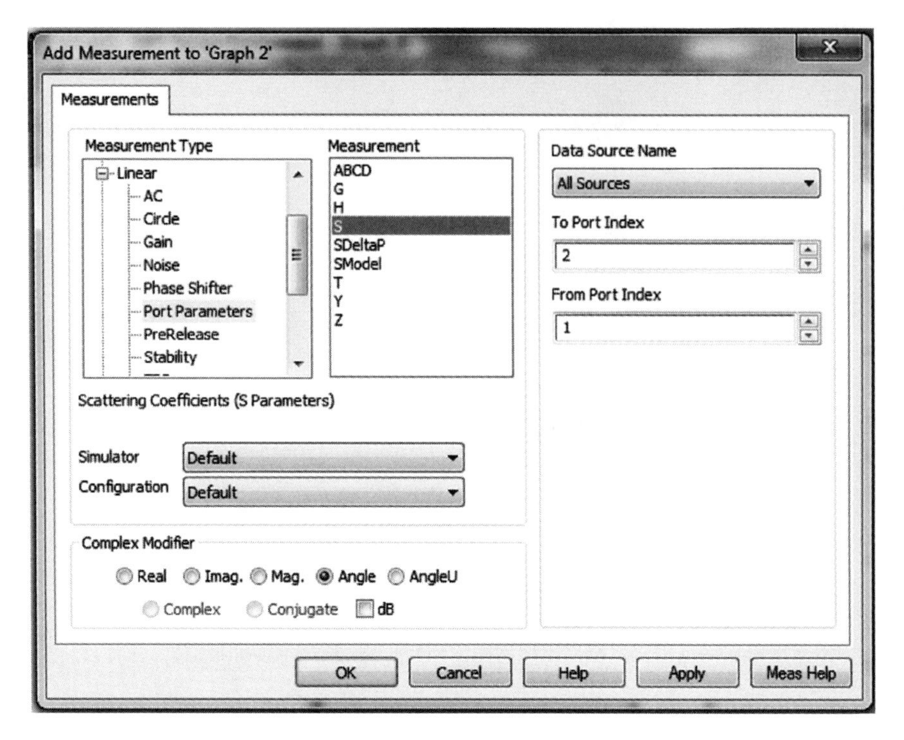

Figure B.19 Plotting the phase of the S-parameters.

Tile the three windows vertically, and then run the simulation to see the results together as shown in Figure B.20. Vary the values of the lengths of the PCB microstrip line transmission lines and observe the effect on the phase response of the filter; then vary the width of the transmission line to create impedance mismatches and observe the effects.

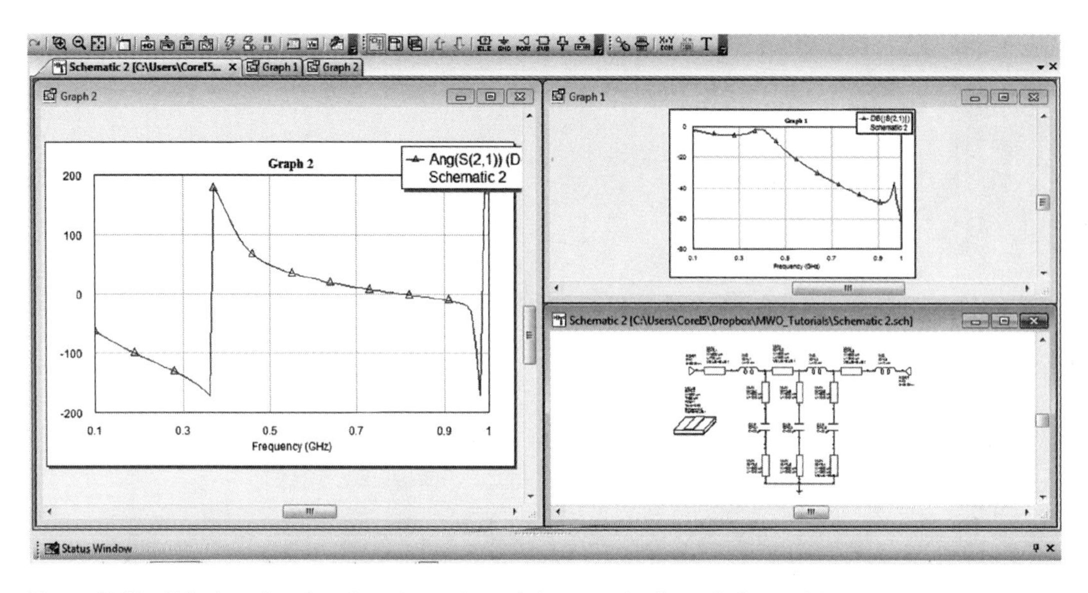

Figure B.20 Window showing the schematic and the magnitude and phase of S_{21}.

APPENDIX C

Using HFSS Tutorial

This tutorial will get you started with HFSS. HFSS is a full-wave electromagnetic field simulator that is based on the finite element method (FEM). It is a commercial software tool that can be used for antenna design, antenna placement analysis, radio frequency (RF) structure performance prediction, and EMC, as well as scattering problems and bio-electromagnetics. This is a quick step-by-step tutorial that is aimed to make learning and using the tool easier for new users. It was made using HFSS V.15, but the same steps can be followed for other versions.

1. We will design a patch antenna with a Pin feed excitation (as opposed to a microstrip feed that is accomplished using a microstrip line with an Edge feed). Figure C.1 shows the geometry of the antenna to be modeled in this tutorial.
2. Open your HFSS program. You will have a window as in Figure C.2.

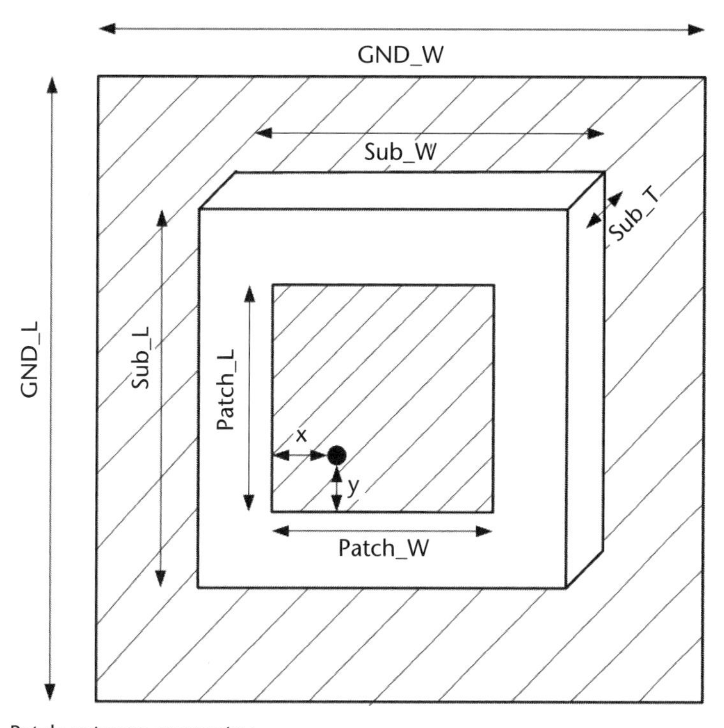

Figure C.1 Patch antenna geometry.

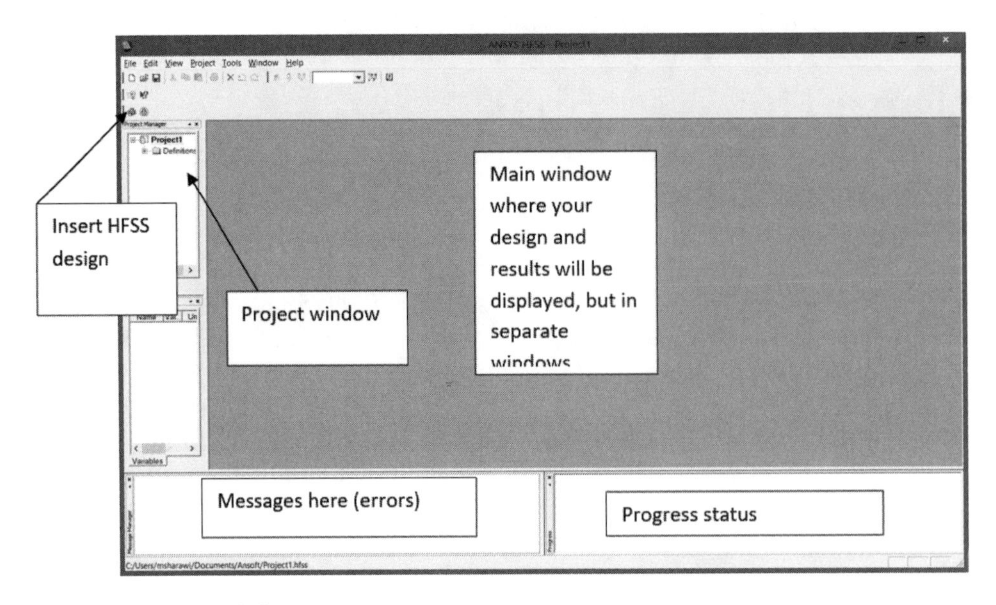

Figure C.2 HFSS workplace.

3. Click the "Insert new design" icon, right-click, and change "Solution Type" to "Driven Terminal" as shown in Figure C.3.
4. If you will be performing some changes on the model geometry (usually the case), it is recommended to define variables and then use them as the dimensions in your design. For this exercise, you will define several variables.

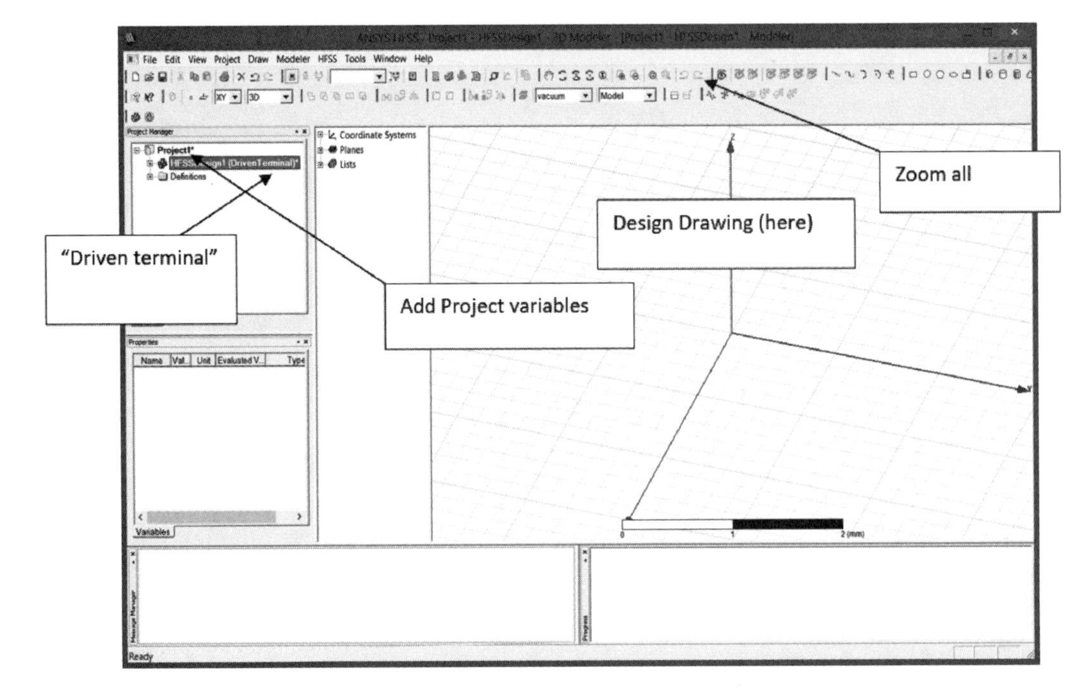

Figure C.3 Adding a new design and solution setup.

5. Right-click the “Project” name tree on the left side of the screen, click on “Project Variables,” and then choose “ADD” to be prompted with a window like Figure C.4. Fill the variables in one after the other for the geometry of interest until all are entered as shown in Figure C.5.

The variables of interest as shown on Figure C.1 are:

GND_L = 90 mm

GND_W = 90 mm

Sub_L = 90 mm

Sub_W = 90 mm

Sub_T = 1.56 mm

Patch_L = 45 mm

Patch_W = 45 mm

X = 11 mm

Y = 0 mm

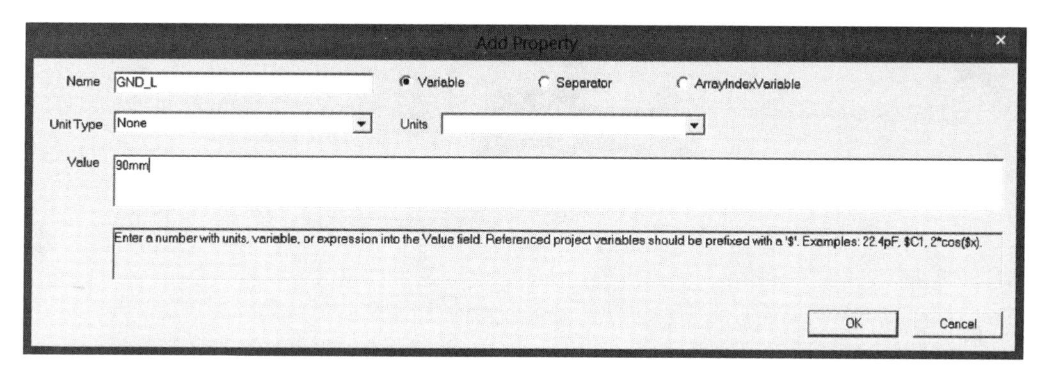

Figure C.4 Variable insertion window.

Properties: Project1

Project Variables | Intrinsic Variables | Constants

Value · Optimization · Tuning · Sensitivity · Statistics

Name	Value	Unit	Evaluated V...	Description	Read-only	Hidden	Sweep
$GND_L	90	mm	90mm		☐	☐	☑
$GND_W	90	mm	90mm		☐	☐	☑
$Sub_L	90	mm	90mm		☐	☐	☑
$Sub_W	90	mm	90mm		☐	☐	☑
$Sub_T	1.56	mm	1.56mm		☐	☐	☑
$Patch_L	45	mm	45mm		☐	☐	☑
$Patch_W	45	mm	45mm		☐	☐	☑
$feed_x	16	mm	16mm		☐	☐	☑
$feed_y	12	mm	12mm		☐	☐	☑

☑ Show Hidden

Add... · Add Array... · Edit... · Remove · OK · Cancel

Figure C.5 Project variables list.

The feed point location (x,y) is not arbitrary, and several iterations are made to make sure the patch is at resonance in the location of the feed. Resonance is identified by having an $|S_{11}|$ value much less than −10 dB for proper matching as well as zero reactive impedence value.

6. Start creating the GND plane by "Add Rectangle" from the top right icon menu, then choose (−45 mm, −45 mm, 0) for x, y, and z at the bottom right corner boxes, and hit enter and then 90 mm, 90 mm, and 0 for dx, dy, and dz, respectively. Then "Zoom all" to obtain Figure C.6. Then change the name of the rectangle to "GND" and add the GND width and length variables to the rectangle shape as shown in Figure C.7.

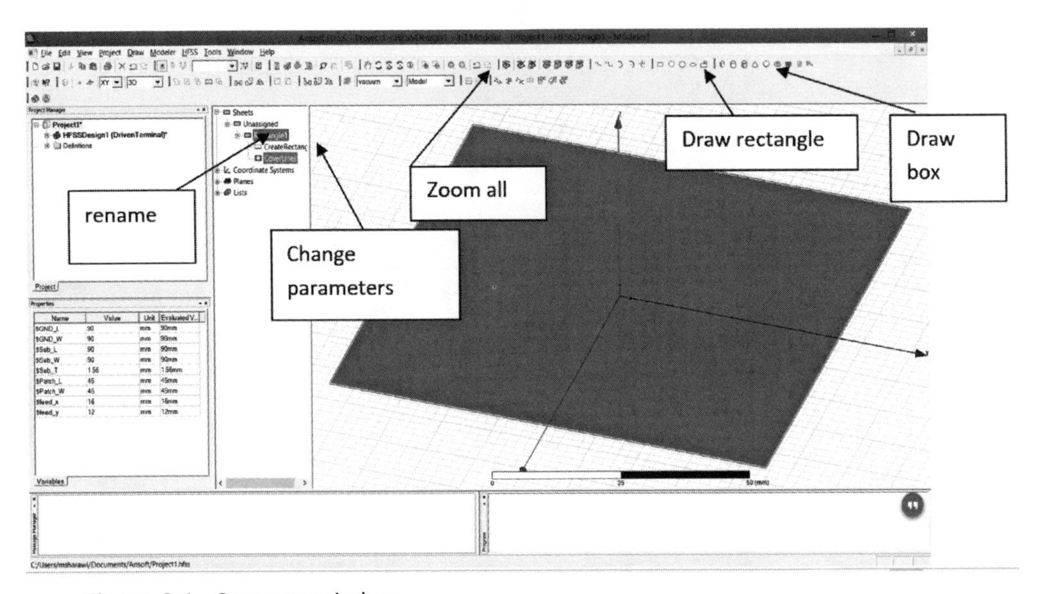

Figure C.6 Geometry window.

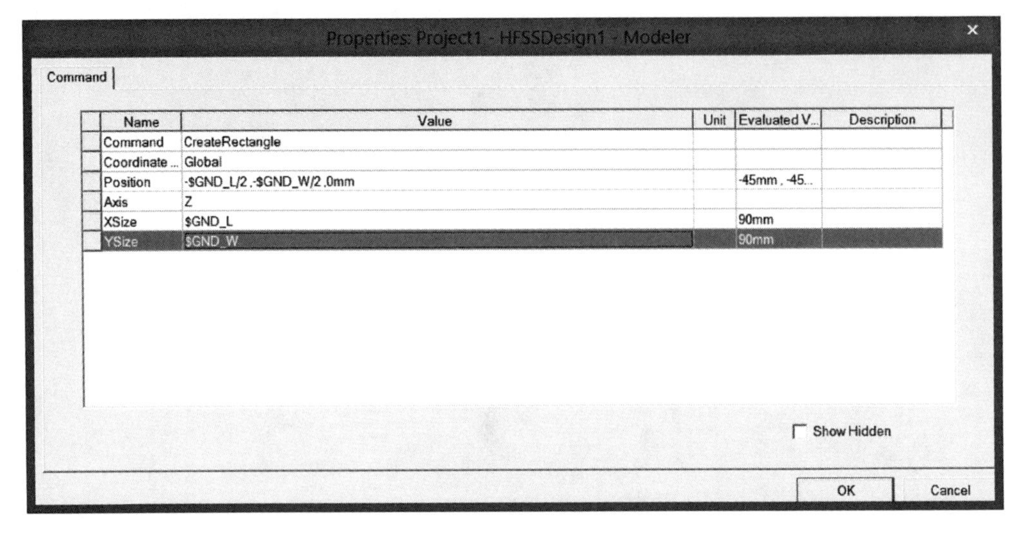

Figure C.7 Rectangle properties and size.

7. Use the "Draw Box" icon (Figure C.6) to draw the substrate rectangular volume needed to hold the patch on its top layer and the GND on its bottom layer. Select (–45 mm, –45 mm, 0) for (x, y, z) initial points, and hit enter and then (90 mm, 90 mm, 1.56 mm) for the changes in these directions to create the rectangular volume. Name it "Substrate" and then change its variables according to the ones on Figure C.1 and the corresponding dimensions as you did for the sheet.
8. Now we will add the radiating patch element. It will be adding a sheet on the top of the substrate, so you need to be careful with the location and dimensions. Add a rectangular sheet "Add rectangle," and then place it at (–22.5 mm, –22.5 mm, 1.56 mm) for initial points, then hit enter, and choose (45 mm, 45 mm, 0 mm) for the increments in each direction (since this is a sheet, the z-value is zero, that is, zero thickness). Change the rectangle name to "Patch" and edit the dimensions with the variables for the patch. You will end up with Figure C.8.
9. To complete the geometry, we need two more things: the feed pin and the radiation boundary. For the feed pin, we need to change the plane to the y-z plane to place a vertical strip representing the feed pin. Change the plane view to "YZ" as shown in Figure C.8. Then add a new rectangle, place it at (11 mm, 0 mm) (here we will have its width to be 0.5 mm, centered at 0.25 mm), choose "Add Rectangle," then choose starting points as (11 mm, 0, 0), and enter lengths as (0, 0.5 mm, 1.56 mm). Hit enter.
10. To add the radiation box, add a box ("Add Box") and specify the dimensions as (–90 mm, –90 mm, –45 mm) and the total size (180 mm, 180 mm, 90 mm); this takes care of the quarter-wave boundary margins. You will end up with the geometry shown in Figure C.9.

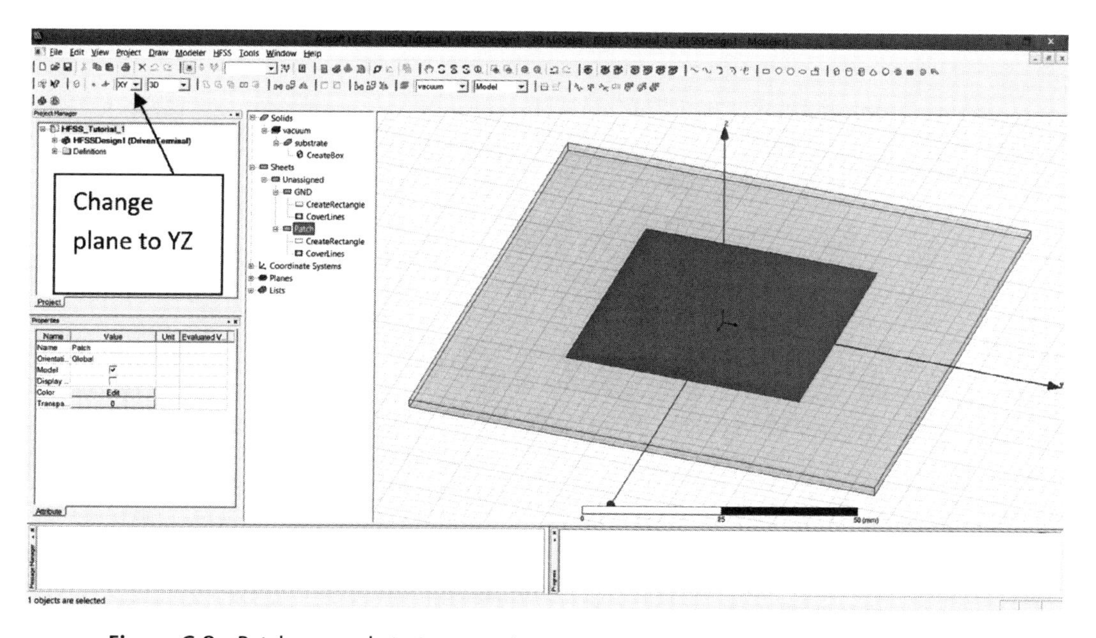

Figure C.8 Patch on a substrate geometry.

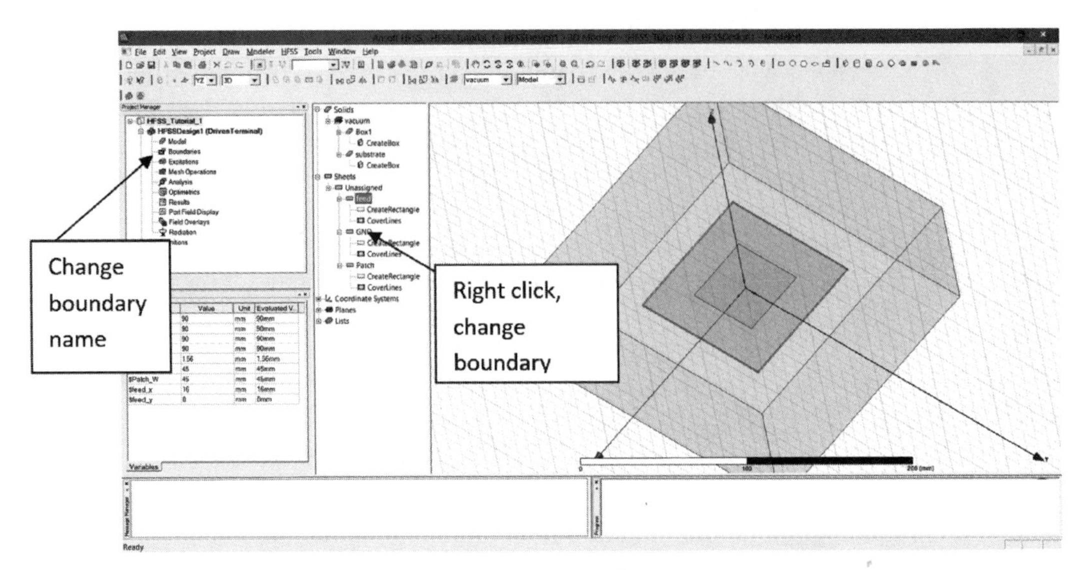

Figure C.9 Patch antenna geometry and its boundary.

11. Now we need to assign materials and boundary definitions to the various parts of the design. Starting with the GND plane, we need to assign a perfect electric conductor (PEC) boundary. To do so, right-click on the GND rectangle sheet, choose "Assign Boundary" and then "Perfect_E," and choose boundary name "GND" as shown in Figure C.9.
12. Repeat this, assign a PEC boundary to the patch, and then assign a radiation boundary to the box. Assign an "Excitation" for the feed and choose a "Lumped Port" with the GND as its reference plane.
13. For the substrate, we need to assign a dielectric material. Right-click on the box, "substrate" (or whatever name you gave it), and then choose "Assign Material"; a box will pop up as shown in Figure C.10. If your material is not listed, you can add it to the library by clicking on "Add Material,"; then add material properties according to Figure C.11 (or the available material that you will use for fabrication). After creating the new material, assign it to the substrate box.
14. The complete setup for the geometry will look like Figure C.12.
15. Now we need we need to set up the analysis.
16. Under the project tree in the Project window, right click on "Analysis" and "Add a Solution Setup." In the "General" tab of the solution setup window, specify the solution frequency that you expect based on your design. In our case it will be 1.5 GHz. Leave the other parameters as for now. Click OK.
17. Expand the analysis tree, right-click on "Setup1," and then choose "Add Frequency Sweep." Here we will sweep over a frequency range to identify the exact frequency of operation and its bandwidth. Choose the following:

    Sweep type: FAST
    Type: Linear Step
    Start: 0.5 GHz

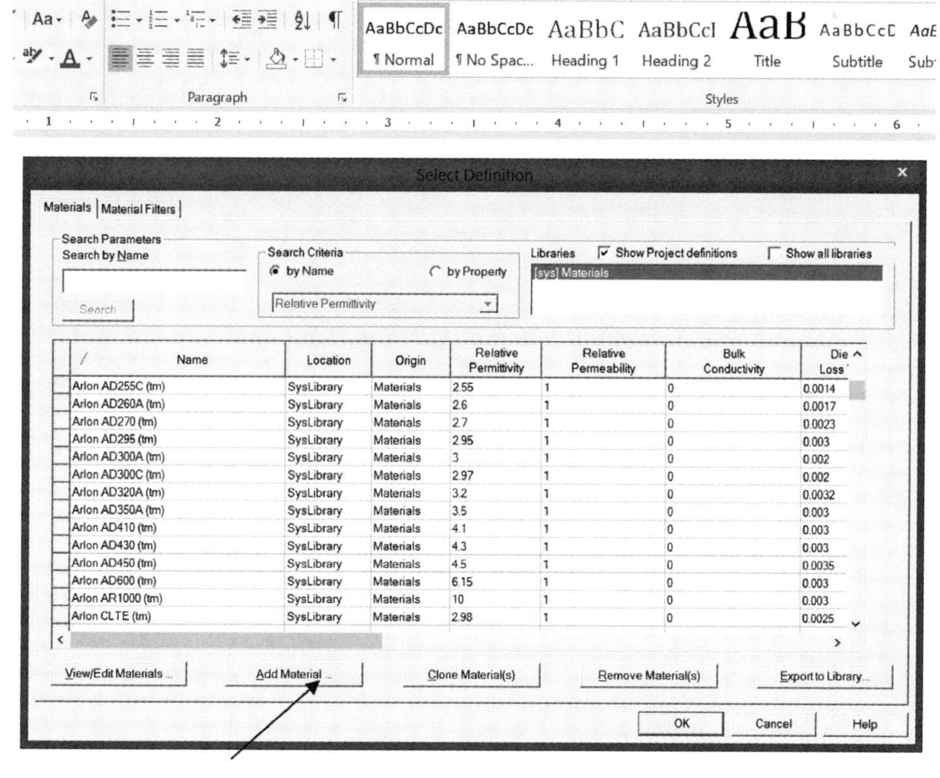

Figure C.10 Material setup window.

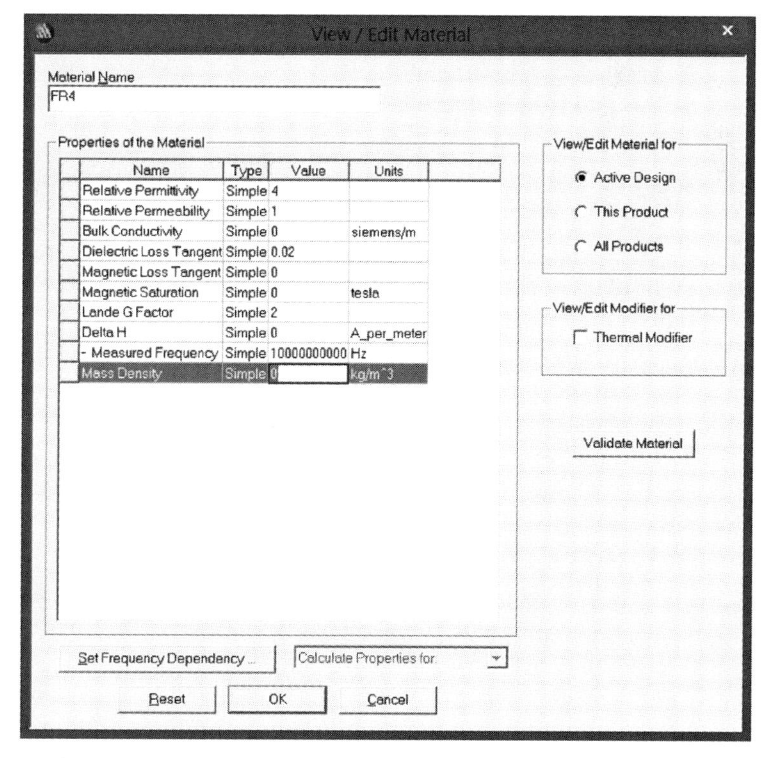

Figure C.11 Edit material window.

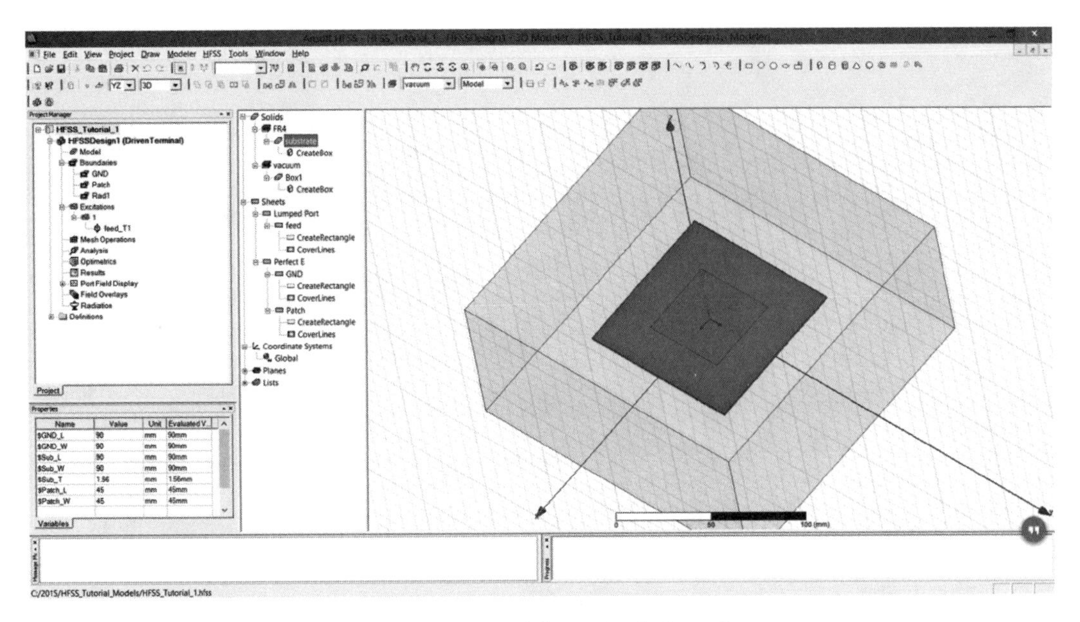

Figure C.12 Complete geometry setup of the probe-fed patch.

Stop: 2.5 GHz

Step size: 0.01 GHz

Hit OK.

18. To be able to see the radiation patterns, right-click on "Radiation" in the project tree, and choose "Insert Far Field Setup" and then "Infinite Sphere." The phi direction will start from 0 and goes up to 360° in 5° steps and same for the theta. This will cover the full sphere (spherical coordinates) of the antenna. Click OK.
19. The final project tree as well as the geometry will look like Figure C.13.
20. Check your design setup and geometry by clicking on the check mark in Figure C.13. You should get all check marks as in Figure C.14.
21. If all is well, now you can run your simulation by clicking on "!" next to the check mark. The analysis should start and wait until its completion. The message for completion will be displayed in the messages and analysis windows.
22. Once the simulation is completed, you can now check the results.
23. To plot the reflection coefficient, or other s-parameters, as well as the input impedance, among many other port parameters, go to "Results" under the main project tree, right-click on it, and then choose "Create Terminal Solution Data Report" and then "Rectangular plot"; you will get the window in Figure C.15. Have the selections highlighted in the figure, and click "New Report." You will get Figure C.16 showing the reflection coefficient, from which you can identify the resonance frequency (point of minimum reflection), and the –10-dB bandwidth. You can use markers to find the locations of the various values required as shown in Figure C.16. Note that the resonance is at 1.6 GHz and the bandwidth is around 34 MHz.

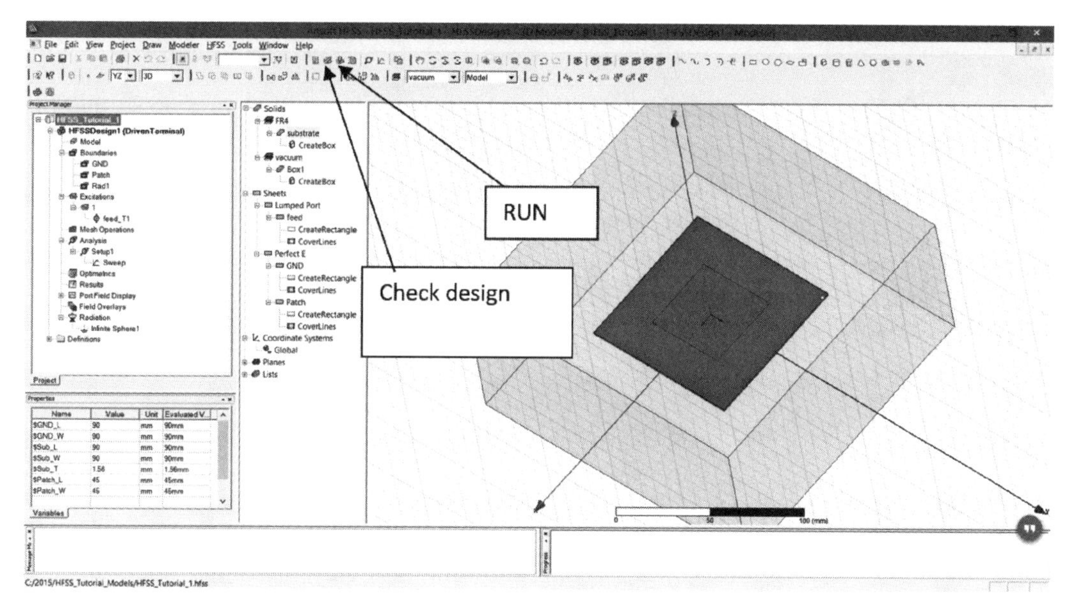

Figure C.13 Simulating the model.

24. To plot the input impedance (real and imaginary), create a new rectangular plot and from Figure C.15 choose "Terminal Z Parameter," then choose "re" for real component and "Create new report," and then in the same window, choose "img" and select "Add Trace." You will have Figure C.17. Note that the resonance is occurring at 1.59 GHz and the real impedance is 56Ω; thus, good matching is already preserved.
25. To plot the radiation gain pattern of this antenna, we need to right-click on "Results" from the project tree, then select "Create Far Fields Report," and choose "3D Polar Plot." The window in Figure C.18 pops up. Choose the values shown in the figure. Then click on the Families tab. There choose the "solution: Setup1 Sweep" and the correct frequency (from Families tab), which is the identified minimum S-parameter frequency (i.e., 1.6 GHz or

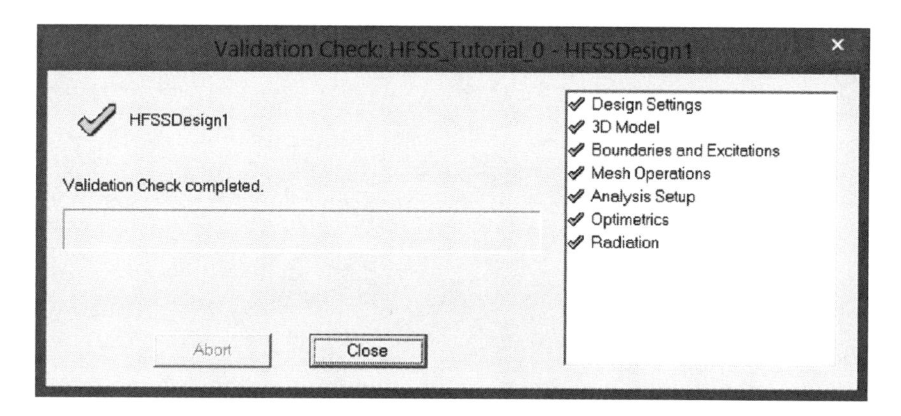

Figure C.14 Model Check window.

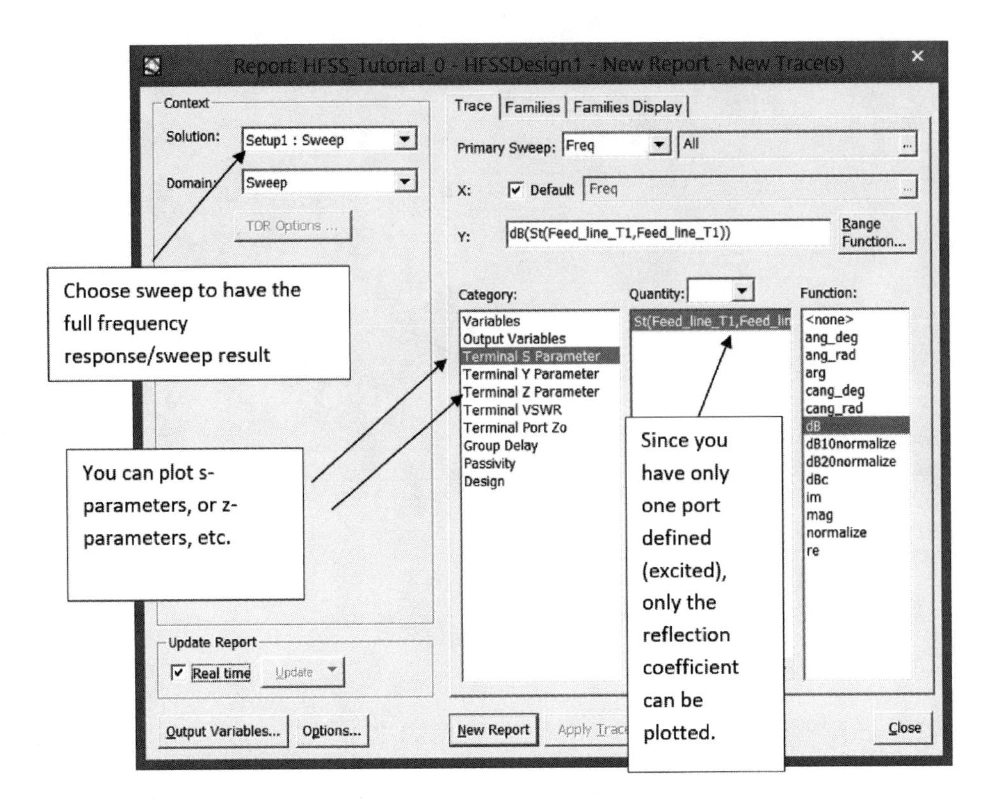

Figure C.15 Results plot window.

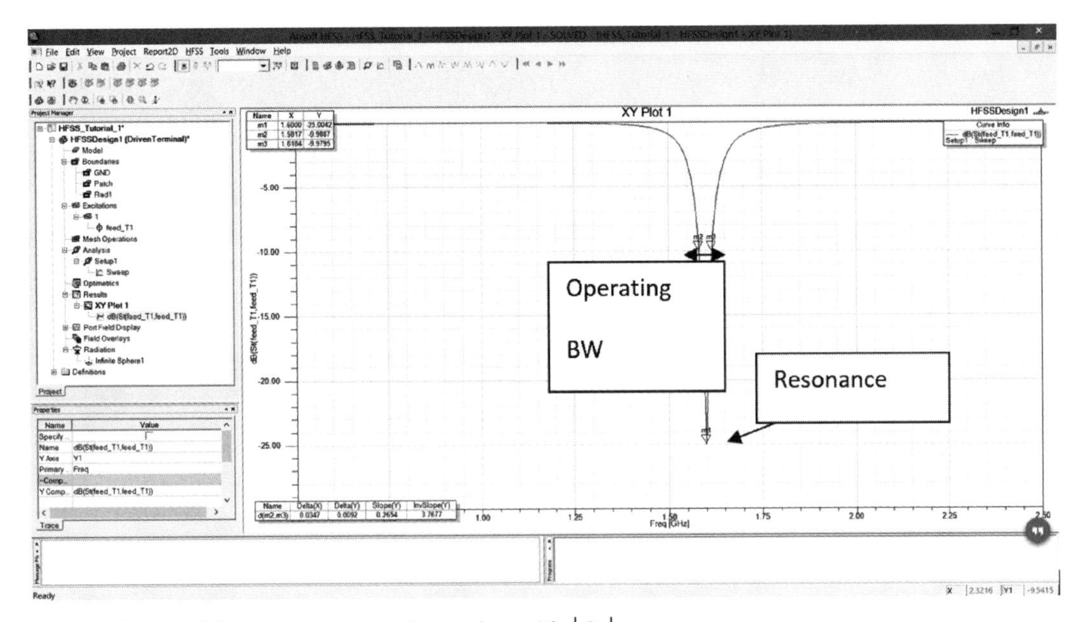

Figure C.16 Rectangular results window with $|S_{11}|$ curve.

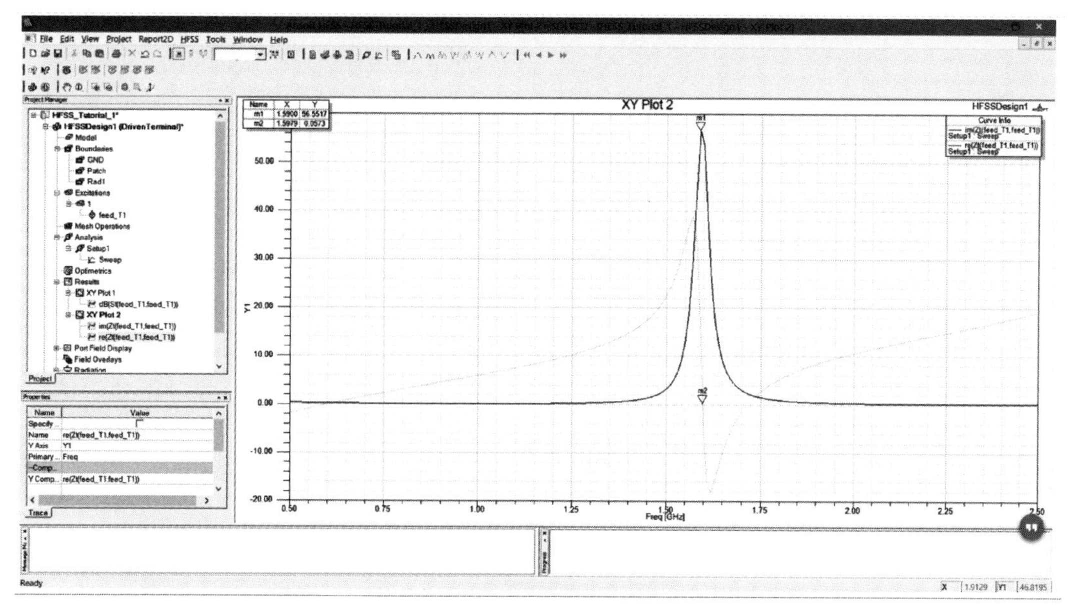

Figure C.17 Impedance curves.

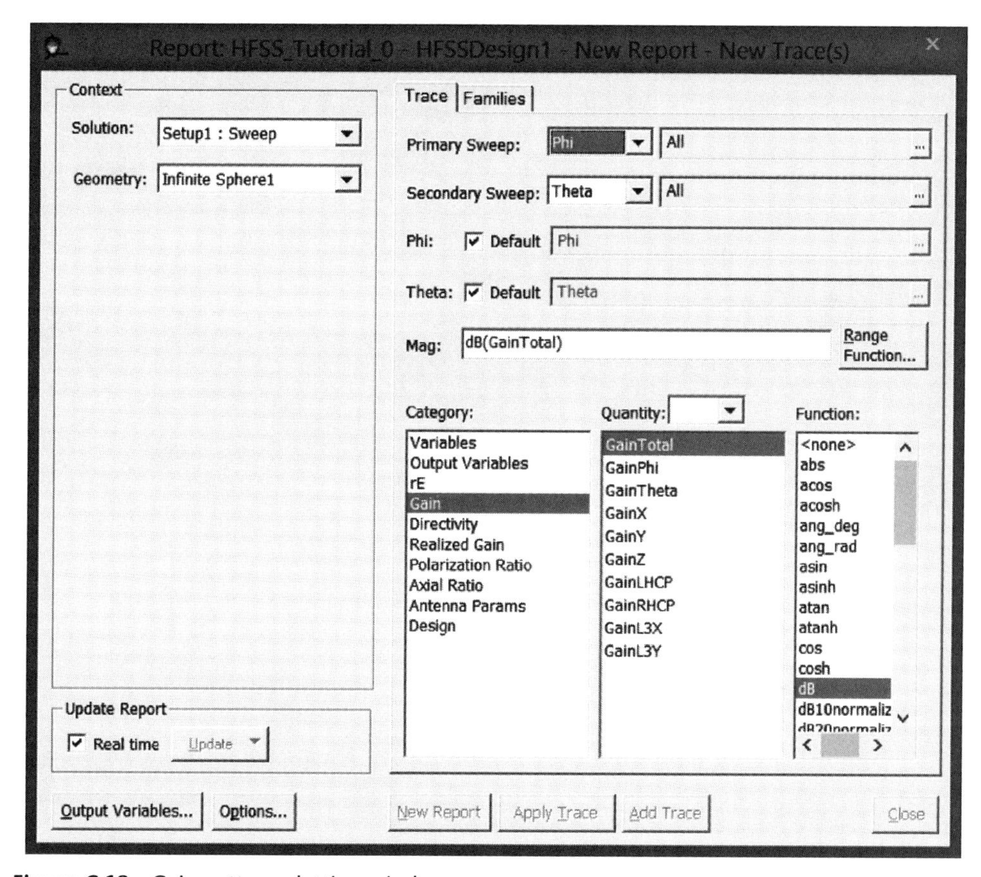

Figure C.18 Gain pattern plotting window.

something close to it). Then click on "New Report." The result will look like Figure C.19 (a directional pattern above the patch).

To check the efficiency of the antenna (i.e., how much does it radiate compared to what it is fed), go to "Radiation" in the main project tree, expand it, right-click on "Infinite Sphere1," and then choose "Compute Antenna Parameters"; then in "Available Solutions" choose "Setup1: Sweep" because we want to check the 1.6-GHz point and not the solution point of 1.5 GHz (default). Then, in the "intrinsic Variation" choose the frequency of interest (i.e., 1.6 GHz), and then click OK. Note the efficiency of the antenna, which is almost 40% (due to high losses in the cheap FR4 substrate with loss tangent of 0.02).

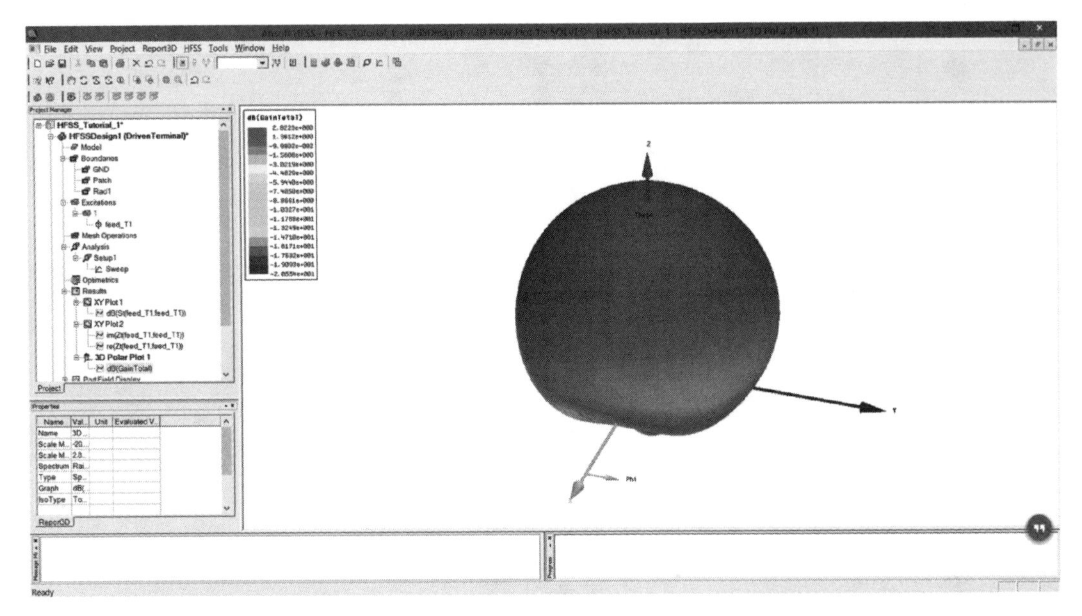

Figure C.19 Three-dimensional gain pattern plot

APPENDIX D
Using CST Tutorial

In this tutorial, CST will be used to model and simulate the response of a patch antenna. Simulation through CST can be divided into the following main steps:

1. Project creation;
2. Variable definition;
3. Model construction;
4. Feeding definition;
5. Simulation setup;
6. Results extraction.

A. Project Creation

- Open CST Studio Suite software and wait for CST main window to open as shown in Figure D.1, in this window you have to choose between creating a new project or loading an old one.
- Click on Create new project, and then choose Microwaves & RF as shown in Figure D.2 and click Next.

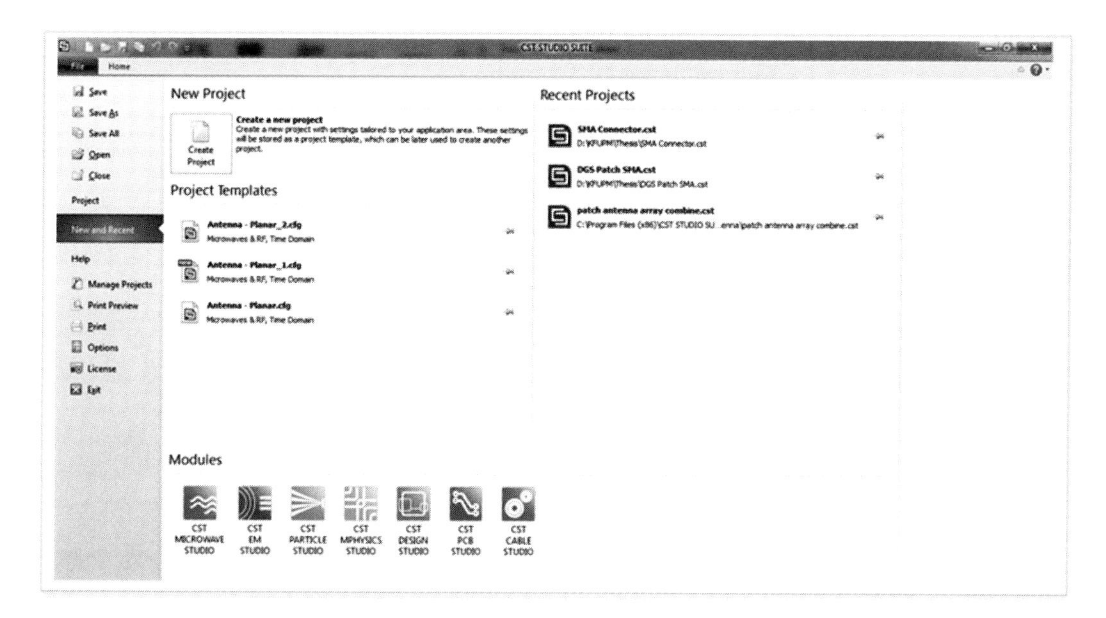

Figure D.1 CST main window.

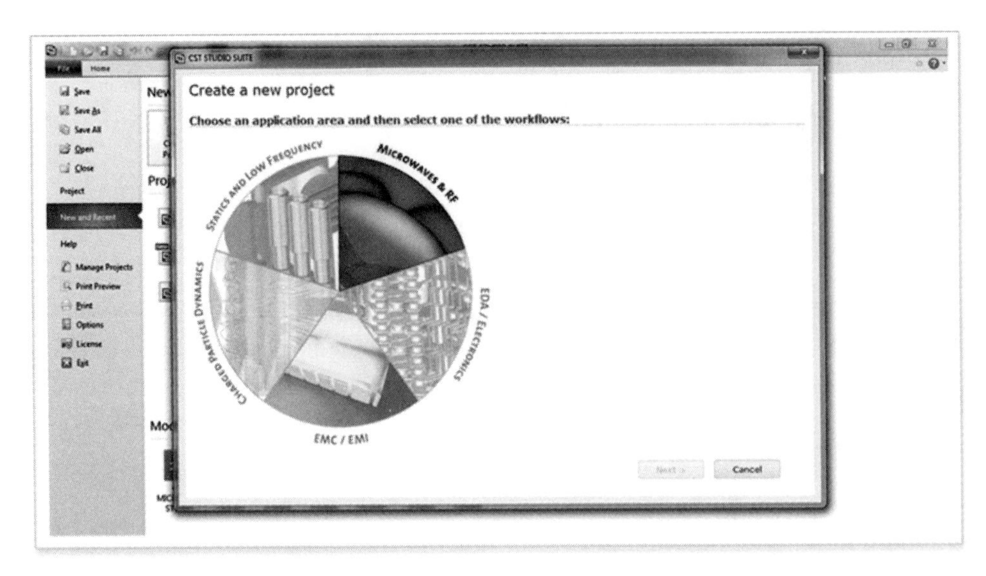

Figure D.2 Project application area.

- For antenna simulaton, click on Antennas as shown in Figure D.3 and then click on Planar to simulate the patch antenna as shown in Figure D.4.
- Then CST will display a list of suitable simulators for the specified application (antennas here) as shown in Figure D.5. You can choose between frequency-domain and time-domain simulators, click on Time Domain (time domain is better to use for wideband antennas, while frequency domain is usually used for narrowband ones).
- The units for the project will be displayed as shown in Figure D.6. For an antenna project, units typically used for dimensions and frequency are millimeters and gigahertz, respectively.

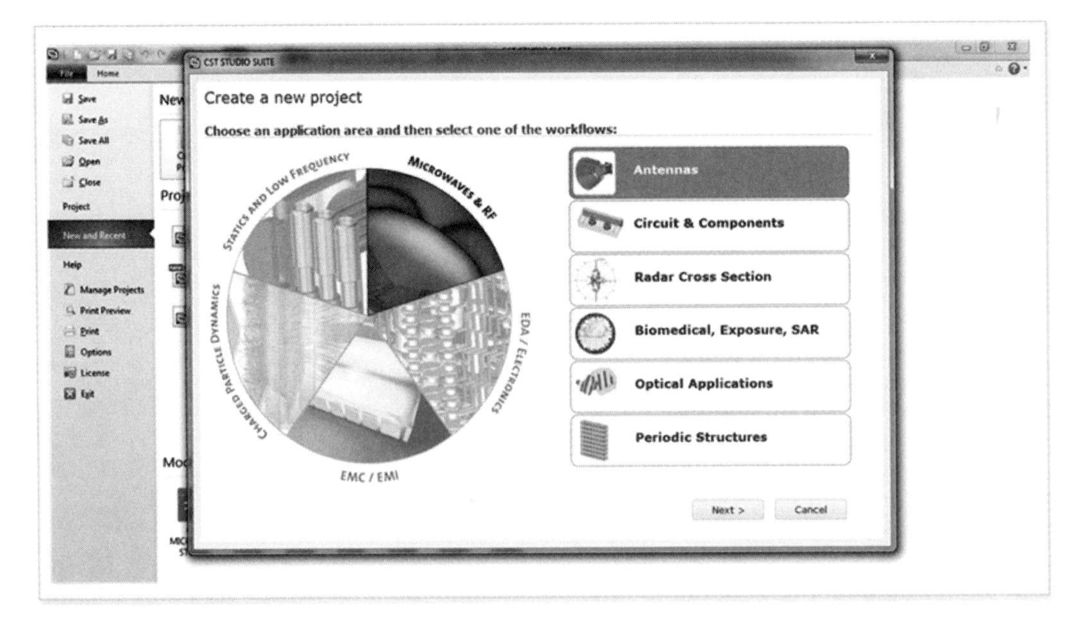

Figure D.3 Project workflows.

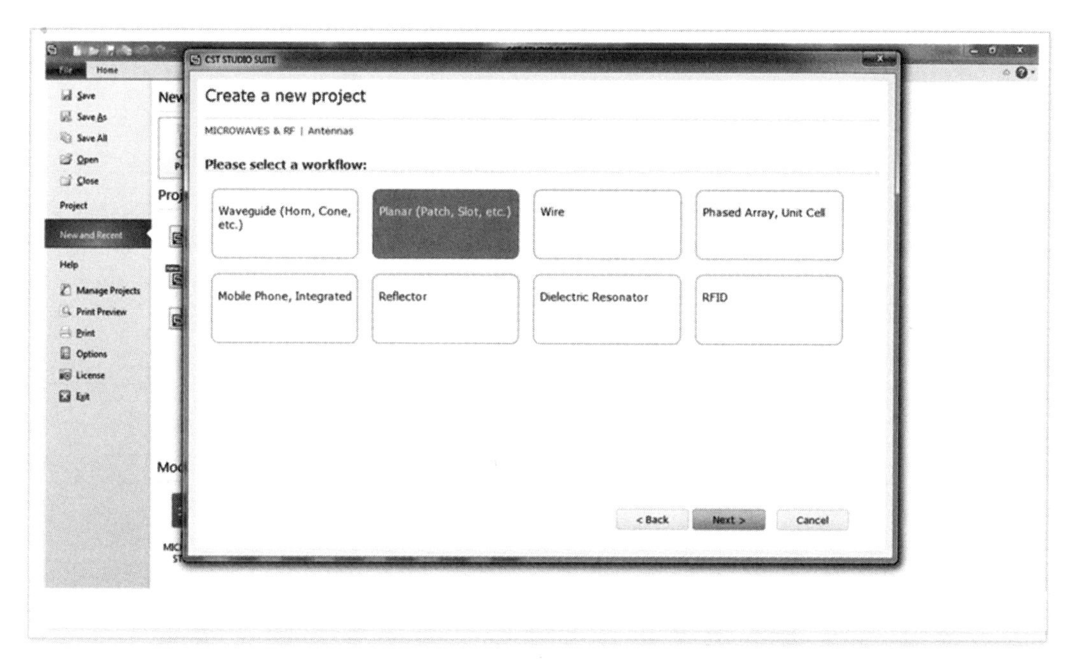

Figure D.4 Antenna categories.

- Now the project setup is completed and the setup summary will be displayed as shown in Figure D.7. Click Finish to proceed to the project main window shown in Figure D.8.
- On the left side of Figure D.8, you will find the navigation tree in which the model components, ports, and results will be found and the parameter list will be in the bottom.

Figure D.5 Simulator selection.

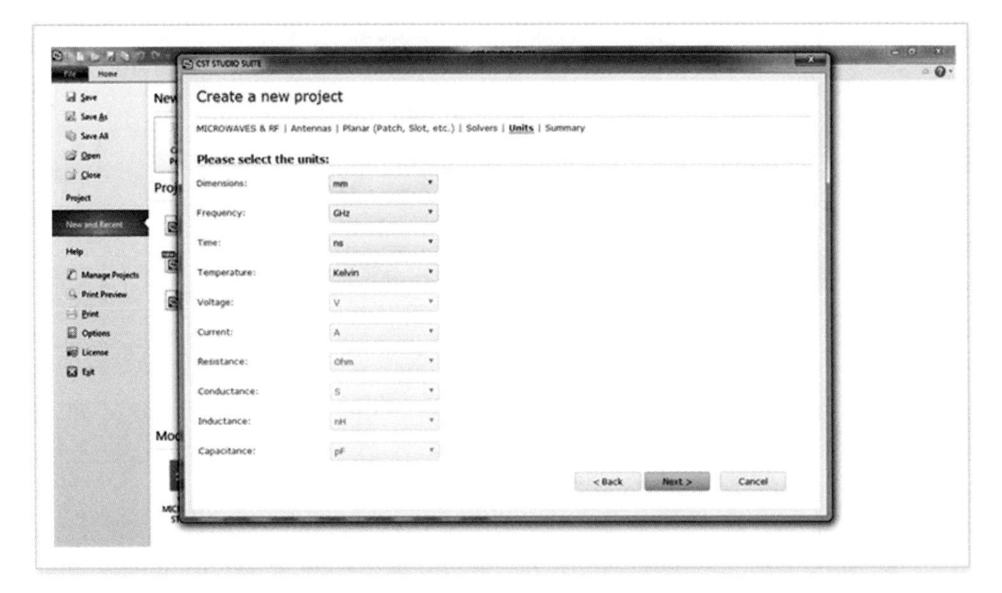

Figure D.6 Unit selections.

B. Variable Definition

- For calculating the dimensions of a rectangular patch with certain design requirements, we can refer to a standard antenna design book, or we can use an online calculator. You can define variables as pw (patch width), pl (patch length), subw (substrate width), subl (substrate length), subthick (substrate height), and tl (transmission line).

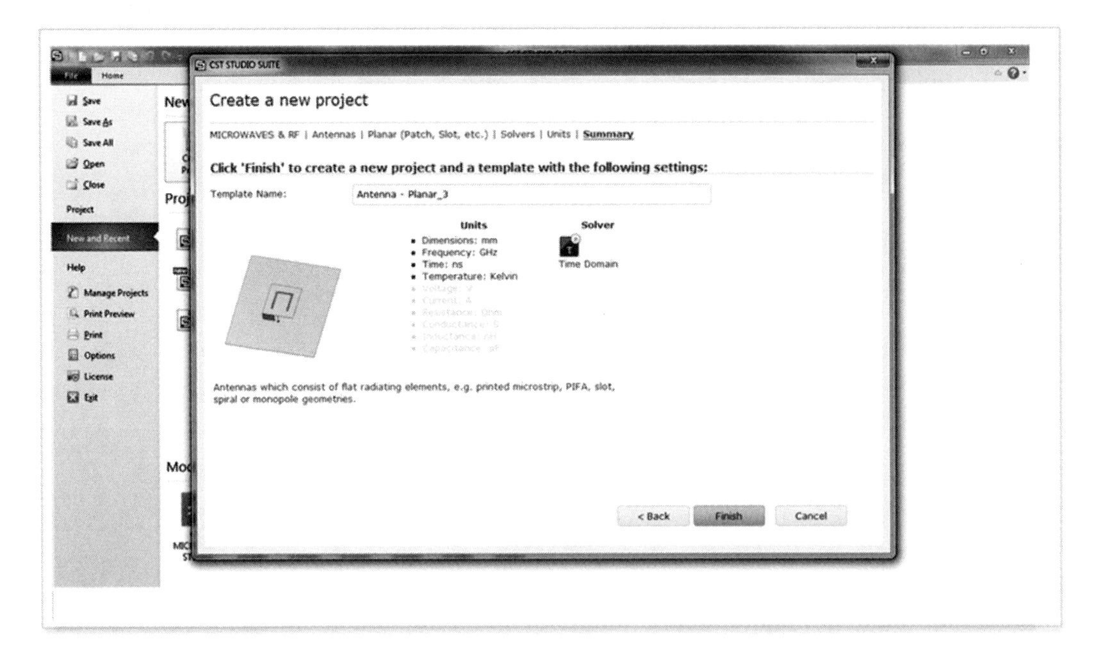

Figure D.7 Project summary.

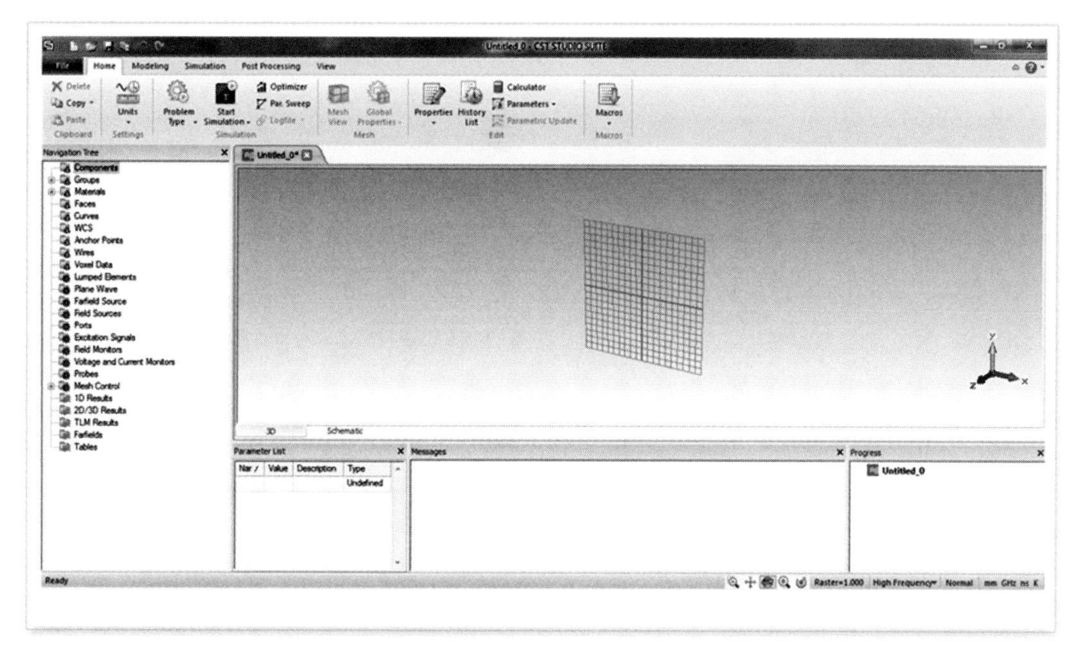

Figure D.8 Project main window.

C. Model Construction

- To create a patch, you can select Brick from the Modeling menu and name its width as pw = 50 mm and length as pl = 40 mm as shown in Figure D.9. Select the material as PEC. You can repeat the same steps for creating the substrate by assigning subw = 70 mm, subl = 70 mm, and subthick = 1.6 mm and then assigning the material as FR-4 (lossy) as shown in Figures D.10 and D.11.

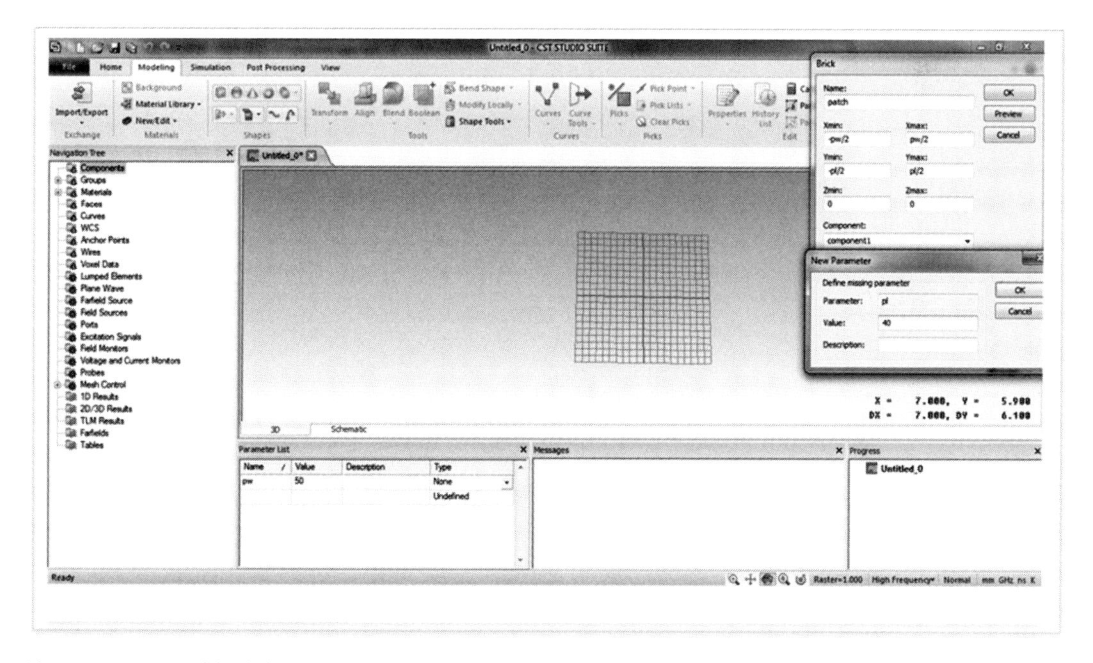

Figure D.9 Variable definitions.

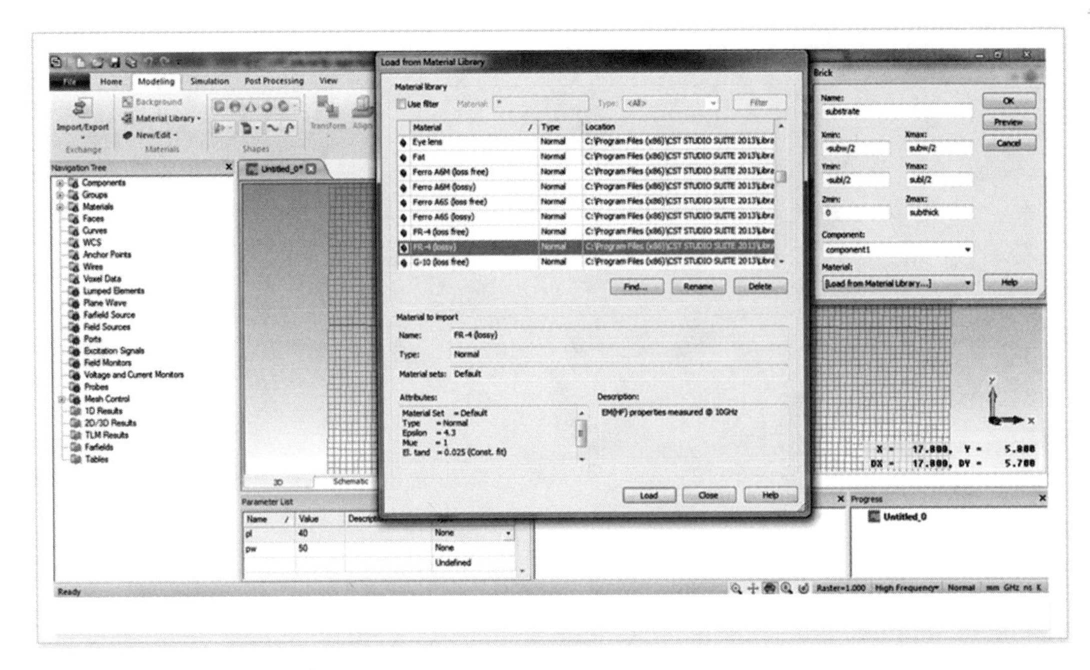

Figure D.10 Choosing a substrate.

- A ground plane is created in the same way as of the patch element and is assigned as a PEC as shown in Figure D.12.
- Local coordinates enable the user to design multiple objects with respect to its own local origin and hence messy calculations are avoided (e.g., here to create a transmission line, we can use the local coordinates by selecting Local WCS from the modeling menu). Select a point for the local coordinates and

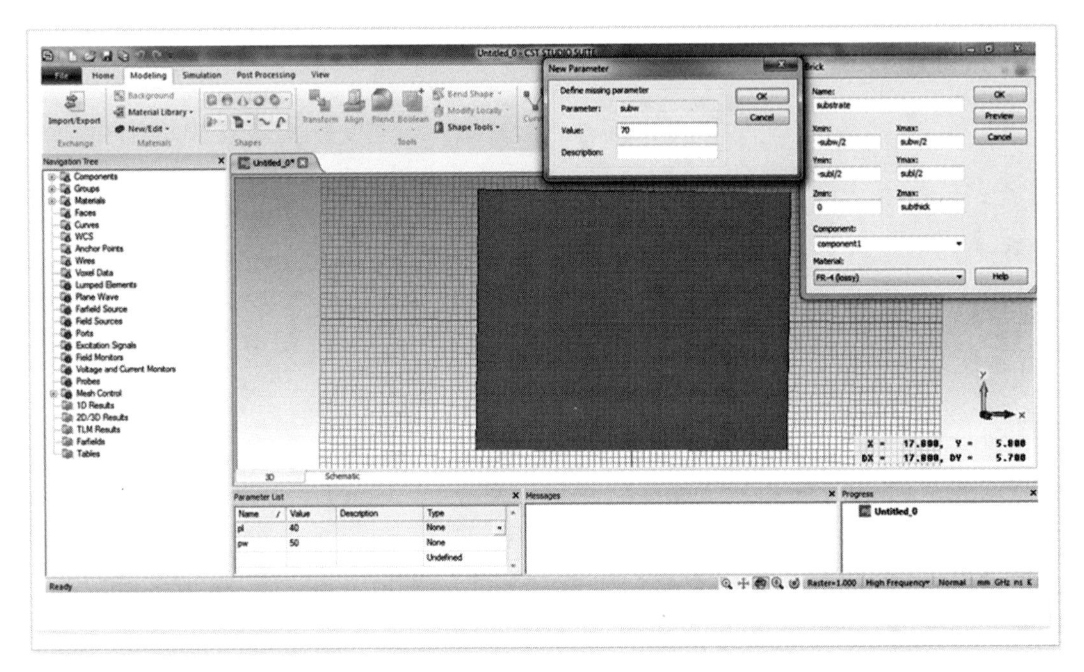

Figure D.11 Designing substrate.

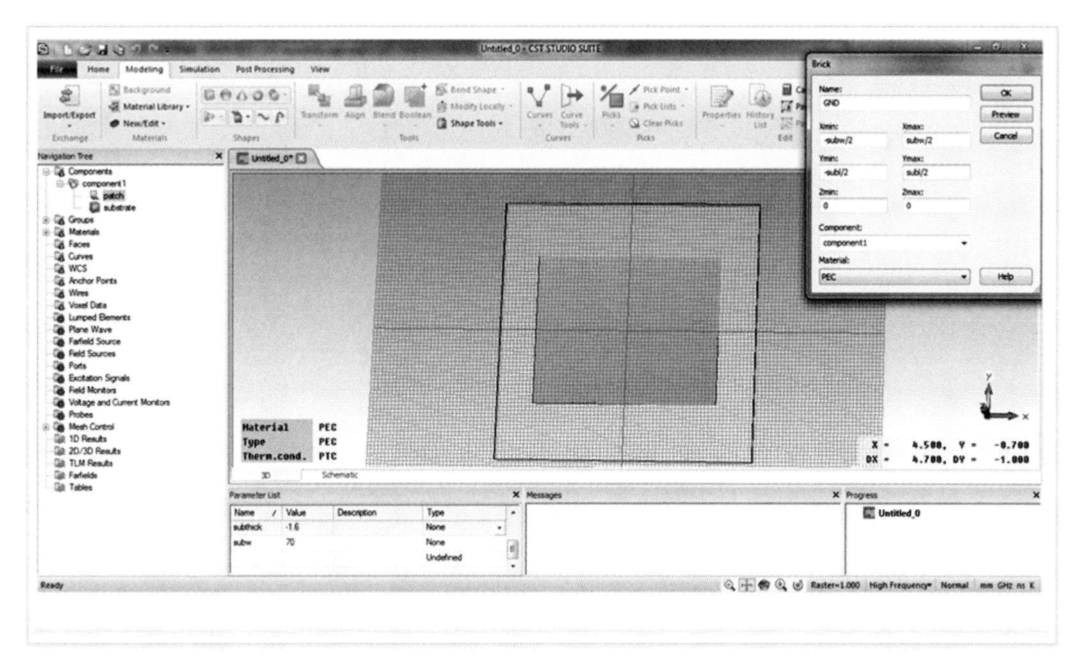

Figure D.12 Defining the ground plane.

you can now assign values and Umin and Umax represent the width of the transmission line, which is 3 mm, so one can set −1.5 to 1.5 mm, similar to the length that you can set −Vmin to Vmax as shown in Figure D.13. Select PEC for transmission line. The length of the transmission line is chosen as 15 mm; you can get the starting point from online calculator and then do further optimization to get good matching.

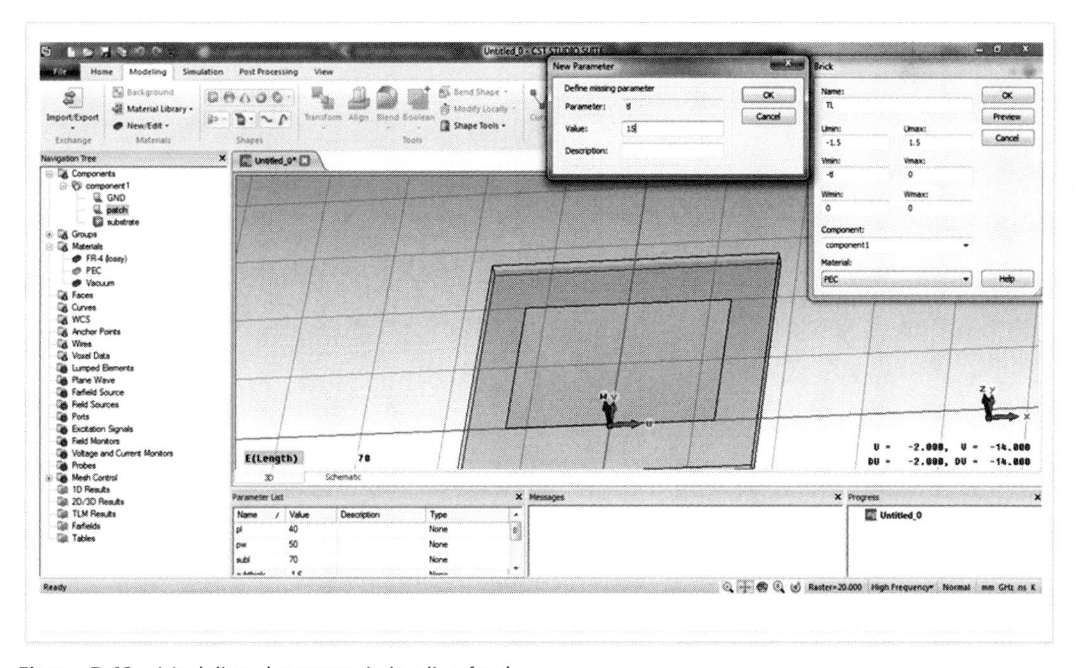

Figure D.13 Modeling the transmission line feed.

- To find the inset point Y_o of the feeding to match the antenna to the transmission line, refer to a standard antenna textbook. Here this distance was found to be 13.8 mm. For simplifying the calculation, you can again select local WCS as we did before and you can now create these two slits (having a width of 1 mm and length of 13.8 mm) as shown in Figures D.14 through D.16. After successfully creating one slit, for simplicity, you can use the Boolean command inside the Modeling menu, as shown in Figure D.16, to create the inset feed clearance.
- After adding the inset feed, the final geometry is shown in Figure D.17.
- To finalize the model, go to Simulation tab and click on Frequency in the left top corner. Specify the frequency range, for instance, as the center frequency is 1.9, so we can select the minimum and maximum values as 1.5 GHz and 2.3 GHz, respectively.
- CST has another nice feature which is calculating the simulation boundary automatically. Click on Boundaries (under Frequency Option mentioned above) and then the boundary conditions window will pop up; click OK to activate the automatic boundary calculation.

D. Feeding Definition

- To define an excitation first you have to pick the brick, go to the Simulation tab and click on Pick edge. Then choose the upper edge of the brick and do the same for the lower edge of the brick as shown in Figure D.18.
- Click on Discrete port as shown in Figure D.18. Then the discrete port setup will pop up, click OK as the edges are already chosen. Now the antenna is excited with a discrete port.

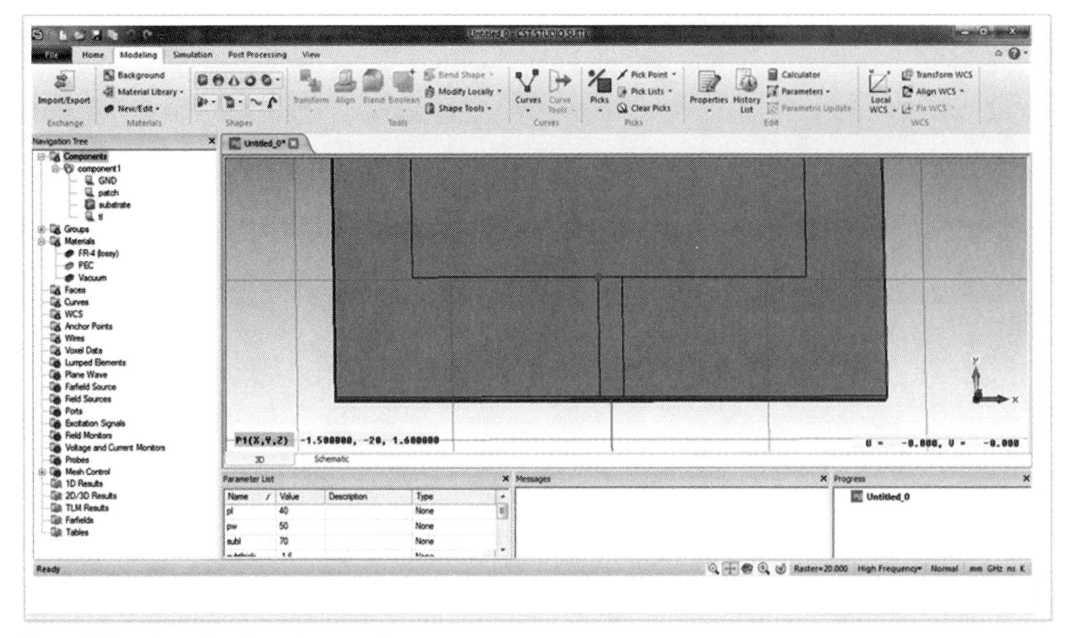

Figure D.14 Inset feed creaton step 1.

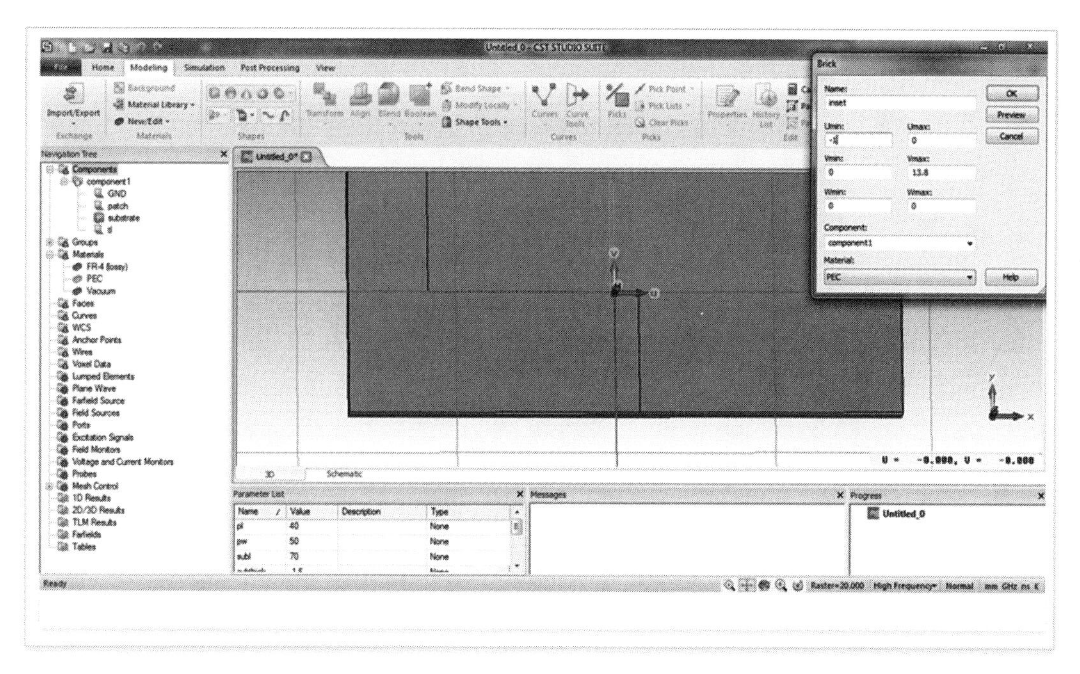

Figure D.15 Inset feed creation step 2.

E. Simulation Setup

- The first step to prepare the antenna simulation is to define a monitor for far fields. First click on Field Monitor in the Simulation tab and then choose Farfield/RCS and click OK.
- Click on Start Simulation wait for the simulation to complete.

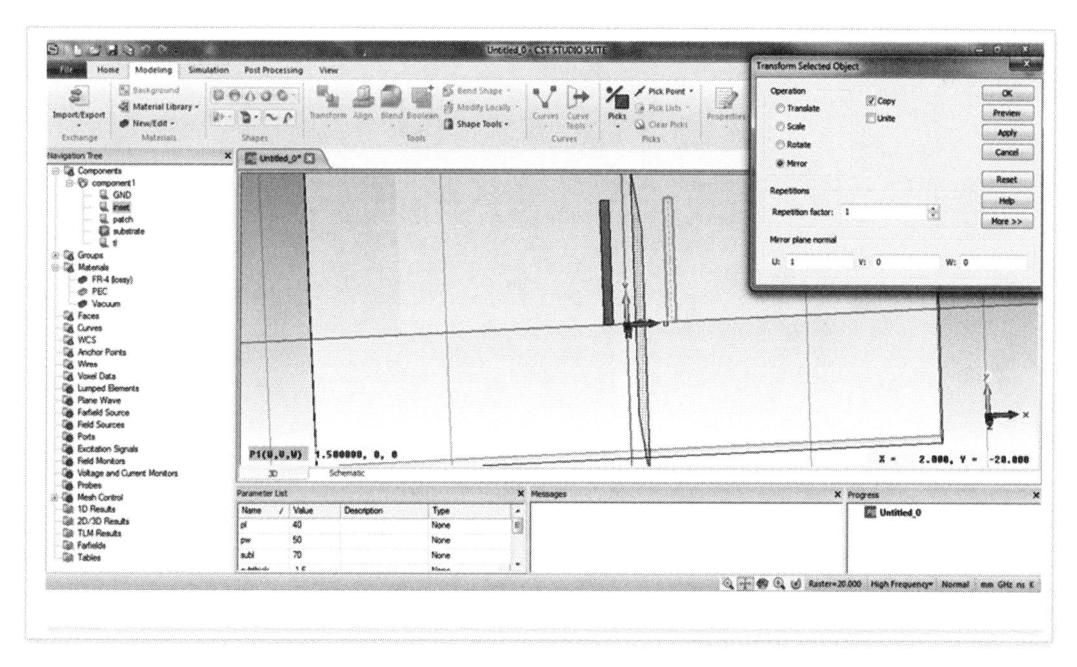

Figure D.16 Inset feed completion.

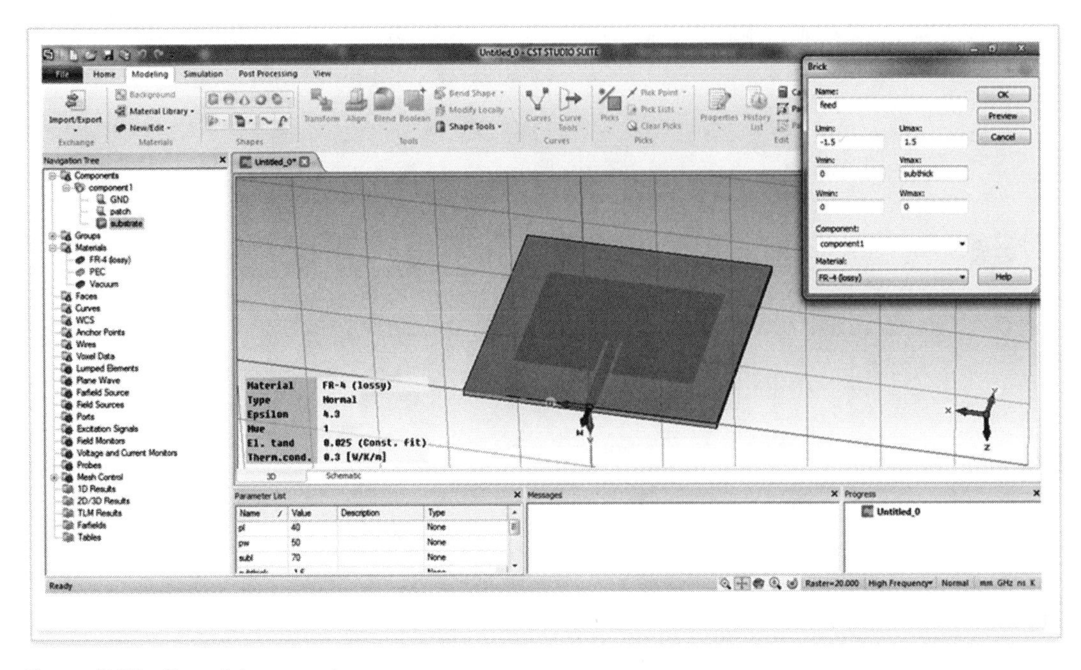

Figure D.17 Complete geometry.

F. Results Extraction

- After the simulation is completed, the results can be found in the Navigation Tree. For example, in one-dimensional results, the *S*-parameters and impedance curves can be found. S11 curve is shown in Figure D.19. The resonance frequency is little shifted to the left, so you need to optimize it.

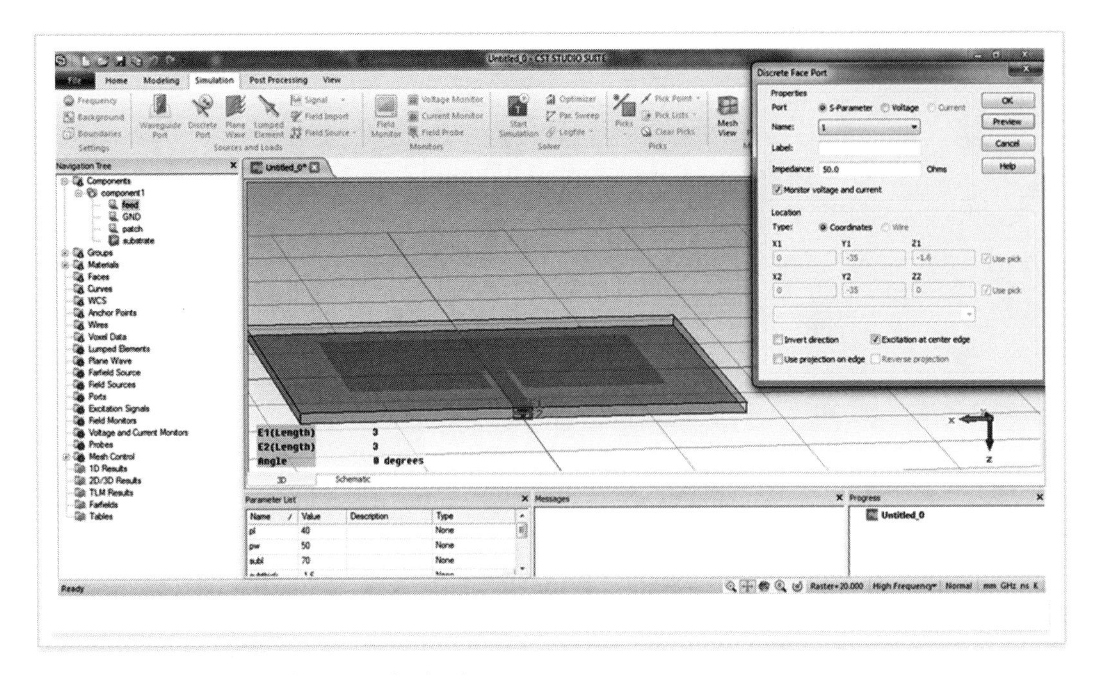

Figure D.18 Excitation addition to the feed.

- The real and imaginary parts of the impedance are shown in Figures D.20 and D.21, respectively. The real impedance is close to 50Ω as expected. Figure D.22 shows the 3-D gain pattern. This antenna has a gain of 4.2 dBi.

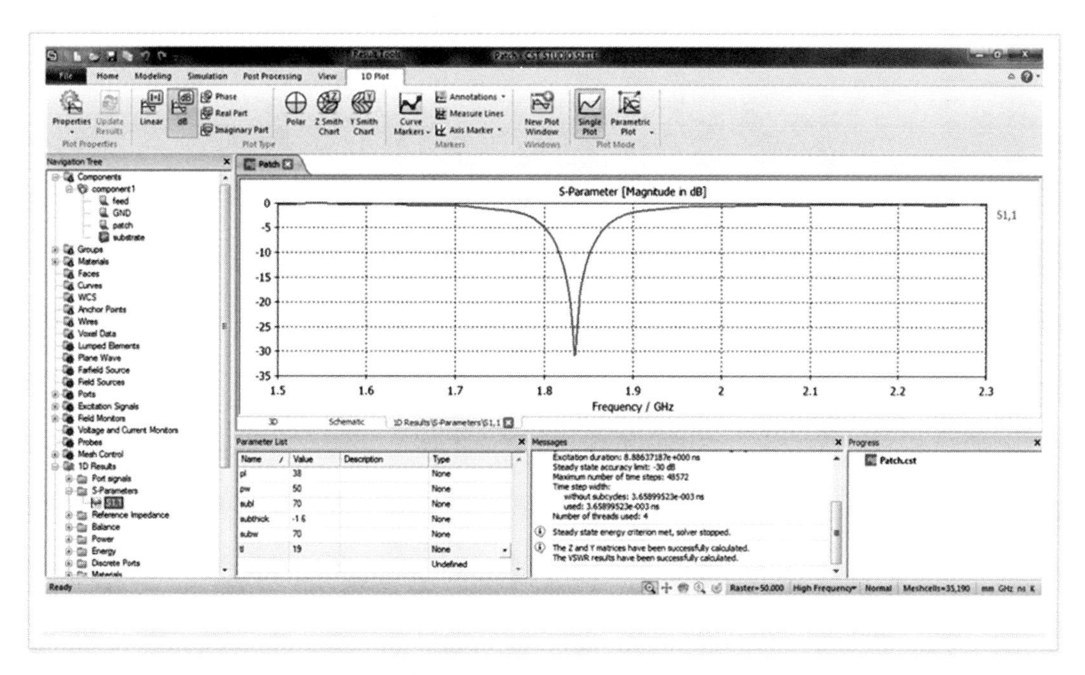

Figure D.19 S11 curve.

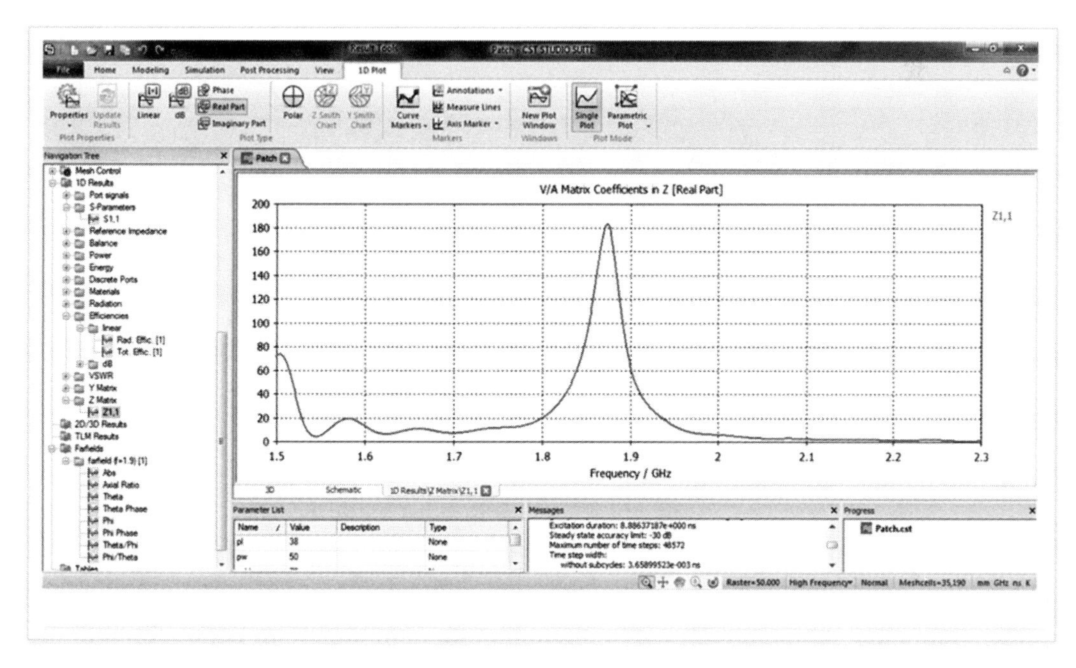

Figure D.20 Impedance, real part.

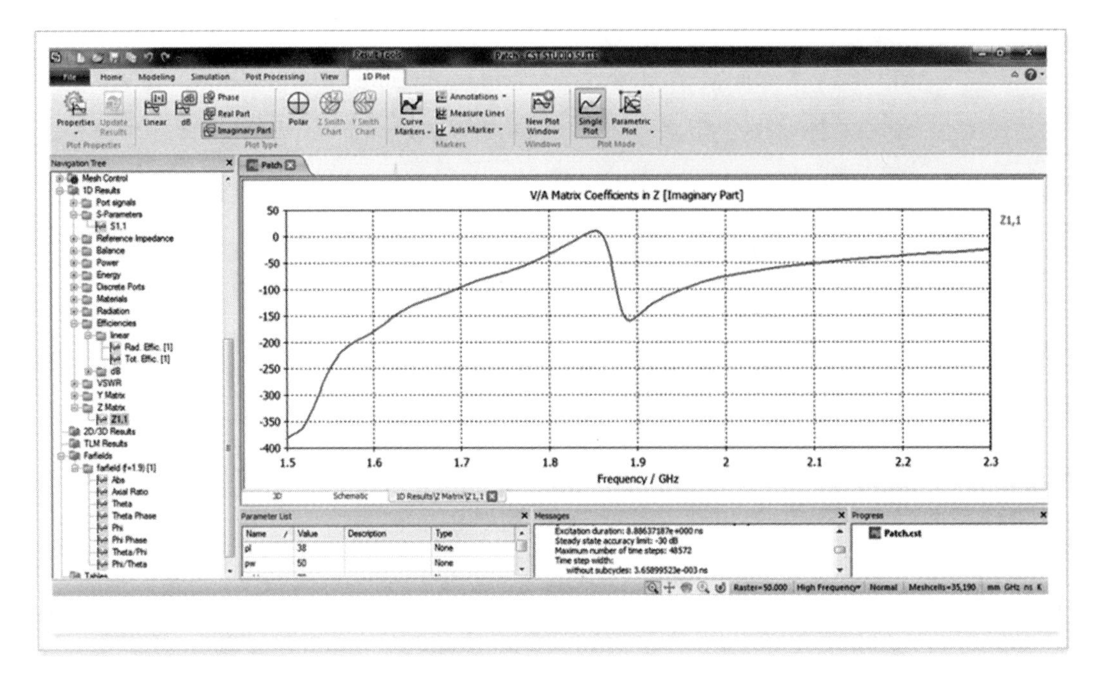

Figure D.21 Impedance, imaginary part.

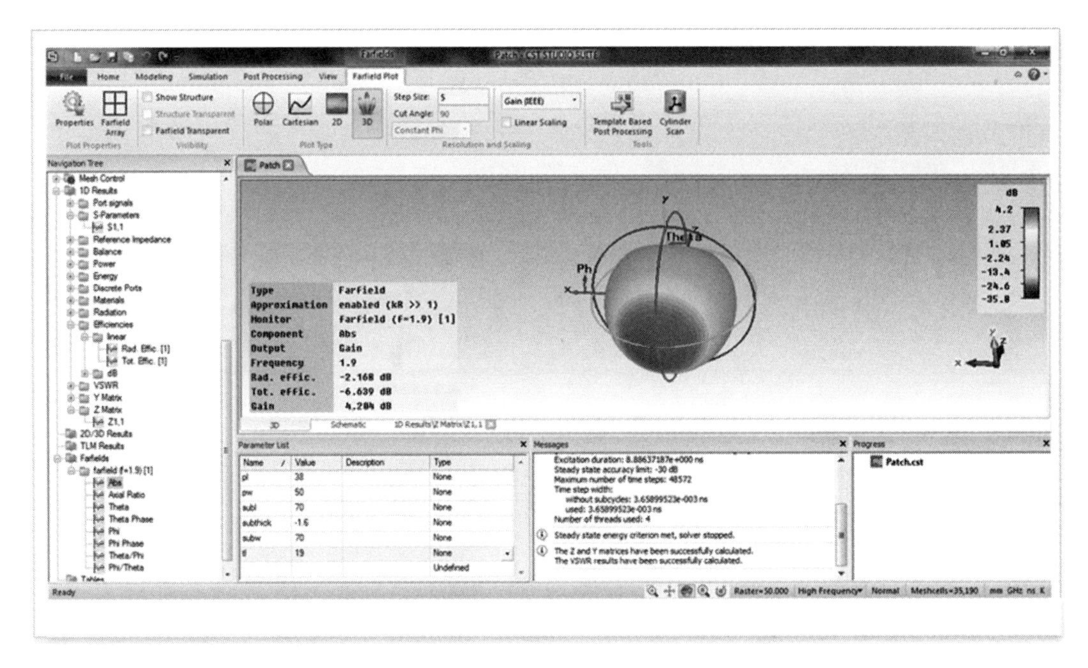

Figure D.22 The 3-D radiation pattern.

List of Acronyms

1G	First generation of the wireless standards
2G	Second generation of the wireless standards
3G	Third generation of the wireless standards
4G	Fourth generation of the wireless standards
5G	Fifth generation of the wireless standards
AC	alternating current
ADC	analog-to-digital converter
ADG	apparent diversity gain
ADS	Advanced Design System (software package)
AF	array factor
AIA	active integrated antenna
AM-PM	amplitude modulation to phase modulation
AMC	artificial magnetic conductor
ATC	air traffic control
BER	bit error rate
BJT	bipolar junction transistor
BPF	bandpass filter
BPR	branch power ratio
BS	base station
BW	bandwidth
CA	carrier aggregation
CAD	computer-aided design
CC	correlation coefficient
CDF	cumulative distribution function
CDMA	code division multiple access
CR	cognitive radio
CSRR	complementary split ring resonator
CST	Computer Simulation Technology (software package)
CW	continuous wave
DAC	digital to analog converter

dB	decibels
dBc	dB carrier or dB for circularly polarized waves/antennas
dBi	dB with respect to an isotropic antenna
DC	direct current
DG	diversity gain
DGS	defected ground structure
DNG	double negative
DPD	digital predistortion
DR	dielectric resonator
DSB	double side band
DSP	digital signal processing
ECC	envelope correlation coefficient
EDG	effective diversity gain
EDGE	enhanced data rates for GSM evolution standard
EIRP	effective isotropic radiated power
EM	electromagnetics
ESA	electrically small antenna
ETSI	European Telecommunications Standards Institute
FDMA	frequency division multiple access
FDTD	finite difference time domain
FEM	finite element method
FET	field effect transistor
FPGA	field programmable gate array
GaAs	gallium arsenide
GND	ground
GPRS	General Packet Radio Service standard
GPS	Global Positioning System
GSM	Global System Mobile
HB	harmonic balance
HEMT	high electron mobility transistor
HFSS	High Frequency Structure Simulator (software package)
HPBW	half-power beamwidth
HSPA	high-speed packet access standard
IC	integrated circuit
IEEE	Institute of Electrical and Electronics Engineers
IET	Institute of Engineering and Technology
IF	intermediate frequency
IIP3	third order input intercept point

IMN	input matching network
IQ	in-phase and quadrature
IRR	image rejection ratio
LINC	linear amplification using nonlinear components
LNA	low noise amplifier
LO	local oscillator
LPF	lowpass filter
LTE	Long-Term Evolution (4G)
LTE-A	LTE-Advanced
MEG	mean effective gain
MESFET	metal-semiconductor field effect transistor
MIC	microwave integrated circuit
MIMO	multiple-input-multiple-output
m-MIMO	massive multiple-input-multiple-output
MMIC	microwave monolithic integrated circuit
MN	matching network
MoM	method of moments
MRC	maximum ratio combining
MTM	metamaterial
MWO	Microwave Office (software package)
NEC	Numerical Electromagnetic Code (software package)
NF	noise figure
NIC	negative impedance converter
OCS	open circuit stable
OFDM	orthogonal frequency division multiplexing
OFDMA	orthogonal frequency division multiple access
OMN	output matching network
OSRR	open split ring resonator
PA	power amplifier
PAE	power added efficiency
PAPR	peak to average power ratio
pHEMT	pseudomorphic high electron mobility transistor
PIFA	planar inverted-F antenna
PLF	polarization loss factor
QAM	quadrature amplitude modulation
RF	radio frequency
RFC	radio frequency choke
RFID	radio frequency identification

RFT	real frequency technique
RIS	reactive impedance surface
RLC	resistor, inductor and capacitor circuit (series or parallel)
SCS	short-circuit stable
SLL	sidelobe level
SMA	SubMiniature version A (connector model)
SMS	short message service
SNR	signal-to-noise ratio
SOM	self-oscillating mixer
SRR	split ring resonator
SSB	single side band
TARC	total active reflection coefficient
TDMA	time division multiple access
TL	transmission line
USB	Universal Serial Bus
UT	user terminal
UWB	ultrawideband
VCO	voltage controlled oscillator
VSWR	voltage standing wave ratio
WCDMA	wideband code division multiple access
WLAN	wireless local area network
XPD	cross-polarization discriminator

About the Authors

Mohammad S. Sharawi is currently a professor of electrical engineering at King Fahd University of Petroleum and Minerals (KFUPM), Dhahran, Saudi Arabia. He was a visiting professor at the Intelligent Radio (iRadio) Laboratory, at the Electrical Engineering Department, University of Calgary, Alberta, Canada, during the summer and fall of 2014. He was a visiting research professor at Oakland University during the summer of 2013. Prior to joining KFUPM in 2009, Professor Sharawi was a research scientist with the Applied Electromagnetics and Wireless Laboratory at the Electrical and Computer Engineering Department, Oakland University, Michigan, United States, during 2008 to 2009. He was a faculty member in the Computer Engineering Department at Philadelphia University, Amman, Jordan, during 2007 to 2008. He served as the acting head of the Computer Engineering Department at the German-Jordanian University in Amman, Jordan during 2006 to 2007. He also served as the organization committee chair of the IEEE Conference on Systems, Signals, and Devices that was held in Jordan in July 2008.

Professor Sharawi obtained his B.Sc. degree in electronics engineering with highest honors from Princess Sumaya University for Technology in Amman, Jordan, in 2000. He received his M.Sc. degree in electrical and computer engineering and Ph.D. in radio frequency (RF) systems engineering from Oakland University, Rochester, Michigan, United States, in 2002 and 2006, respectively, with a focus on microwave and antenna systems design. He was an exchange student during the academic year of 1997 and 1998 at the University of Illinois at Urbana-Champaign. He was an intern with Silicon Graphics Inc. (SGI), Mountain View, California, United States, during the summers of 1998, 2001, and 2002, where he worked on various circuit design and interconnect modeling projects. Then he became a member of technical staff, design engineer, with SGI during 2002 to 2003, where he was responsible for modeling, design, and verification of high-speed PCBs and circuits for various high-performance computing systems.

Professor Sharawi is the founder and director of the Antennas and Microwave Structure Design (AMSD) Laboratory at KFUPM. His research interests include printed antennas and antenna arrays, MIMO antenna systems, millimeter-wave antennas, reconfigurable antennas, applied electromagnetics, microwave electronics, and microwave system integration. Professor Sharawi has brought more than 10 million SAR in research funding (approximately $2.5 million in U.S. dollars) in 8 years while with KFUPM. He has organized several special sessions in international IEEE conferences in the area of printed, MIMO, and millimeter-wave antenna systems and has served on the technical program committees (TPC) at APS, EuCAP, APCAP, ICCE, and APWC among other international conferences.

Professor Sharawi is the author of the book *Printed MIMO Antenna Engineering* (Artech House, 2014) and the lead author of 8 book chapters on antennas and RF system design. He has published more than 200 refereed journal and conference papers and has 14 issued patents and 15 pending patents with the U.S. Patent Office. Professor Sharawi is a Senior Member in the IEEE and a Fellow IET.

Oualid Hammi received his B.Eng. degree from the École Nationale d'Ingénieurs de Tunis, Tunis, Tunisia, in 2001, his M.Sc. degree from the École Polytechnique de Montréal, Université de Montréal, Montréal, Quebec, Canada, in 2004, and his Ph.D. degree from the University of Calgary, Calgary, Alberta, Canada, in 2008, all in electrical engineering. Since 2015, he has been an associate professor of electrical engineering at the American University of Sharjah, Sharjah, United Arab Emirates. He has also been an adjunct associate professor with the Electrical and Computer Engineering Department, Schulich School of Engineering, University of Calgary, Calgary, Alberta, Canada, since 2015. From 2010 to 2015, he was a faculty member with the Department of Electrical Engineering, at King Fahd University of Petroleum and Minerals (KFUPM), Dhahran, Saudi Arabia.

He coauthored one book, *Behavioral Modeling and Predistortion of Wideband Wireless Transmitters* (John Wiley & Sons, 2015), and one chapter, "Digital Predistortion for Microwave Transmitters," for the *Wiley Encyclopedia of Electrical and Electronics Engineering*, in 2014. His research activities led to over 100 technical papers in refereed journals and international conferences, 10 patents awarded by the U.S. Patent and Trademark Office (USPTO), and one university spin-off company. He is a regular reviewer for numerous *IEEE Transactions* and several other journals.

His teaching interests are in the area of electronics, RF circuits and systems, microwave techniques, and innovative approaches to learning engineering through projects. His research interests include the characterization, behavioral modeling, and linearization of RF power amplifiers and transmitters, the design of energy-efficient RF circuits for 5G wireless communications, Internet of Things (IoT), and satellite applications.

Index

A

B

C

D

E

F

G

N

O

S

T

U

V

W

Recent Titles in the Artech House Antennas and Electromagnetics Analysis Library

Christos Christodoulou, Series Editor

Adaptive Array Measurements in Communications, M. A. Halim

Advanced Computational Electromagnetic Methods and Applications, Wenhua Yu, Wenxing Li, Atef Elsherbeni, Yahya Rahmat-Samii, Editors

Advances in Computational Electrodynamics: The Finite-Difference Time-Domain Method, Allen Taflove, editor

Advances in FDTD Computational Electrodynamics: Photonics and Nanotechnology, Allen Taflove, editor; Ardavan Oskooi and Steven G. Johnson, coeditors

Analysis Methods for Electromagnetic Wave Problems, Volume 2, Eikichi Yamashita, editor

Antenna Design for Cognitive Radio, Youssef Tawk, Joseph Costantine, and Christos Christodoulou

Antenna Design with Fiber Optics, A. Kumar

Antenna Engineering Using Physical Optics: Practical CAD Techniques and Software, Leo Diaz and Thomas Milligan

Antennas and Propagation for Body-Centric Wireless Communications, Second Edition, Peter S. Hall and Yang Hao, editors

Antennas and Site Engineering for Mobile Radio Networks, Bruno Delorme

Analysis of Radome-Enclosed Antennas, Second Edition, Dennis J. Kozakoff

Applications of Neural Networks in Electromagnetics, Christos Christodoulou and Michael Georgiopoulos

The Art and Science of Ultrawideband Antennas, Second Edition, Hans G. Schantz

AWAS for Windows Version 2.0: Analysis of Wire Antennas and Scatterers, Antonije R. Djordjević, et al.

Broadband Microstrip Antennas, Girsh Kumar and K. P. Ray

Broadband Patch Antennas, Jean-François Zürcher and Fred E. Gardiol

CAD of Microstrip Antennas for Wireless Applications, Robert A. Sainati

The CG-FFT Method: Application of Signal Processing Techniques to Electromagnetics, Manuel F. Cátedra, et al.

Computational Electrodynamics: The Finite-Difference Time-Domain Method, Third Edition, Allen Taflove and Susan C. Hagness

Design and Applications of Active Integrated Antennas, Mohammad S. Sharawi and Oualid Hammi

Electromagnetics and Antenna Technology, Alan J. Fenn

Electromagnetic Modeling of Composite Metallic and Dielectric Structures, Branko M. Kolundzija and Antonije R. Djordjević

*Electromagnetic Waves in Chiral and Bi-Isotropic Media,*I. V. Lindell, et al.

Electromagnetics, Microwave Circuit and Antenna Design for Communications Engineering, Peter Russer

Engineering Applications of the Modulated Scatterer Technique, Jean-Charles Bolomey and Fred E. Gardiol

Fast and Efficient Algorithms in Computational Electromagnetics, Weng Cho Chew, et al., editors

Frequency-Agile Antennas for Wireless Communications, Aldo Petosa

Fresnel Zones in Wireless Links, Zone Plate Lenses and Antennas, Hristo D. Hristov

Handbook of Antennas for EMC, Thereza MacNamara

Handbook of Reflector Antennas and Feed Systems, Volume I: Theory and Design of Reflectors, Satish Sharma, Sudhakar Rao, and Lotfollah Shafai, editors

Handbook of Reflector Antennas and Feed Systems, Volume II: Feed Systems, Lotfollah Shafai, Satish Sharma, and Sudhakar Rao, editors

Handbook of Reflector Antennas and Feed Systems, Volume III: Applications of Reflectors, Sudhakar Rao, Lotfollah Shafai, and Satish Sharma, editors

Introduction to Antenna Analysis Using EM Simulators, Hiroaki Kogure, Yoshie Kogure, and James C. Rautio

Iterative and Self-Adaptive Finite-Elements in Electromagnetic Modeling, Magdalena Salazar-Palma, et al.

LONRS: Low-Noise Receiving Systems Performance and Analysis Toolkit, Charles T. Stelzried, Macgregor S. Reid, and Arthur J. Freiley

Measurement of Mobile Antenna Systems, Second Edition, Hiroyuki Arai

Microstrip Antenna Design Handbook, Ramesh Garg, et al.

Microwave and Millimeter-Wave Remote Sensing for Security Applications, Jeffrey A. Nanzer

Mobile Antenna Systems Handbook, Third Edition, Kyohei Fujimoto, editor

Multiband Integrated Antennas for 4G Terminals, David A. Sánchez-Hernández, editor

New Foundations for Applied Electromagnetics: The Spatial Structure of Fields, Said Mikki and Yahia Antar

Noise Temperature Theory and Applications for Deep Space Communications Antenna Systems, Tom Y. Otoshi

Phased Array Antenna Handbook, Third Edition, Robert J. Mailloux

Phased Array Antennas with Optimized Element Patterns, Sergei P. Skobelev

Plasma Antennas, Theodore Anderson

Polarization in Electromagnetic Systems, Second Edition, Warren Stutzman

Printed MIMO Antenna Engineering, Mohammad S. Sharawi

Quick Finite Elements for Electromagnetic Waves, Giuseppe Pelosi, Roberto Coccioli, and Stefano Selleri

Radiowave Propagation and Antennas for Personal Communications, Third Edition, Kazimierz Siwiak

Reflectarray Antennas: Analysis, Design, Fabrication and Measurement, Jafar Shaker, Mohammad Reza Chaharmir, and Jonathan Ethier

RF Coaxial Slot Radiators: Modeling, Measurements, and Applications, Kok Yeow You

Solid Dielectric Horn Antennas, Carlos Salema, Carlos Fernandes, and Rama Kant Jha

Switched Parasitic Antennas for Cellular Communications, David V. Thiel and Stephanie Smith

Ultrawideband Antennas for Microwave Imaging Systems, Tayeb A. Denidni and Gijo Augustin

Ultrawideband Short-Pulse Radio Systems, V. I. Koshelev, Yu. I. Buyanov, and V. P. Belichenko

Understanding Electromagnetic Scattering Using the Moment Method: A Practical Approach, Randy Bancroft

Wavelet Applications in Engineering Electromagnetics, Tapan Sarkar, Magdalena Salazar Palma, and Michael C. Wicks

For further information on these and other Artech House titles, including previously considered out-of-print books now available through our In-Print-Forever® (IPF®) program, contact:

Artech House
685 Canton Street
Norwood, MA 02062
Phone: 781-769-9750
Fax: 781-769-6334
e-mail: artech@artechhouse.com

Artech House
16 Sussex Street
London SW1V HRW UK
Phone: +44 (0)20 7596-8750
Fax: +44 (0)20 7630 0166
e-mail: artech-uk@artechhouse.com

Find us on the World Wide Web at: www.artechhouse.com